COLLECTIVE BEINGS

Contemporary Systems Thinking

Series Editor: Robert L. Flood
Monash University
Australia

ᵔder Plan is available for this series. A continuation order will bring delivery of each new volume
ᴵblication. Volumes are billed only upon actual shipment. For further information please contact

COLLECTIVE BEINGS

Gianfranco Minati
Eliano Pessa

 Springer

Gianfranco Minati
Italian Systems Society
Milan, Italy

Eliano Pessa
University of Pavia
Pavia, Italy

Library of Congress Control Number:

ISBN-10: 0-387-35541-3 (HB)   ISBN-10: 0-387-35941-9 (e-book)
ISBN-13: 978-0387-35541-2 (HB) ISBN-13: 978-0387-35941-0 (e-book)

Printed on acid-free paper.

Printed in the United States of America.

9 8 7 6 5 4 3 2 1

springer.com

Contents

Preface

Why introduce a new concept such as *Collective Beings*, which is the subject of this book? The motivations are strongly rooted in the very history of Systemics. When its founding fathers, including Von Bertalanffy, Ashby, Boulding, Von Foerster, for the first time recognized the potential associated with the concept of "system" and understood that there were new possibilities for controlling and managing complex systems, physical as well as social, they were forced to rely on the achievements of systems theory existing at that time. This produced a great disparity between the available tools, on the one hand, and the complexity of the problems to be dealt with on the other. It was this disparity which limited the effectiveness of the application of Systemics to a whole range of domains, nothwithstanding the revolutionary importance of its ideas.

Over the years, however, the situation has changed and new contributions have increased the power of systemic tools: theories of self-organization, phase transitions, collective behaviors and of emergence have allowed a better understanding of many problems connected to extremely complex systems. Nevertheless, these new achievements were reached within specific disciplinary domains, such as physics, computer science or biology. In a number of cases this prevented a complete understanding and a correct assessment of their systemic value by researchers accustomed to different cultural traditions. Now is the time for a profound reflection upon these topics and for ascertaining whether the new tools will allow Systemics to make a qualitative leap towards dealing with more complex systems, which we call *Collective Beings*.

Roughly speaking, the concept of Collective Being embodies the main features of complex systems, such as social ones, made up of a number of agents, each endowed with a cognitive system, and belonging simultaneously to various subsystems. The introduction of this concept takes us beyond the traditional conception of systems in which each element is associated with a list of features with fixed and invariant structures. Namely, in a social system every element (it would be better to call it an *agent*) can change its role with time or play, at one and the same time, different roles: for instance, within modern society I can play, at the same time, the role of consumer, worker, member of a family and so on.

Clearly, new conceptual tools are required for describing Collective Beings, as well as a critical revisitation of all previous achievements of Systemics, even technical ones. But the effort in this field is necessary if we want, in the near future, Systemics to play a more important role in managing complex social systems.The aim of this book, therefore, is to introduce *new* ideas and approaches, based upon the most advanced research achievements in self-organization and emergence, with the prospect of applying them to Collective Beings, so as to promote their use in various disciplinary contexts, improving our ability to design and manage systems.

In order to give the reader a preliminary idea of the contents of the book, as well as its perspective, here we provide a bird's-eye view of the topics dealt with in the individual Chapters.

Starting with Chapter 1, we summarize information about the basis of General Systems Theory and of the more general cultural field of *Systemics*. A brief reference is also made to historical circumstances. Some outstanding contributions to dynamical systems theory (related to the concepts of equilibrium, limit cycles, chaos) are presented in some detail, even of a technical nature.

Chapter 2 discusses the fundamental role of the observer in modern science, mostly with reference to emergence. This entails an assessment of the role of uncertainty principles in science.Chapter 2 also introduces one of the new core ideas proposed in the book, the one of the *DYnamic uSAge of Models* (DYSAM). It applies to situations commonly occurring when dealing with complex systems, in which we cannot, in principle, resort to a unique model (the *correct* one) to describe the system being studied. Thus we are forced to allow for a multiplicity of different models (and modeling tools), all related to the same system. In this context DYSAM refers to the ability to *systemically use* the available models by *resorting to*:

- their results in a *crossed* way, that is when one is considered as a function of another, by *using* what are assumed to be *mistakes* (according to an objectivistic view) and not just *avoiding* them;
- learning processes developed on pre-processed frameworks, on past, current or expected contextual information;

- context-sensitiveness, depending on the global context, without reference to the specific decision or selection to be made;
- behavioral strategies adopted for any reason;
- any kind of recorded information to be considered for making decisions;
- affective-cognitive processes such as those regarding emotional activity, affection, attention, perception, inferencing and language.

An amusing computer simulation of a DYSAM-like behavior is presented briefly at the end of the Chapter.

Chapter 3 is devoted to a preliminary discussion of the concept of emergence. In this context we introduce the second new idea presented in the book, the one of Collective Beings. As previously outlined, this expression denotes systems emerging from interacting subsystems, in turn consisting of interacting agents simultaneously belonging to various subsystems. *Belonging* in this case means that the *same* components make different systems *emergent*. Nevertheless these roles may be played **at different times or simultaneously.** We see how such a level of complexity in emergence is related to the ability of autonomous agents to use the *same* cognitive models over time. This concept is obviously related to the topic of collective behaviors. On this point, we introduce a distinction between collective behaviors occurring within systems such as flocks, swarms and herds and those characterizing the emergence of Collective Beings. Namely, to use the metaphor of the flock, Collective Beings emerge from interactions among different flocks constituted over time by the same agents. Such a level of complexity is strongly related to the ability of the agents to use cognitive models. Human social systems are typically constituted in this way. We also refer to the idea that in each of these situations, the emergent subsystem, though it has the same components, may show different behaviors. These behaviors can be so different that, even though they have emerged from the same components, conceptually it is as if we are dealing with completely *different subsystems*. For example, the subsystem of a "crowd", emerging from people, may show the behavior of a riot, of people panicking or simply shopping, attending a concert or waiting for something to happen. Also, through *collective components* one subsystem may affect the other. If people have the experience of contributing to a riot, this experience may be remembered when they interact within another subsystem, such as a family or a company. When trying to model these situations we cannot rely on traditional models alone and we need an *architecture of usage* of different models, **a dynamic use of them,** as introduced with DYSAM. We can apply different models depending upon the context, the time, the kind of emergent behavior to be modeled or on the observer's purposes. The classical approach based upon the usage of single models one at a time, searching for the most appropriate, the most effective, is not sufficient to manage, control and induce the emergence of Collective Beings. The classical approach based upon the **non-dynamic use of models** is quite adequate when interactions take place among inanimate particles, such as electrons, or agents following *single* behavioral models, but it is not suitable when interactions take place among agents using different **cognitive models**.

Examples of Collective Beings based upon components giving rise to subsystems having different behaviors according to their aggregation and interactions, include the stock exchange, families of financial operators, political parties, sports teams, as well as marketing centres, families, companies, clubs and financial companies acting on the same market, or within the framework of a global network engaging in many activities (as in the case of *Industrial Districts*), or within a virtual company.

Chapter 4 introduces a detailed reference to the theoretical tools available for the problem of modelling emergence, by focussing only on traditional methods. The latter methods were the first to be used at the beginning of Systemics and are characterized by the search for exact, deterministic results through rigorous analytical methods. Any reference to noise, fluctuations, uncertainty, probability, fuzziness is therefore banned from the start. An important part of this Chapter, is a short, although somewhat technical, exposition of the main findings of Dynamical Systems Theory, upon which all traditional methods are based. The topics dealt with include concepts such as stability, bifurcation, chaos and hierarchical systems. Ample space is devoted to the theory of Prigogine's Dissipative Structures, as well as to the theory of solitary waves and of models of pattern formation based upon differential equations. It is shown how all these theories are unable to describe intrinsic emergence, owing to the fact that they lack the tools for taking into account the role of fluctuations and of uncertainty. Nevertheless, they show remarkable potential for describing pattern-formation phenomena at a macroscopic level.

Chapter 5 is, obviously, devoted to a presentation of non-traditional models. It includes topics such as Synergetics, phase transitions, Quantum Field Theory, Symmetry Breaking, to cite only the idealized models. Even a bird's-eye view of non-idealized models is included, with references to topics such as neural networks, Artificial Life, cellular automata and fuzzy sets. The Chapter even contains a discussion about possible relationships between idealized and non-idealized models. It is shown how these models effectively allow a description of intrinsic emergence, even though in most cases the latter is too simple to account for emergence phenomena observed within biological and social systems. We also discuss the possibility of generalizing these tools to account for biological emergence and, as a final goal, for the emergence of Collective Beings.

Chapter 6 introduces the third new idea proposed in the book, related to the usage of ergodicity to detect emergence and to manage Collective Beings. The idea is based on the fact that during the process of the emergence of collective behaviors within a Collective Being the agents are not assumed, of course *with reference to an observer*, to all have the same behavior nor to be distinguishable from one another due to the fact that each one plays a different and well distinguishable role (as in organizations), which is constant over time. In emergence processes of this kind the agents can take on the *same roles at different times*, and *different roles at the same time*. Namely, as we are dealing with autonomous agents, the latter may behave by deciding at any given time which cognitive model to use or by using the same model at different times, that is with different parameters. This is why the possibility of an index for measuring the dynamics of usage of cognitive models may be very helpful for detecting emergence processes and the establishment of Collective Beings. The concept of ergodicity is proposed for such a purpose. In the Chapter we discuss how this concept, initially proposed within statistical mechanics to describe the behavior of physical systems consisting of identical particles, can be generalized to more complex systems, such as Collective Beings.

Chapters 7 and 8 deal with the application to Social Systems of the concepts and tools previously introduced. The topics considered include:

- *Growth, Development and Sustainable Development.* We discuss the problem of representing development, of describing development as a process of emergence, and of development for a Collective Being.
- *Ethics.* This topic includes the emergence of ethics, its crucial role for inducing and maintaining the emergence of social systems, its relationships with quality as well as its crucial role in keeping strategic corporate profitability.
- *Virtual systems.* Here the focus is on all those company devices (virtual and non-virtual, as specified in the text) which process virtual goods, such as money, shares, stocks, insurance products and information. Real goods are processed in these companies through "abstractions, information and images" as happens in e-commerce and in *Industrial Districts.*
- *Knowledge.* This topic deals with the fact that the production, management, distribution, application, transmission, approachability, classification, memorization, protection, representation and marketing of knowledge as well learning technologies will become more and more the core business of the Post-Industrial Society. The subject of knowledge

also refers to knowledge representation, knowledge management and organizational learning.

- Finally we discuss *Industrial Districts* as a typical example of emergence of Collective Beings in the economy.

Finally, the concluding Chapter 9 is devoted to a discussion about the application of the concepts introduced above to the study of cognitive systems (the essential feature of each autonomous agent belonging to a Collective Being) and, more in general, of Cognitive Science. The systemic approach sketched above entails overcoming the traditional computational approach used so far within this domain. The subject of this Chapter, therefore is mainly related to the conceptual tools available for going Beyond Computationalism.

The book also includes two Appendices. Appendix 1 lists brief descriptions of and information about some crucial theoretical concepts dealt with in the book. Appendix 2, on the other hand, discusses some general theoretical questions, supported by typical real-life examples, with proposed answers, related to the ideas, methods and concepts introduced in the book.

From this short description, it can be seen how the main scope of the book is to introduce new ideas for the control and management of human systems, in turn based on a lot of technical material, deriving from different approaches and different disciplines. It is evident how such a synthesis has been a very difficult enterprise. While aware of the intrinsic limitations of our effort, we hope that in some way it will be useful to all those rethinking in a critical way the very foundations of the systemic approach. Dealing with the problems related to Collective Beings is certainly necessary, but we consider that it cannot be the subject of a particular discipline (even though disciplinary knowledge is essential), owing to its transdisciplinary nature. We would consider this book as successful if the readers could include it within the category of transdisciplinarity.

Gianfranco Minati

Eliano Pessa

Foreword

In their impressive book, Gianfranco Minati and Eliano Pessa, introduce the concepts of Collective Beings in order to propose a conceptual tool to study collective behaviors. While such behaviors play a fundamental role in many scientific and technical disciplines, the authors focus their attention on socio-economic applications to be used, for instance, to increase corporate profitability and productivity. To this end, they provide the reader with deep insights into modern conceptual tools of economy based on systemics.

Collective behaviors are shown by systems composed of individual parts or agents. Minati and Pessa illuminate the interplay between system and individual in a remarkable new way: their collective beings give rise to systems having quite different behavior, namely that of the group and that of the individual agent. The authors always take all the various aspects into account - for instance the components of one system may be simultaneously the components of another system as well. They stress the importance of cognitive models used by the interacting agents, of the role of the observer and of the dynamic use of many cognitive models. They deeply penetrate into the concept of being virtual in contradistinction to being actual. They underline the role of ethical agreements, jointly with a careful discussion of different concepts of ethics as well as their relation to quality. I fully agree with their view on systemics as a cultural framework crossing disciplines. Their book is coined by a profound humanitarian attitude, for instance when they state that often technological solutions are designed for problems rather than for people having these problems. I have found the two appendices on systemics characteristics and on various critical questions concerning systemics, e.g. on systemics as a discipline, on the possibility of its falsification, etc. highly informative.

I have read this book not only with the utmost interest, but also with great delight and I can recommend it highly to all those who are interested in the modern and fascinating field of systemics.

Hermann Haken

Center of Synergetics
University of Stuttgart

Chapter 1

THE BACKGROUND TO SYSTEMICS

1.1 Introduction

This book is devoted to the problems arising when dealing with emergent behaviors in complex systems and to a number of proposals advanced to solve them. The main concept around which all arguments revolve, is that of Collective Being. The latter expression roughly denotes multiple systems in which each component can belong simultaneously to different subsystems. Typical instances are swarms, flocks, herds and even crowds, social groups, sometimes industrial organisations and perhaps the human cognitive system itself. The study of Collective Beings is, of course, a matter of necessity when dealing with systems whose elements are agents, each of which is capable of some form of cognitive processing. The subject of this book could therefore be defined as "the study of emergent collective behaviors within assemblies of cognitive agents".

Needless to say, such a topic involves a wide range of applications attracting the attention of a large audience. It integrates contributions from Artificial Life, Swarm Intelligence, Economic Theory, but also from Statistical Physics, Dynamical Systems Theory and Cognitive Science. It concerns domains such as organizational learning, the development or emergence of ethics (metaphorically intended as *social software*), the design of autonomous robots and knowledge management in the post-industrial society. Managers, Economists, Engineers, as well as Physicists, Biologists and Psychologists, could all benefit from the discoveries made through the trans-disciplinary work underlying the study of Collective Beings.

From the beginning, this book will adopt a *systemic* framework. The attribute "systemic" means that this framework fits within *Systemics*, a thinking movement which originated from *General Systems Theory*, proposed by Von Bertalanffy and from *Cybernetics*, introduced by Wiener and developed by Ashby and Von Foerster (Von Foerster, 1979). A systemic framework is characterized by the following features:

- Focus is placed upon the global, holistic properties of entities qualified as *systems*, which, in general, are described in terms of elements and of their interactions;
- the role and the nature of the *observer*, as well as the *context*, are taken into account, as far as possible, within the description and the modelling of each and every phenomenon;
- the goal is not that of obtaining a *unique, correct* model of a given behavior, but rather of investigating the *complementary* relationships existing between *different* models of the same phenomenon.

In order to better specify the domain under study, we will introduce, distinctions (which could even be considered as hierarchical) between different kinds of systems:

- *simple systems*, where each component is associated (in an invariant way) with a single *label* (which could even be a number), specifying its nature and allowed operations ; a limiting case of simple systems is given by *sets*, in which the individual components cannot perform operations, but only exist;
- *Collective Beings* (Minati, 2001), where each component is associated (in a variable way) with a *set of possible labels*; the association between the component and the labels depends upon the *global behavior* of the system itself and can vary with time; a typical case is a flock of birds, within which each bird can be associated with a single label, specifying both its relative position within the flock and the fact that its operation consists only of flying in such a way as to keep constant its distance with respect to neighboring birds. Such an association, however, holds as

long as the flock behaves like a flock, that is like a single entity; as soon as the flock loses its identity, a single bird becomes associated with a set of different labels, specifying different possible operations such as flying, hunting, nesting and so on; this new association can define a different Collective Being, such as a bird community;

- *Multi-Collective Beings*, characterized by the existence, not only of different components, but even of different levels of description and of operation; each component and each level is associated (in a variable way) with a set of possible labels; the forms of these associations depend upon the relationships existing between the different levels. Examples of multi-Collective Beings include the human cognitive system and human societies.

This book principally considers the study of Collective Beings and, in addition to a review of the existing approaches for modeling their behaviors, we will introduce a general methodology for dealing with these complex systems: the DYnamic uSAge of Models (DYSAM) (Minati, 2001). The latter will be applied to cases in which it is manifestly impossible, in principle, to fully describe a system using a single model.

This chapter will introduce the reader to some fundamental concepts of Systemics, by starting from a short history of Systemics and of the associated evolution from the concept of 'set' to that of 'system'. Several examples will help the reader in this introductory approach. The distinctions between sets, structured sets, systems and subsystems will allow the reader to better understand new theoretical concepts, introduced in subsequent Sections of this book, such as those of Collective Beings, of DYnamic uSAge of Models (DYSAM) and of Ergodicity, within the context of the tools used to detect emergence.

In the second part of this chapter reference will be made to some technical tools of Systemics both to complete the historic overview and because they serve as an introduction to Chapter 4, where we will deal with the problems of managing emergence.

1.2   What is Systemics ?

The father of Systemics was Ludwig von Bertalanffy (1901-1972). He was one of the most important theoretical biologists of the first half of the Twentieth Century. His interdisciplinary approach (researcher in comparative physiology, biophysics, cancer, psychology, philosophy of science) and his knowledge of mathematics allowed him to develop a kinetic theory of stationary open systems and General Systems Theory. He was one of the founding members and Vice-President of the Society for General Systems Research, now renamed as the International Society for Systems Sciences (ISSS). The "Society for General Systems Research" (SGSR) was formally established at the 1956 meeting of the American Association for the Advancement of Science (AAAS), founded in 1848. The SGSR was born under the leadership of Ludwig von Bertalanffy, the economist Kenneth Boulding, the neurophysiologist Ralph Gerard, the anthropologist Margaret Mead, the psychologist James Grier Miller and the mathematician Anatol Rapoport.

Von Bertalanffy held positions, to mention but a few, at the University of Vienna (1934-48), the University of Ottawa (1950-54), the Mount Sinai Hospital (Los Angeles) (1955-58), the University of Alberta (1961-68) and the State University of New York (SUNY) (1969-72).

A collection of his essays was published in 1975, three years after his death.. This collection (Von Bertalanffy, 1975) included forewords written by Maria Bertalanffy (his wife) and Ervin Laszlo. The latter added the following considerations about the term *General Systems Theory*:

> "The original concept that is usually assumed to be expressed in the English term *General Systems Theory* was Allgemeine Systemtheorie (or Lehre). Now "Theorie" or Lehre, just as Wissenschaft, has a much broader meaning in German than the closest English words *theory* and *science*."

The word *Wissenschaft* refers to any organized body of knowledge. The German word *Theorie* applies to any systematically presented set of concepts. They may be philosophical, empirical, axiomatic, etc. Bertalanffy's reference to *Allgemeine Systemtheorie* should be interpreted by understanding a new perspective, a new way of *doing science* more than a proposal of a *General Systems Theory* in the dominion of science, i.e. a *Theory of General Systems*.

In this book, instead of using terms such as *Theory of General Systems* or *General Systems Theory* we will use the word *Systemics*, widely used in English language systems literature (see par. 1.3, point i), keeping in mind the distinction mentioned above, and emphasizing that the reference is not only to the scientific domain, which is the topic of this book, but to an overall, general approach towards understanding phenomena in an interdisciplinary manner. The meaning adopted for the word Systemics, therefore, will be that specified in the introduction to this chapter, with the proviso that such an approach to the study of scientific questions will need the design of suitable methodologies and technical tools, which will be described in this book.

1.3   A short, introductory history

In this chapter a short introductory history of systems thinking will be outlined. The reader, by using some of the keywords and consulting a history of philosophy and science and encyclopaedic sources, some of which are listed in the bibliography, will be able to reconstruct a disciplinary framework adequate for his/her interest and background. Information about the history of systems thinking evolution is available in the literature in many books and papers (see, for example, Von Bertalanffy, 1968; Umpleby and Dent, 1999). References and key concepts are also described in Appendix 1.

a) The concept of System as a **mechanism** and as a **device**. From the idea of system as a configuration of assembled components, producing a *working* mechanism, based on the concept of *machine*, in its turn based on many concepts of classic physics, it is possible to extrapolate the powerful abstraction of *device*. The latter concept still makes reference to assemblies of components working as a whole, but having non-mechanical relationships among the components themselves; typical examples are given by electronic devices or software programs. In these cases we may refer to abstract entities, such as procedures, and within this context we will deal with *systems control, automata theory, control techniques*. This context is known as *Cybernetics*, a term coined from the Greek "pilot of the boat" (Ashby, 1956). This approach provided the basis for modern **systems engineering** (Porter, 1965).

b) **Cybernetics** has been very important in the process of establishing systems thinking. It has been defined as the science of behavior, communication, control and organization in organisms, machines and societies. One of its salient features was the introduction of the concept of *feedback*, viewed as a sort of self-management or self-regulation. Cybernetics as a scientific discipline was introduced by **Norbert Wiener** (1894-1964) in the Forties (Wiener, 1948; 1961), with the goal of studying the processes of control and communication in animals and machines. Initially, (Ashby, 1956; Heims, 1991) it was identified with information theory. A very well- known stereotyped example of a cybernetic device, often used in a metaphorical way, is Watt's centrifugal regulator designed for steam engines (see Figure 1.1): it is based on a feedback process able to keep constant the angular velocity of a steam engine. As can be seen in

Figure 1.1 the base R of the regulator moves upward or downward, its direction of motion depending on the rotation speed of the shaft A. If the base R of the regulator is connected to a regulating valve, the device is able to self-regulate by keeping the shaft rotation velocity constant.

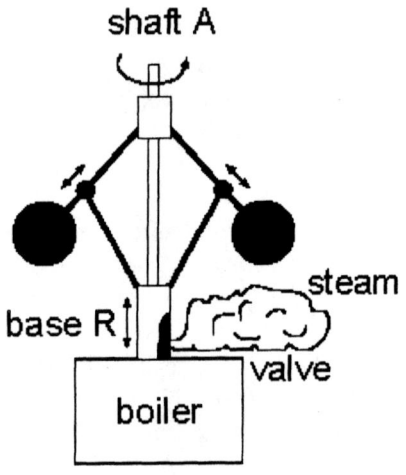

Figure 1-1. Watt's centrifugal regulator.

The behavior of Watt's centrifugal regulator can be easily described in mathematical terms, through the equation of motion:

$$m\ddot{\phi} = mn^2 \sin\phi\cos\phi - mg\sin\phi - b\dot{\Phi} \qquad (1.1)$$

where ·
   - ϕ is the rotation angle of the axis,
   - m is the mass of the revolving pendulum,
   - n is the transmission ratio,
   - Φ is the rotation speed of the motor axis,
   - g is the gravitational constant,
   - b is the dissipation constant depending on the viscosity of the pivot.

Cybernetics allowed the creation of relationships among regulation models operating in different fields, such as those describing the operation of animal sense organs, where self-regulation processes are identifiable. One example is the eye which, when hit by light, automatically reduces the aperture in the iris thus regulating the amount of light entering the eye.

Another example may help to characterize the domain in which the concepts of Cybernetics can be applied.

The problem of computing the trajectory of an artillery shell, starting from the knowledge of all initial factors determining the shell's motion, cannot be considered as a cybernetic problem. On the contrary, it becomes cybernetic when the missile itself is capable of continuously correcting its trajectory, as a function of the information about the nature of the trajectory and the position of the target.

Other approaches to Cybernetics were introduced by:

- **Warren McCulloch** (1898-1968), neuro-physiologist, introduced the mathematical model of Neural Networks and considered cybernetics as the study of the communication between observer and environment;
- **Stafford Beer** (1926-), researcher in management, considered cybernetics as the science of organization (Beer, 1994);
- **Gregory Bateson** (1904-1980), anthropologist, introduced a distinction between the usual scientific approach, based on matter and energy, and cybernetics, dealing with models and forms (Bateson, 1972).

c) **System Dynamics (SD)**, in which a system is identified with a configuration of regulatory devices. The expression "System Dynamics" actually denotes a methodology introduced by **Jay W. Forrester** (1918-) in 1961, in his book "Industrial Dynamics" (Forrester, 1961) to study and implement systems of feedback loops (an example of a single feedback loop involving two elements A and B is shown in Figure 1.2), associated with configurations of interacting elements. A system consisting of interacting (through feedback) elements can exhibit global emergent behaviors, not reducible to those of the single individual elements nor to the feedback among them. Such behaviors, for instance, occur within electrical networks and traffic flows. This approach was assumed to be the most suited to describe the interactions among industrial departments which emerge within companies.

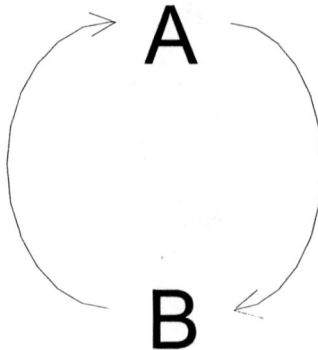

Figure 1-2. System of feedback loops

To summarise, Systems Dynamics deals with conceptual networks of elements interacting through feedback loops. This approach is mainly used for software simulations of corporate dynamics and social systems (Forrester, 1968), but also to model organized social systems (Meadows *et al.*, 1993).

d) **The theory of dynamical systems**

System Dynamics (SD) must be not confused with *Dynamical Systems Theory*. In the mathematical literature often a *continuous dynamical system* in an open interval *w* is described by an autonomous (i.e. whose right hand members are time independent) system of ordinary differential equations which hold for a vector of dependent variables x:

$$dx/dt = F\ (x) \tag{1.2}$$

The theory of dynamical systems, implemented on the basis of the fundamental intuitions of **H. Poincaré** (1854-1912), showed the coexistence of ordered and chaotic behaviors in the study of almost any kind of system which can be represented in mathematics and physics. Simple systems, such as a pendulum or the Moon moving along its orbit, can be described by using the equations of motion of classical mechanics. A dynamical system is associated with two kinds of information:

- One which deals with the representation of the system's state and with basic information about the system itself;
- Another specifying the dynamics of the system, implemented through a rule describing its evolution over time.

The time evolution of a dynamical system may be geometrically represented as a graph in a multidimensional space, the so-called *phase space.* It should be noted that by looking only at the form of the orbits in the phase space, we do not describe the geometrical movement of the system, but only the relationships among its independent variables (see Appendix 1).

e) **Gestalt Psychology** introduced an important new approach, related to systems thinking. It originated in Germany, in 1912, under the name of *Gestaltpsychologie.* In the same period in the U.S., *Behaviorism* (Skinner, 1938; 1953), holding an opposite view of psychology, was born.

The German term *Gestalt* refers to a structure, a schema, a configuration of phenomena of different natures (psychological, physical, biological and social) which are so integrated as to be considered an indivisible whole having different properties from those of its component parts or of a subset of them.

Gestalt Psychology is part of the anti-mechanistic and anti-reductionistic movement deriving from the crisis of positivism.

According to this approach it is not possible to reduce psychological phenomena to a chain of stimulus-response associations as, on the contrary, is held by the Behaviorist approach.

Gestalt Psychology triggered a thinking movement, which led to the establishment of systemic psychology.

f) The term **organicism** refers to a view which, in contrast with positivism, assumes that a living system is a finalistic organized whole and not simply the mechanical result of the sum of its component parts. This conception, of biological origin, has been more generally expressed, for instance, by (Whitehead, 1929) who used the word 'organicism' to denote his general philosophical conception. In the field of sociology Compte and Spencer adopt this approach.

g) The term **vitalism** refers to conceptions according to which the phenomena of living beings are so peculiar as to make their reduction to physico-chemical phenomena of the inorganic world impossible. In the second half of the 18th century vitalistic doctrines opposed mechanicism, by hypothesizing, to explain the phenomena of life, a force acting as an organizing principle at the molecular level, separated from the soul or spiritual values. This force, called from time to time 'life force', or 'life surge', is the key aspect of such a conception. This concept began to wobble with the synthesis of urea, representing the birth of organic chemistry, and Darwinism.

Recent advances in genetics and molecular biology have reduced interest in the confrontation between vitalists and mechanicists.

Even today a *theory of living* is still lacking, even though very important progress in the *physics of living matter* has been made (Vitiello, 2001).

h) The term **complexity** relates to problems and conceptual tools (Flood and Carson, 1988) which have chiefly emerged from physics. Brownian motion provides an important historic example. In this case random fluctuations are directly observable, and this circumstance gave rise to the basic concepts of complexity. Brownian motion is the irregular, disordered, unpredictable motion of a speck of pollen in water. The motion is caused by interactions with water molecules, moving in their turn with thermal energy. As a consequence, it becomes impossible to build a deterministic model of this phenomenon. According to classical physics the reason why deterministic models of phenomena like this are not available is because of an incomplete knowledge of all physical features of the components of the involved systems.

On this point, there are two conflicting views:

- The *mechanistic* view, based on the so-called *strong deterministic hypothesis*, according to which we can reach a presumed infinitely precise knowledge of position and speed of all components of a physical system and this knowledge, in principle, could give rise to a deterministic theory of the system itself. Among the holders of this view we can quote Newton (1643-1727) and Laplace (1749-1827). Faith in this conception was shaken for the first time by the failure of classical mechanics to solve the so-called *Three-Body Problem* (Barrow-Green, 1997), tackled by mathematicians including Eulero (1707-1783), Lagrange (1736-181), Jacobi (1804-1851), Poincarè (1854-1912), (see *chaotic* in Appendix 1). Another key principle of the mechanistic view is that of Descartes, according to which *the microscopic world is simpler than the macroscopic one.*

- The view based on the theory of complexity, according to which a complex system can exhibit behaviors which cannot be reduced to those of its component parts, even if it were possible to know with absolute precision their positions and velocities. This view also acknowledges that in most cases it is practically impossible to obtain complete information about microscopic positions and velocities, hypothesized by Newton and Laplace. This circumstance has been proven by many experiments, whose explanation needs more effective conceptual tools. With the term *complexity* reference is made to themes such as (see Appendix 1) *deterministic chaos, role of the observer* (Chapter 2), *self-organization, science of combined effects* or *Synergetics* (Haken, 1981). A typical example of a complex system, containing a huge quantity of elements and interconnections among them, is the **brain**.

i) **Systems thinking** may be dated back to cultural frameworks of different natures, all oriented towards recognizing *continuity* and unity in a reality fragmented and desegregated into different disciplines, languages, approaches and conceptions (Checkland, 1981; Checkland and Scholes, 1990; Emery, 1969; Flood and Jackson, 1991). Thousands of references relating to different domains are available (introductory ones include, Bohm, 1992; Boulding, 1956; 1985; Briggs and Peat, 1984). References to approaches, currently denoted as *systemic*, may even be found in the Biblical theme of the confusion of languages in the story of the *Babel tower*; in the Talmudic way of thinking in Hebrew culture, as told by S. Freud [1]; in Heisenberg's autobiography with the original German title "The share and the whole" (Heisenberg, 1971); and in many others cases quoted by Capra (Capra, 1996), where systems thinking and its birth are discussed. The expression *General Systems Theory* refers to the fundamental work by Von Bertalanffy (Von Bertalanffy, 1968). Von Bertalanffy states in that book (in which, by the way, a sound introduction to the history of systems thinking is presented) that he introduced the idea of a General Systems Theory for the first time in 1937 during a philosophy conference in Chicago. Around the concept of 'system', suitable for generalizing concepts previously formulated within different contexts, intense research activity has grown. The goals of the latter include both the study of invariant system features and the search for conceptual and methodological (Churchman, 1968; 1971) application to different disciplinary contexts, General Systems Theory (Rapoport, 1968; Sutherland, 1973) was introduced to describe a system as a phenomenon of **emergence** (see Chapter 3) (Von Bertalanffy, 1950; 1952; 1956; 1968; 1975). As introduced at the beginning of this chapter, this expression refers to a general cultural approach more than to a real theory. Actually, *theory* is a very strong word in science (Kuhn, 1962). Following the approach proposed by Popper it must be possible to falsify a scientific theory, if we want to adopt a scientific and not a *pseudo*-scientific attitude (see Appendix 2). The hypotheses on which a theory is based must be validated. It must be possible to design an experiment, a validation test which, if a given result is obtained, would confute the hypothesis on which the theory itself is based.

Usually people speak of
• Systemic approach, with reference to a methodological framework;

---

[1] S. Freud (1908), to Abraham, May 8th , in Correspondence (1907-1926), Paris, Gallimard, 1969

- Production Systems, a term having different meanings in management science and in a logic-mathematical context;
- System analyst, which is a profession in the field of Computer Systems;
- Electronics and telecommunications systems;
- Systemic therapy, in a psychotherapeutic context and so on.

General Systems Theory looks like a cultural framework, a set of disciplinary meanings extrapolated from theories sharing the topic of systems. A structured and formalized organization of this approach may be found in Klir (Klir 1969; 1972; 1991).

For the reasons presented in the previous section, the term **Systemics** (**Systémique** in French, **Sistemica** in Italian and Spanish) has thus been introduced. The term is used not only in academic literature for referring to holistic concepts (Smuts, 1926), but also with reference to other conceptual extensions of the word 'System'. Systems Research Societies, such as the International Society for Systems Sciences (ISSS), as well as a number of national societies, use this term. It is also used in modern expressions when referring to applications in various disciplines, in order to emphasize the complexity, the web of relations, the interdependency between components. Typical cases are net-economy, software development, organizations, medical applications, pharmacology, electronics, biology, chemistry and so on.

At this point it is important to make a fundamental clarification of the terms introduced so far. This will avoid ambiguities and serious conceptual mistakes in assuming Systemics as referring to a traditional scientific domain (the so called "hard" sciences, such as mathematics, physics, biology, chemistry, etc.) rather than to a general cultural approach.

As introduced at the beginning of this chapter, the term Systemics refers to a cultural framework which crosses various disciplines. Disciplinary applications of this crossing within the scientific context, although particularly important, are only a part of the possible outcomes. The systemic contributions from various disciplines are fundamental for the emergence of Systemics. In its turn, Systemics is a source of innovative approaches within each particular discipline.

However, the term Systemics does not mean a particular disciplinary context in which this approach takes place, but a general strategy for approaching problems, emphasizing the need for a generalised view of events, processes and complex entities in which they are interrelated (see Appendix 2). This is not a trivial observation such as: *arithmetic is applicable to apples, people and trains*. The difference is that, when Systemics is applied within a given context, a model designed for the latter is enriched with new disciplinary concepts and becomes a *systemic invariant* (i.e. a concept, an approach which can be used within other contexts) As such, it allows the use of approaches and strategies designed in other contexts. *Systemic invariants* cannot qualify single elements but the behavior of the whole emerged system. General examples of *systemic invariants* identified within individual disciplines are listed in Appendix 1 and *Systems Archetypes* are discussed in Chapter 7. The concept of systemic *openness and closeness* applies for instance to biology, physics and economics. Moreover, even in *multidisciplinary* fields such as **Cognitive Science,** *when science studies itself, its own processes,* there is a continuous enrichment among applications of mathematical models, computer processing techniques based on Neural Networks, psychological experimental activities, modeling, language research and representation. The same circumstance occurs in domains, which are multidisciplinary *in principle*, such as **Environmental Science** which combines physics, chemistry, biology, economy and engineering.

Thus, the important relationships between *interdisciplinarity* and Systemics can be emphasized. To summarise:

- *Mono-disciplinary* approaches take place when specific domains are studied by designing specific tools. Different fields of interest deal with individual disciplines, such as mathematics, arts, economics. Education is usually *fragmented* into individual disciplines.
- *Multidisciplinary* approaches require the use of several different disciplines to carry out a project. For instance a project in telecommunications needs individual engineering, economic, legal, managerial competences working in *parallel*. Implementation of projects requires more and more multidisciplinarity. Multidisciplinary education means teaching one discipline while discussing another, i.e. language and history, mathematics and economy and so on.
- *Interdisciplinary* approaches involve problems, solutions and approaches (and not just tools) of one individual discipline being used in another following from systemic concepts such as those listed in Appendix 1. This is different from just using the *same tools*, such as mathematical ones. For instance, the use of the systemic concept of openness in physics, economics and biology allows scientists to deal with corresponding problems, solutions and approaches even using the same tools.
- *Trans-disciplinary* approaches are taken when problems are considered between, across and *beyond* disciplines, in a *unitary* view of knowledge. In this case the interdisciplinary approach is reversed: it is not a matter of an inter-crossing, cooperative use of disciplinary approaches looking for conceptual invariants using the same concepts in different

disciplines, but of finding disciplinary usages of the same trans-disciplinary knowledge. Trans-disciplinarity refers to something beyond individual disciplinary meanings and effects. It refers to the multiple levels and meanings of the world, the multiple levels of descriptions and representations adopted by the observer. While disciplinary research concerns *one disciplinary level*, trans-disciplinary research concerns the *dynamics between different levels of representation* taking place at the same level of description. Examples include multi-dimensional education focusing on the development of different, simultaneous, cognitively and ethically related disciplinary interests (Gibbons *et al.*, 1994; Nicolescu 1996) and in the approach to phenomena by *simultaneously* using different representations, descriptions, languages and models. These aspects are introduced in the DYnamic uSAge of Models (DYSAM) in Chapter 2.

The most significant contribution of the systemic approach is its ability to demonstrate that a strategy based only on the identification and study of the behavior of single, isolated components is ineffective and unsuitable for problems carrying the complexity of emergent processes and systems. At one level, this approach, in systems engineering, is based upon using, designing and controlling input-output and feedback-controlled devices as considered by System Dynamics (Forrester, 1968). In other words, societies, corporations, biological systems, the human mind and even a magnet or a superconductor can not be studied as if they were made up of individual component parts such as: pendula, levers and bolts. The machine paradigm, in short, is adequate only for machines and it is useless and ineffective in all other cases. To recognize this idea implies a very profound conceptual revolution, given that our scientific tools (mathematical, physical, biological, medical, economical models), as well as our legal and social frameworks have all been designed for a world where the machine concept and model is a fundamental element, within a more deterministic than probabilistic context. The mechanistic view is used as a touchstone to represent and design any other kind of operational device. Systemics, on the other hand, produces conceptual tools, devices and methodologies to deal with situations in which classical mechanistic approaches are ineffective.

Attempting to explain everything by using the available conceptual tools is an understandable and unavoidable human attitude. The following story may be illuminating. *In a dark room only one corner is lit up by a small bulb hanging from the ceiling. The light falls on a disordered set of objects. In this narrowly lit area a person is desperately searching for something. A friend arrives and asks: are you looking for something? Yes, the other says, I have lost my keys. The friend asks if he can help. Sure! After some fruitless searching the friend says: "There is nothing here! Are you sure you lost them here?" "No", says the other, "I lost them over there". "Then why are you searching here?". "Because here there is light".*

We must underline that the systems approach is a scientific, cultural approach and not an **ideological** one. The systemic approach is very appropriate and effective, *today*. Systems scientists must not only be ready for the emergence of a new paradigm, but also welcome it, search for it (Minati and Pessa, 2002).

Due to a systemic understanding of social processes it is now possible to realize how often technological solutions have been designed for *functions* more than *uses*, that is, designed for **problems rather than for people having the problems**. This process has led to very poor human-machine interfacing, as well as the prevalence of non-learning systems.

In the same way it becomes clear how solutions to problems have been designed with no awareness that **solutions to old problems may generate more complex and more difficult new problems**, requiring new approaches, such as the use of drugs, technologies for food conservation, energy production, temperature control and so on.

It should also be recalled here that a complete *history* of Systemics is not yet available in the literature. One might say that the field is too young to have a *history*. We believe that the reason for this is not related only to chronological events, nor to contributors to Systemics including those quoted above. The point is to **recognize** cultural events which form part of Systemics. In order to decide whether an element belongs to a set or not we need a rule and, in the same way, a very important first step is to have such criteria for the recognition of systemic thinking in processes where there was no *awareness* of it. The reason for finding such criteria and events is to establish general paths and trends in system thinking, to prepare for going *beyond* Systemics, by designing the *logistic curve* (see Figure 1.3 in Section 1.4.4) of this approach.

1.4 Fundamental theoretical concepts

Before starting a short exposition of some fundamental concepts of Systemics, a general point should be made about its history. Until the early 1980s, advances were made using concepts deriving from classical physics, according to which a knowledge of general, abstract laws and rules was sufficient to account for the whole range of phenomena observed in the real world. After this point in time, however, it major questions were posed in the description and treatment, of processes of emergence observed in ecological (and not abstract) systems, that is in the presence of noise, disturbances, context-dependent constraints together with the influence of the observer. Thus, the need arose for new methodologies and tools able to deal with real-life questions, avoiding the pitfalls of the abstract descriptions. These methodologies were not seen as being opposed to the previously used ones, but rather as generalizing them. This chapter presents a description of some of the tools traditionally used within the first phase of the development of Systemics, which can be denoted as *pre-emergence Systemics*. A knowledge of these tools could be considered as a sort of pre-requisite for designing a systemic approach. The methodologies introduced in *post-emergence Systemics* will be described in Chapter 4.

1.4.1   Set theory

As we will see in the following sections, various approaches and definitions of systems have been based on set theory.

This requires a clarification of exactly what constitutes a set of elements and how their relationships can be defined, detected or controlled. A number of different approaches are possible. The first, and most popular of them, is based on a mathematical theory, known as *Set Theory*. Within this theory the concept of *Set* is taken as *primitive*, together with that of *belonging to a set*.

Why is it usually assumed that a system may be established by interacting components *belonging to a set* and not just to an aggregate lacking any rules of membership?

The concept of set has been studied and analysed in the fields of logic, philosophy and mathematics by authors including Cantor, Dedekind, Russell, Zermelo, to mention only a few). For our purposes it is sufficient to recall that a set is characterized by elements having *at least one common property* (as we will see it is not possible to leave aside the role of the observer *detecting* the property, and having a *cognitive model* of it). A *set* may be defined as a group of elements for which there is a criterion, a rule, allowing one to *decide* whether an element belongs to it or not.

The reason for starting from the consideration of a *Set* and not just of an aggregate of elements is based on the logical power of the definition of a set. Within Set theory (Cantor, 1884), introduced by **Georg Cantor** (1845 - 1918), a set is defined when a rule of membership is defined. In such a way it is always possible to decide whether an element belongs to a set or not. In this theory sets are collections of objects in which ordering has no significance. Intuitively, the purpose was to have a theory with which it would be possible to state properties which, when valid for one element, were valid for the entire set and, when valid for the entire set, were automatically valid for *each* element.

The rule of membership assures a logical homogeneity which is very useful for the observer. This rule may even be represented by a list.

Set theory has been of great relevance, as a conceptual framework, for mathematics and science in general.

Briefly, it should be mentioned how its logical power has been *limited* by the discover of antinomies in the concept of set, especially when applied to the set of (all) sets and when dealing with the concept of infinite. Two classic antinomies can be mentioned:

1) Russell's Antinomy

The set of all sets having more than k elements is an element of itself. The set of all books is not a book. Consider the set of all sets which are not elements of themselves, as in this example. If this set is an element of itself, then it is not an element of itself by definition. If it is not, then it is so.

2) Cantor's Antinomy

Referring to the concept of *power of a set*, expressed by its *cardinality*, that is the possibility of a one-to-one correspondence of its elements with the elements of another set, many questions arose, such as: Can a set having an infinite number of elements have more or less elements than another set having an infinite number of elements? Is a part of an infinite set finite or infinite?

Consider the following example.

In the triangle ABC

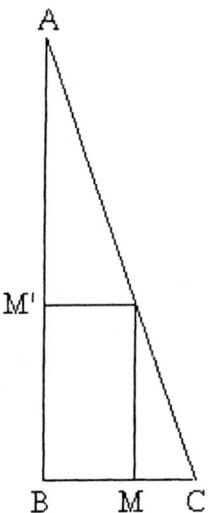

it is possible to identify on AB a point corresponding to any point of the segment BC. This does not mean that they have the same length, even though they have *same number of points*. Actually, the same may be done with another triangle having BC in common where AB has a different length.

The length has nothing to do with the number of points. It makes no sense to refer to the *number of points*, but to the *power of a set*.

A set is a <u>finite</u> or <u>infinite</u> collection of objects in which order has no significance, and multiplicity is also generally ignored (unlike a <u>list</u> or <u>multi-set</u>). Members of a set are often referred to as <u>elements</u> and the notation $a \in A$ is used to denote that a is an element of set A.

Symbols used to operate on sets include ⌐(which means "<u>and</u>" or <u>intersection</u>) and ∪(which means "<u>or</u>" or <u>union</u>). The symbol $\varnothing$ is used to denote the set containing no elements, called the *empty set*.

Let E, F and G be sets. Then the operators ∩ and ∪ fulfil the following properties:

1) Commutative $E \cap F = F \cap E$, $E \cup F = F \cup E$
2) Associative $(E \cap F) \cap G = E \cap (F \cap G)$, $(E \cup F) \cup G = E \cup (F \cup G)$
3) Distributive $(E \cap F) \cup G = (E \cup G) \cap (F \cup G)$, $(E \cup F) \cap G = (E \cap G) \cup (F \cap G)$

1.4.2   Set theory and systems

One of the first *definitions of system* was proposed by A. D. Hall and R. E. Fagen in 1956. According to them, a system is defined as "a set of objects together with relationships between the objects and between their attributes" (Hall and Fagen, 1956).

Other more structured definitions of systems based on set theory have since been introduced.

Mihajlo D. Mesarovic introduced his mathematical approach in defining systems (Mesarovic, 1968, Mesarovic, 1972). In his research two lines of development have been followed:

a)   via abstraction (Mesarovic and Takahara, 1989)

In this approach starting from well-established, specific theories (examples include automata and systems of differential equations) it is possible to embrace the common features of the initial assumptions. The main problem for Mesarovic is that "the new concepts are not general enough, so that one is bogged down by technical problems of minor conceptual importance." (Mesarovic, 1972).

b)   via formalization (Mesarovic, 1968; 1972)

The idea was to first define concepts verbally, then define them axiomatically by using the *minimal mathematical structure*.

Let us recall that the *Cartesian Product*, after René Descartes (1596-1650), of two sets A and B is a set containing all possible ordered combinations of one element from each set: A x B = {(a, b) | a in A, b in B}. This definition can be extended to products of any number of sets.

Obviously, A x B # B x A. When the elements of sets A and B are considered as points along perpendicular axes in a 2D space then the elements of the *Cartesian Product* are the *Cartesian Coordinates* of points in this 2D space.

In this approach the general system S is the starting notion, defined as: $S \subset V_1 \times ... V_n$

The components of the relation, V_i, are the objects belonging to the system. The set V_i is to be intended as the "totality of alternative ways in which the respective feature is observed or experienced." (Mesarovic, 1972). In this way a system may be defined as a set (of relations, for instance).

There are two ways of specifying systems, by distinguishing a relationship with another made on the same objects:
a) the input-output approach, also referred to as stimulus-response;
b) the goal-seeking approach, also referred to as problem solving, decision-making.

During the same period, **George Klir**, starting from a compilation of definitions of system, introduced a definition of system at five levels (Klir, 1969). First of all Klir introduces some basic definitions:
- *The ST-Structure (State transition state)*: "The complete set of states together with the complete set of transitions between the states of the system."
- *The UC-structure (Structure of the universe and couplings)*: "A set of elements together with their permanent behaviors and with a UC-characteristic".
- *The space-time resolution level* is the feature of the observations or measurements (quantities to be observed, accuracy of measurements, frequency).
- *Permanent (real) behavior*: is defined by the relationships linking given quantities at the resolution level.

The five definitions introduced are:
1. *Definition by a set of external quantities and the resolution level*: The System S is a given set of quantities regarded at a certain resolution level.
2. *Definition by a given activity*: The system S is a given ensemble of variations in time of some quantities under consideration.
3. *Definition by permanent behavior*: The system S is a given time-invariant relation between instantaneous and/or past and/or future values of

external quantities; every element of the relation may, but need not be, associated with a probability of its occurrence.

4. *Definition by real UC-structure*: The system S is a given set of elements, together with their permanent behaviors, and a set of couplings between the elements on the one hand, and between the elements and the environment on the other.

5. *Definition by real ST-Structure*: The system S is a given set of states together with a set of transitions between the states; every transition may, but need not be, associated with a probability of its occurrence.

Klir (Klir, 1969) presents a complete formalization of his approach which was very innovative for its time.

He has since developed his views of systems theory in many publications of fundamental importance for Systemics (see, for example, Klir, 2001).

Another way to define systems, based on the concept of set, was introduced by **James Grier Miller** (1916-2002). In his book *Living Systems* (Miller, 1978) he introduced the following definition: "A system is a set of interacting units with relationships among them. The word set implies that the units have some common properties." (p. 16).

He distinguished between different levels, such as conceptual systems, concrete systems, abstract systems, structure and process. In basic concepts he made reference to emergence: "I have stated that a measure of the sum of a system's units is larger than the sum of that measure of its units.... Because of this, the more complex systems at higher level manifest characteristics, more than the sum of the characteristics of the units, not observed at lower level. These characteristics have been called *emergent*" (p. 28).

This way of introducing the concept of system is based on the concept of set, but without using the formalization introduced in set theory.

The simultaneous usage of the concepts of *relationship* and *interaction,* as well as the reference to *emergence* is particularly interesting.

The concept of set has been metaphorically used in other approaches and ways of introducing systems. The metaphorical usage refers both to the common properties of the elements considered and to their grouping, considered as components having relationships and interactions between them. In this case the formalization introduced in set theory is not used to describe, nor for making inferences about, systemic properties. Examples are sets of values, laws, organizational schemas and strategies, which we assume to become systems when common features, relationships and interactions among them are also considered.

This book will mainly consider the formalizations of the concept of system based on set theory. **This is the path followed in *systems theory*, and allows the reusability of concepts, approaches and results. This reusability is not based on metaphors or analogies, but on modeling and simulating activities grounded within conceptual frameworks, in turn granting high levels of coherence, robustness and validity in specific corresponding domains.**

In any case, it will be possible to extrapolate general concepts and metaphorical knowledge suitable for approaches based on *other kinds of knowledge*, such as artistic, religious, political, managerial and so on.

It is also possible to consider concepts of systems not based on the classic concept of set even within a scientific framework. One example is the introduction of *fuzzy sets, fuzzy logic and fuzzy systems* by **Lofti Zadeh** (Zadeh *et al.*, 1996). Fuzzy set theory is a subject of current scientific discussion (see Fuzzy Sets and Systems, An International Journal in Information Science and Engineering, Official Publication of the International Fuzzy Systems Association (IFSA)). Lofti Zadeh, in 1991, also introduced the concept of *soft computing* thus highlighting the emergence of computing methodologies focused upon exploiting a tolerance for imprecision and uncertainty to achieve tractability, robustness and low solution costs (Klir and Yuan, 1995).

Another approach, introduced by the theoretical biologist **Robert Rosen** (1934-1998), should also be mentioned. He introduced the concept of *anticipatory system* (Rosen, 1985): "An anticipatory system is a system containing a predictive model of itself and/or of its environment, which allows it to change its state at an instant in accord with the model's prediction pertaining to a later instant."

Formally, an anticipatory system is a system X whose dynamical evolution is governed by the equation: $X(t + 1) = F(X(t), X^*(t + 1))$ where $X^*(t + 1)$ is X's anticipation of what its state will be at time $(t + 1)$.

In this theory Rosen uses a mathematical approach known as "category theory" (Rosen, 1978).

The relation between systems theory and category theory is based on the original definition of systems introduced by Bertalanffy: "A system is a set of units with relationships among them" (Von Bertalanffy, 1956). The notion of category is very suitable for use with this definition.

Category Theory is a mathematical domain which unifies a large part of mathematics by developing a general theory of relations and structures. Eilenberg and MacLane (Eilenberg and MacLane, 1945) introduced this approach to transform difficult problems of Topology into more accessible problems of Algebra.

Consider a graph constituted of vertices linked by arrows. In an arrow *a:* $V \rightarrow V_1$, V is the vertex representing the domain, and V_1 the target, represents the co-domain. Several arrows may have the same source, the same target, and may even be closed (when source is equal to target). A category is a graph combining different arrows in such a way that each pair of consecutive arrows, such as *a:* $V \rightarrow V_1$, *b:* $V_1 \rightarrow V_2$, is associated with a composite arrow *ab* from V to V_2 satisfying the conditions of:

- Associativity. States that paths a(bc) and (ab)c equal. There is a unique composite for a path of length n, whatever n is;
- Identity. There is a closed arrow from V to V for each vertex V. This arrow is called the *identity* of V whose composite with any arrow beginning or ending in V is this other arrow.

An arrow a is said to be an *isomorphism* if there exists another arrow b such that ab and ba are identical; b is unique and is the inverse of a.

Category theory is a stratified or hierarchical structure without limits, which makes it suitable for modeling the process of modeling itself.

The conceptual approach of *anticipatory systems* has been and is applied in different fields, such as biological, engineering, computing, cognitive, physical, economical, fuzzy systems (see the proceedings contained in the International Journal of Computing Anticipatory Systems, published by CHAOS, Center for Hyperincursion and Anticipation in Ordered Systems, Institute of Mathematics, University of Liège, Belgium).

1.4.3 Formalizing systems

The mathematical biologist **Ludwig von Bertalanffy** (1901 – 1972), considered the father of *General Systems Theory*, described a system S, characterized by suitable state variables Q_1 , Q_2 , . . . , Q_n ,, whose instantaneous values specify the state of the system. In most cases the time evolution of the state variables is ruled by a system of *ordinary differential equations*, such as:

$$\begin{cases} \dfrac{dQ_1}{dt} = f_1(Q_1,Q_2,...,Q_n) \\ \dfrac{dQ_2}{dt} = f_2(Q_1,Q_2,...,Q_n) \\ .... \\ \dfrac{dQ_n}{dt} = f_n(Q_1,Q_2,...,Q_n) \end{cases} \qquad (1.3)$$

The system (1.3) specifies how the change in value of a given state variable, affects all other state variables. This is **interaction**. An interaction between elements is said to take place when *the behavior of one affects the behavior of another*. For instance, the behaviors of the elements belonging to a given system may be mutually affected by physical interactions due to collisions, magnetic fields and energy exchanges. When the interacting elements are autonomous agents, assumed to have a cognitive system, interactions can take place through *information processing*, in order to *decide* which behavior to follow.

In the case of a system characterized by only one state variable, the previous system becomes a single differential equation:

$$\frac{dQ}{dt} = f(Q) \tag{1.4}$$

It should be noted that, in general, interactions can take place only among elements possessing suitable *properties*. For instance, in order to have magnetic interactions among elements they should have the property of being able to act as magnetic poles.

When interactions are based on *cognitive processing* and *information exchange*, single elements are assumed to have *cognitive systems* and be able to interact using specific *cognitive models*.

Before proceeding some terminological definitions will be made, starting from the frequently used word **Reductionism**, an approach which attempts to describe every entity as a set, or a combination, of other simpler, smaller, basic component entities. Within this approach each feature of a given entity is considered as *reducible* to the features of its component parts. As will be seen below, this approach can be classified as anti-systemic, as it excludes the concept of *emergence* (see Appendix 1).

Modern reductionism, however, is no longer based on the naïve assumption that the whole is the sum of its parts and that the whole can be understood by studying only its components. Modern reductionism, (see, for instance, Crick, 1994), states that " … while the whole may not be the simple sum of its separate parts, its behavior can, at least in principle, be *understood* from the nature and behavior of its parts plus the knowledge of how all these parts interact".

Another important term is **emergence**, which roughly denotes the occurrence of properties of the whole which are not deducible from those of its components. *A system is considered to be such when, acting as a whole, it produces effects that its parts alone cannot.*

It follows that the occurrence of an interaction among components is a **necessary condition** in order to describe a system. *Many* different interactions may occur simultaneously within a given system. In some cases the interactions may influence every component, whereas in other cases they may affect only certain specific kinds of components. In the latter case, the system of differential equations (1.3) should be rewritten by taking into account that each variable is affected only by other specific variables. The usefulness of this remark will appear later, when the concept of **multiple-systems** is introduced, in which different interactions affect different kinds of interacting components (see Section 3.4).

Moreover, the occurrence of interactions among components is not a **sufficient condition** for the appearance (or *emergence*, as we shall see later) of a system. A sufficient condition, on the other hand, is when the interactions among components give rise to suitable cooperative effects endowing the system with properties that the individual components do not possess.

In general cooperative effects can occur when different components:
- play different roles (e.g., within electronic devices, a living body, a company), or
- perform the same tasks in certain specific ways, depending upon the phenomenon being considered (typical cases are flocks, traffic, laser light).

In both cases we observe the appearance of a new entity possessing specific *emergent* properties. Here, *interaction* is the key word. The concept of system is based upon the assumption that, during interaction, components have:
- distinguishable roles and functions, even though endowed with the same characteristics, as in an anthill,
- cooperative, synergic behaviors, as in ecosystems.

A system may in turn interact with other systems contributing to the appearance of another higher-order system. In this case the systems are considered as **subsystems**, like departments in a company or organs in a living system.

One further observation: Systemics develops within a cultural context (as mentioned below) which takes into account the key role played in the modeling activity by observation, in addition to the observer's classical role of relativity and noise generator. When it is stated that the appearance (emergence) of a system requires an observer, the reference is not only to a device able to physically *detect* the appearance itself. It also refers to the need for a *cognitive model* suited to *acknowledge* rather than *detect* the appearance, i.e., for processing the related information. Lack of knowledge, of attention, of interest may make the agent completely insensitive to the presence of a system: in this case there is no observer even though the phenomenon physically exists.

1.4.4 Formalizing Systemics

In this paragraph we will introduce some approaches to the problem of formalizing a framework for Systemics, in order to develop models of systems suitable for simulations and for understanding their theoretical features. In this context we will introduce several models which have played a crucial role in the conceptual evolution of systems thinking

The cultural and scientific basis of Systemics came from studies in various disciplinary domains, as mentioned in 1.3. The unifying concepts of Systemics introduced by Ludwig von Bertalanffy allowed recognition and conceptual organization of the systemic meaning of the different studies and approaches (see Appendix 1).

The most important of these, for the formalization of Systemics and associated modeling activity, came from disciplines including physics, engineering, mathematics and biology.

Classical **thermodynamics**, for instance, implicitly uses a systemic view of physical problems. This discipline deals with the study of the exchange of energy between a system and its external environment and, in particular, most attention was attracted to the transformation of heat into work within machines.

The more advanced topics of thermodynamics refer to subjects such as: equations of state, phase equilibrium and stability, molecular simulations, statistical mechanics, statistical thermodynamics, stability in multi-component systems, energy and energy analysis of open systems, entropy, irreversibility, general thermodynamic relationships, external-field effects, low temperature thermodynamics, irreversible thermodynamics, energy conversion.

The problems dealt with by thermodynamics are inherently *systemic*, such as that which constitutes the subject of Statistical Thermodynamics: the relationship between macroscopic thermodynamic variables and the microscopic variables describing the dynamics of the individual components of a macroscopic system such as, for instance, the molecules of a gas. Statistical Thermodynamics deals with this problem by considering only systems consisting of very large numbers of particles and neglecting the details of their individual behaviors.

The generalizations of Thermodynamics take place conceptually when trying to apply its concepts and laws to generic systems, for instance when considering that the problem of energy exchange with the environment may be studied in the same way within different contexts such as a star, a living system, a machine. Classical thermodynamics is based upon two laws:

a) First law: "It is impossible to design a machine that operates in a cyclical manner and performs work without the input of energy to the machine itself" (impossibility of perpetual motion) or "Energy can neither be created nor destroyed but it can be converted from one form to another." (Energy conservation);

b) Second law: "It is not possible to design a thermal machine integrally transforming heat into work"

There have been different formulations of the second law, a circumstance which underlines its great physical importance. For instance: "Elements in a closed system tend to seek their most probable distribution", or "in a closed system entropy always increases."

Heinz von Foerster (1911-2002) collected the following reformulations:

1. Clausius (1822-1888): It is impossible that, at the end of a cycle of changes, heat has been transferred from a colder to a hotter body without at the same time converting a certain amount of work into heat.

2. Lord Kelvin (1824-1907): In a cycle of processes, it is impossible to transfer heat from a heat reservoir and convert it all into work, without at the same time transferring a certain amount of heat from a hotter to a colder body.

3. Ludwig Boltzmann (1844-1906): For an adiabatically enclosed system, the entropy can never decrease. Therefore, a high level of organization is very improbable.

4. Max Planck (1858-1947): A perpetual motion machine of the second kind[2] is impossible.

5. Caratheodory (1885-1955): Arbitrarily near to any given state there exist states which cannot be reached by means of adiabatic processes.

The great importance of the second law is related to its implications on *irreversibility*. William Thomson (Lord Kelvin) introduced the second law in 1852, in a publication on the journal *Philosophical Magazine* entitled "On the universal tendency in nature to the dissipation of mechanical energy".

Ilya Prigogine (1917-2003) introduced a very innovative view of thermodynamics. In his book "From being to becoming" (Prigogine, 1981)

---

[2] A perpetual motion machine of the *first order* is one which produces power without energy uptake. A perpetual machine of the *second order* is one undergoing a cyclic process that does nothing but convert energy into work (i.e. without *any* dissipation or *other* use of the energy).

he states that the second law seems more a program than a principle. This is the main reason why thermodynamics focused upon the problem of equilibrium. In advanced thermodynamics he introduced a total change of perspective by looking at irreversible processes in a new way, particularly for systems far from equilibrium. Irreversible processes were considered for their constructive aspects more than for their usual degenerative ones. He introduced the very powerful notion of *dissipative structures* having profound consequences for our understanding of biological systems (see Appendix 1).

Thermodynamics has been studied by contemporary physics with reference to new problems such as the relationships between order and disorder, reversibility and irreversibility, static and dynamic stability, the role of bifurcation, symmetry-breaking, self-organization. In this way scientists deal with new theoretical problems within different frameworks, and this domain is usually referred to as the study of **complexity,** as introduced in Section 1.3. The term "complexity" does not denote the topic of a single discipline, but rather a category of conceptual problems. In other words, it refers to the need to use conceptual approaches and theoretical tools having in common the ability to use and deal with different levels of description, representation, modeling using different computational approaches and simulation tools, taking into account different observational scales, the observer being an integral part of the problem.

In short, generally, the common theoretical feature of complexity studies is the need to deal with **emergence**, a notion introduced above which will be discussed in detail throughout this book as the key theoretical concept.

An example for illustrating the switch from classical physics to the physics of complexity is the **Three Body Problem**, i.e., the problem of computing the orbits resulting from the mutual gravitational interaction among three separate masses. This problem is surprisingly difficult to solve, even in the case of the so-called *Restricted Three-Body Problem*, corresponding to the simple case of the three masses moving in a common plane. It is an example of the lack of adequacy of the classical mathematical models used in science to deal with an aspect of complexity, i.e., many body interactions. The strategy of looking for a deeper, more complete knowledge of the single components was absolutely inadequate for dealing with the behavior of many body systems.

The study of dynamical systems described by evolution equations is the subject of *dynamical systems theory*, dealing mostly with non-linear and chaotic systems. The problem of carrying out theoretical analyses on the structural properties of a system independently from its practical implementation, was dealt with by resorting to different approaches, all based on suitable mathematical descriptions. Mathematical modeling allows the exploration of the structural features of the asymptotic state of a system on the basis of the description of the structural properties of its evolution, independently from particular interpretations of the state variables.

One of the first domains of application of dynamical systems theory was the mathematical description of **Population Growth**.

In 1798 **Thomas R. Malthus** proposed in his essay, "An Essay on the Principle of Population" a simple mathematical model of population growth for modeling the time evolution of biological populations.

Malthus observed that, without environmental or social constraints, human populations doubled every twenty-five years, regardless of the initial population size. The principle was that each population increased by a fixed proportion over a given period of time and that, in the absence of constraints, this proportion was not affected by the size of the population.

In the Malthusian model of population growth the only hypothesis is therefore that a population grows at a rate proportional to itself. The model equation is

$$dP/dt = k\,P \qquad\qquad\qquad (1.5)$$

where P (which is a function of time, t) stands for the population density at time t and k is a proportionality factor.

Malthus's model is an example of a model with one *variable* and one *parameter*.

One of the first models of population growth which took into account some social constraints was that describing the interactions between predator and prey. It was proposed in the Twenties by the US biophysicist Alfred Lotka and the Italian mathematician Vito Volterra. To be precise, the Lotka-Volterra (LV) model was originally introduced in 1920 by A. Lotka (Lotka, 1920) as a model for oscillating chemical reactions. It was only later applied by V. Volterra (Volterra, 1926) to predator-prey interactions. Like the Malthusian model, the **Lotka-Volterra** model is based on *differential equations*.

The Lotka-Volterra model describes interactions between two species in an ecosystem, one acting as a predator and the other as a prey. Since it deals with two species, the model involves two equations, one describing the time evolution of the density of individuals belonging to the prey population, and the other describing the time evolution of the density of predators. The explicit form of the model equations is:

$$dx/dt = ax - cxy$$

$$dy/dt = -bx + cxy \qquad (1.6)$$

where x is the density of prey individuals, y the density of predators, a is the intrinsic rate of prey population increase, b the predator mortality rate, and c denotes both predation rate coefficient and the reproduction rate of predators per prey eaten.

The model of Lotka and Volterra is not very realistic. It does not consider any competition among prey or predators. As a result, prey population, in the absence of predators, may grow infinitely without any resource limits. Predators have no saturation: and their consumption rate is unlimited. The rate of prey consumption is proportional to prey density. Thus, it is not surprising that model behavior is unnatural, showing no asymptotic stability. However, there exist many variants of this model which make it more realistic.

As the model lacks asymptotic stability, it does not converge towards an attractor (i.e., it does not "forget" initial conditions) and its solutions consist of periodic behaviors, whose amplitude depends on the initial conditions. An example is given in Figure 1.3.

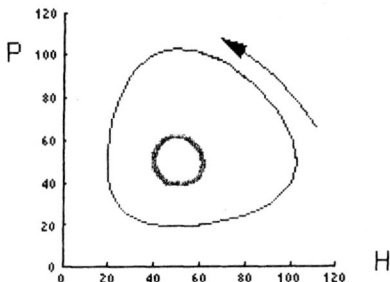

Figure 1-3. Periodical solutions of the Lotka-Volterra equations.

The LV model has become a classic example of a nonlinear dynamical system (Hernandez-Bermejo and Fairén, 1997; Minorsky, 1962; Verhulst, 1990) and has been applied to very different kinds of problems, including population biology (Tainaka, 1989), epidemiology (Roussel, 1997), neural networks (Noonburg, 1989), and chemical kinetics (Noszticzius *et al.*, 1983).

Another, earlier model was introduced by the Belgian mathematician **P. Verhulst** (1804-1849) with reference to the study of population growth within the context of limited resources. The mathematical solution of this model is given by the well known *logistic curve* (Figure 1.3).

To illustrate the Verhulst model, we start from the definition of system introduced in Section 1.2.1 and refer to the particular case of a single state variable, described by equation (1.4). Introducing a Taylor series development of the right-hand member of this equation around $Q = 0$ (assumed to be an equilibrium point of this equation), we obtain:

$$\frac{dQ}{dt} = a_1 \cdot Q + a_{11} \cdot Q^2 + \dots \tag{1.7}$$

Keeping only the first order term $a_1 Q_1$, system growth will be directly proportional to the actual value of Q and the growth will be exponential, the dynamical evolution equation having the form

$$\frac{dQ}{dt} = a_1 \cdot Q \tag{1.8}$$

In this case the solution will take the form $Q = Q_0 e^{a_1 t}$, where Q_0 is the number of elements at time $t = 0$. This is the *exponential law* used in many contexts. Now, if terms up to the second order in Eqn. 1.7 are kept we have:

$$\frac{dQ}{dt} = a_1 Q + a_{11} Q^2 \tag{1.9}$$

A solution of (1. 9) is $Q = \dfrac{a_1 C e^{a_1 t}}{1 - a_{11} C e^{a_1 t}}$, the equation of the so-called

logistic curve (see Figure 1.4).

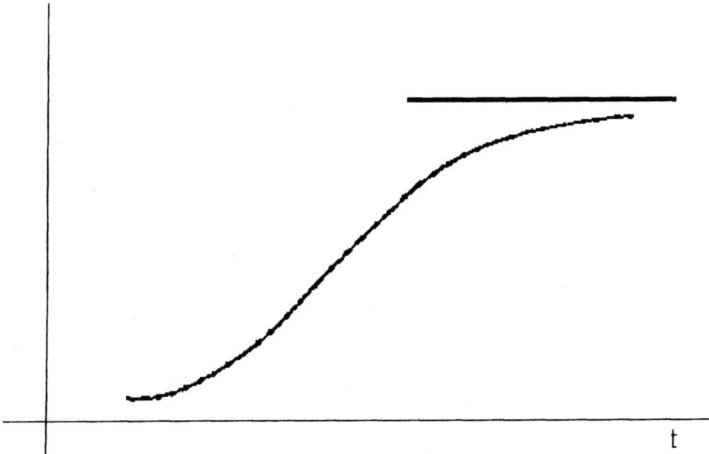

Figure 1-4. An example of a logistic curve

A different class of models was introduced by **Balthasar Van der Pol** (1889–1959), who introduced the concept of *limit cycle*. The latter term denotes a self-sustained stable periodic oscillation which occurs, after a certain transient behavior which, together with frequency and amplitude, are independent of the initial conditions themselves.

Van der Pol was the initiator of modern experimental laboratory dynamics during the 1920s and 1930s. He studied electrical circuits employing valves and found that the behavior of the state variables characterizing them showed stable oscillations, corresponding to the *limit cycles* mentioned above.

In 1927 Van der Pol and his colleague Van der Mark published a paper (Van der Pol and Van der Mark, 1927) in which they reported that in electronic valve circuits an "irregular noise" was observed at certain driving frequencies, differing from the natural entrainment frequencies. We now know that they had experimentally discovered *deterministic chaos*.

The mathematical model introduced by Van der Pol regarded electronic circuits, but it is also very suitable for describing general features of some kinds of nonlinear dynamical systems. Namely, the very nature of the solutions of the equations describing the model can change as a function of the value of a parameter contained within those equations. As a matter of fact, the appearance or not of a limit cycle is determined by whether these values fall within a given interval. This aspect was absent from previous models, such as those of Lotka-Volterra and of Verhulst.

Behavior of this kind, in which a structural change of behavior sets in only when the value of a parameter crosses a given *critical point*, occurs very often in physical, biological, and socio-cognitive phenomena. Let us consider, for example, *fluid mechanics* systems. Generally, a fluid is *viscous*. This means that a source of energy is necessary to sustain its motion. Consider a layer of fluid heated from below. In this case the fluid at the bottom becomes warmer, less dense and tends to move upwards. When a sufficiently large temperature difference between the bottom and the top layer occurs, *convection rolls* develop. If the temperature difference is high enough to allow convection rolls, the velocities are time-independent. This fluid motion is called a *steady flow*. In general, a very large number of variables (as large as the number of its constituent molecules) is necessary to describe the state of the system, but a steady flow may be represented as a *point* in this large, ideally infinite-dimensional phase space. When the temperature difference is increased, periodic time-dependent patterns appear. On further increasing the temperature difference, intricate time-dependent motion emerges. This sequence of changes is called the *onset of turbulence*.

In the case discussed above, the difference of temperature plays the role of a *control parameter* and the change of its values can give rise to profound changes in the nature of the fluid behavior. However, an understanding of the effect of changes in control parameter values is difficult in complicated mathematical models, such as that describing fluid motion. The importance of the Van der Pol model stems from the fact that it happens to be one of the simplest dynamical system models in which the role of a control parameter value can be fully understood. The explicit form of the Van der Pol model is described by the following pair of first-order equations:

$$dx/dt = y + ax^3 - x$$

$$dy/dt = -x \qquad\qquad\qquad\qquad (1.10)$$

where *a* is the control parameter. When *a* is negative, the system is "trapped" in an attractive fixed point. When *a* is positive but not too large a *limit cycle appears*. When the value of *a* crosses the zero value, from negative becoming positive, the structural change of system behavior (a limit cycle suddenly appears) is called *Hopf bifurcation*.

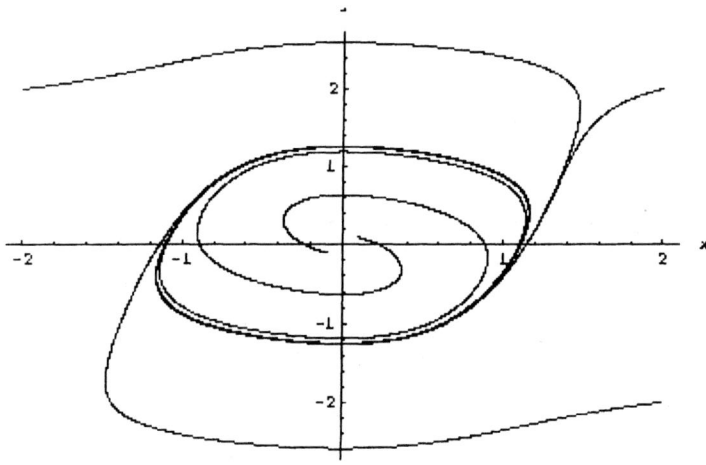

Figure 1-5. Some phase curves for the Van der Pol equation with $a=1$.

reactants, A and B, leading to products C and D, formed through unstable Figure 1.5 shows graphically that the Van der Pol differential equations exhibit a limit cycle, i.e., there is a closed curve in the phase plane corresponding to a periodic behavior.

Another interesting class of models are those formulated in terms of partial differential equations. In this case, besides the usual time independent variable, there are other independent variables, such as the spatial positions. Generally systems of this kind allow for an infinite number of solutions, whose general form cannot be discovered through the introduction of suitable parameters. Namely, the difference between two solutions of a partial differential equation is given, in general, by a completely arbitrary function. However, even within these models it is possible to have locally stable solutions, attracting the behavior of the entire system. This circumstance allows the description of self-organization phenomena. The simplest and most celebrated of such models is the so-called **brusselator** (from its origin in Ilya Prigogine's group in Brussels) introduced by Prigogine and Lefever in the Seventies to model tri-molecular reactions (*cfr* Nicolis and Prigogine, 1977; Babloyantz, 1986).

This model was found to be very useful for describing the most-studied self-organizing chemical reaction-diffusion system, known as the *Belousov-Zhabotinsky reaction*, discovered by the Russian chemist Belousov in 1958 and later studied by Zhabotinsky and many others (see Appendix 1). The Brusselator models the reaction between two intermediates X and Y. The model equations are:

$$d\phi/dt = D_1 \Delta_2 \phi + A - (B + 1)\, \phi + \phi^2\, \psi$$

$$d\psi/dt = D_2 \Delta_2\, \psi + B\phi - \phi^2\, \psi \qquad\qquad\qquad (1.11)$$

where:
- ϕ and ψ are to be interpreted as concentrations of appropriate chemical substances;

- D_1, D_2, A, B are control parameters of the model;
- Δ_2 is the *Laplace operator* which, in the case of three spatial dimensions has the explicit form $\Delta_2 f = d^2 f/dx^2 + d^2 f/dy^2 + d^2 f/dz^2$.

The *Laplace equation* $\Delta_2 f = 0$ was introduced by the French mathematician and astronomer Pierre-Simon Laplace (1749-1827); the operator appearing within it was named the *Laplace operator* by James Clerk Maxwell in his *Treatise on Electricity and Magnetism*, published in Oxford in 1873. The solutions of the Laplace equation are called *harmonic functions* and have wide applications in physics in the description of *gravitational, electrostatic* and *magnetic potentials*.

Since the reactants (A and B) are maintained at constant concentration, the two dependent variables are the concentrations of X and Y. The behaviors of some numerical solutions of this system of equations are plotted in Figure 1.6.

The interesting feature is that, whatever the initial concentrations of X and Y, the system settles down into the same periodic variation of concentrations. The common trajectory is a limit cycle, and its period depends on the values of the rate coefficients.

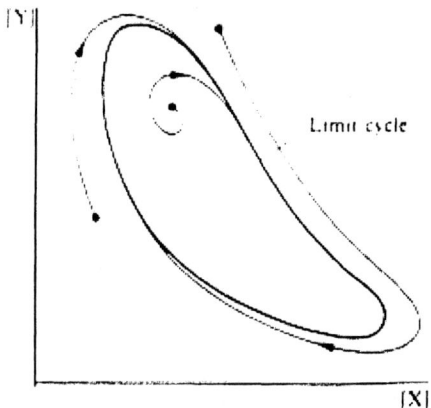

Figure 1-6. Some oscillating reactions approach a closed trajectory whatever their starting
conditions.

The Brusselator has been used to study a range of phenomena. One, relevant for Systemics, is the formation of so-called *dissipative structures*. The term was introduced by Prigogine (Prigogine, 1978; Nicolis and Prigogine, 1977) to denote self-organizing structures in non-linear systems far from equilibrium (*i.e., whirlpools* existing for as long as they are continuously fed by a running fluid).

Another model based on a system of ordinary differential equations, the so-called *Lorenz equations*, is of special interest to scientists studying the occurrence of deterministic chaos (the latter can be identified with an apparently random motion stemming from deterministic equations). The Lorenz equations were discovered by **Ed Lorenz** in 1963 (Lorenz, 1963) as a very simplified model of convection rolls in the upper atmosphere. Later these same equations appeared in studies of lasers, bacteria, and in a simple chaotic waterwheel (Sparrow, 1982; Zwillinger, 1997).

The Lorenz equations are:

$$x = \sigma(y - x)$$

$$y = rx - y - xz$$

$$z = xy - bz \tag{1.12}$$

where r, b, and σ are control parameters.

The Lorenz *butterfly* attractor refers to the so-called "Butterfly Effect", or more technically the "sensitive dependence on initial conditions", which is the essence of deterministic chaos. In simple terms, this means that, in a situation where deterministic chaos occurs, the difference, at any point in time, between two behaviors, associated with two different initial conditions, grows exponentially with time, *however small* the initial value of that difference. Therefore, if we perturb, even very slightly, a chaotic deterministic behavior, the effect of this perturbation, instead of fading away, will grow in a disordinate manner, thus reaching macroscopic levels. In other words, when a butterfly beats its wings in one part of the globe, this circumstance alone could ultimately give rise to a hurricane in another.

The Lorenz attractor is a solution to the Lorenz equations displaying some rather remarkable behavior and represents one of the landmarks in the field of Chaos. These equations were originarily designed to describe the 2D flow of a fluid in a simple rectangular box heated from below. This simple model was intended to simulate medium-scale atmospheric convection.

By plotting the behavior of its numerical solution three-dimensionally we obtain, instead of a simple geometric structure or even a complex curve, a structure which weaves in and out of itself.

Projected onto the X-Z plane (Figure 1.7), the attractor looks like a butterfly; on the Y-Z plane, it resembles an owl mask. The X-Y projection is often useful for glimpsing the three-dimensionality of the attractor.

As the Lorenz Attractor is plotted, a strand will be drawn from one point, and will start weaving the outline of the right butterfly wing. Then it swirls over to the left wing and draws its centre. The attractor will continue weaving back and forth between the two wings, its motion seemingly random, its very action mirroring the chaos that drives the process.

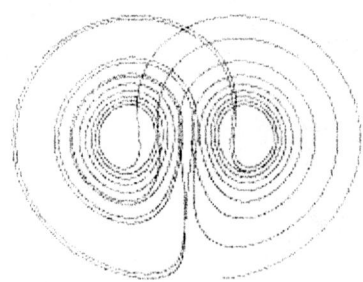

Figure 1-7. The Lorenz attractor projected onto the XZ plane.

A new approach in studying complex systems was introduced by **Hermann Haken** (Haken, 1983a; 1983b; 1987). This approach, named

Synergetics, focuses on the emergence of order from chaos. Synergetics is an interdisciplinary field of research founded by Hermann Haken in 1969 (Graham and Haken, 1969). Synergetics deals with complex systems where interacting components give rise to self-organized functional, spatial or temporal, structures. It searches for general principles of self-organization in different disciplinary fields, such as physics (lasers, fluids, plasmas), meteorology, chemistry (pattern formation by chemical reactions), biology (morphogenesis, evolution theory), economics (financial markets), brain activities, computer sciences, sociology (e.g. urban growth).

Synergetics deals with systems in which *cooperation* among subsystems creates organized structures on macroscopic scales (Haken 1993). Problems dealt with by Synergetics are, for instance, the study of *phase transitions,* (see Chapter 3 and Haken, 1983), of convective instabilities, of bifurcations, of coherent oscillations in lasers, of nonlinear oscillations in electrical circuits, of social and economic behaviors, of population dynamics, etc.

One of the many important new ideas introduced by Synergetics is related to the concept of *order parameter.* When complex systems undergo phase transitions, a special type of ordering occurs at the microscopic level. Instead of addressing each of very large number of atoms of a complex system, Haken (1988) has shown, mathematically, that it is possible to address their fundamental *modes* by means of *order parameters.* The very important mathematical result obtained using this approach consists in drastically lowering the number of degrees of freedom to only a few parameters. Haken also showed how *order parameters* guide complex processes in self-organizing systems.

When an *order parameter* guides a process, it is said to *slave* the other parameters, and this slaving principle is the key to understanding self-organizing systems. Complex systems organize and generate themselves at far-from-equilibrium conditions:

> *"In general just a few collective modes become unstable and serve as 'order parameters' which describe the macroscopic pattern. At the same time the macroscopic variables, i.e., the order parameters, govern the behavior of the microscopic parts by the 'slaving principle'. In this way, the occurrence of order parameters and their ability to enslave allows the system to find its own structure".* (Graham and Haken, 1969, p. 13).

"In general, the behavior of the total system is governed by only a few order parameters that prescribe the newly evolving order of the system" (Haken, 1987), p. 425.

1.5   Sets, structured sets, systems and subsystems

In this Section several practical examples are presented with reference to what has been discussed above.

a) Sets

As introduced above, this concept refers to collections of elements having a common property, defined through the rule of membership.

It is, for example, possible to consider the set of *all* electronic components as well as the set of electronic components within a box. In the latter case the rule of membership makes reference to more then one property: being an electronic component *and* being contained within a box (tested components, components selected by the designer for a project, etc.).

b) Structured sets

The next step is based on the consideration of elements having a *suitable* (for an observer) structure, an organization. In this way order, synchronization, hierarchy may be established. The single elements have relationships amongst them: one may come before another, one be heavier than another, one may have a particular position in a configuration, etc. Relationships amongst elements allow actions upon them to be more effective, by amplifying and multiplying their properties, as may occur, for example, when we scan a structured database in order to retrieve a particular piece of information.

Regarding the example of the electronic components which are, in this case, assembled, structured on a circuit board: they are structured, arranged in a given configuration.

We have moved from the concept of *set* to the concept of *structured set*.

c) Systems and subsystems

Another conceptual step takes place when it is not so much the structure of relationships among elements which characterizes the new entity, but rather the *interactions* among them.

In this case the circuit board is supplied with power and the components interact.

From the interacting elements a new entity having its own characteristics, non-deducible from those of component elements, *may* **emerge**: a system.

Due to interactions, sets of elements become *something else* both with reference to the elements themselves and with reference to the structured set. Most of the situations we are dealing with relate to **emergent** phenomena, i.e. systems created by interacting elements through cooperative, synergistic effects and based upon the creative role of the observer who recognizes, realizes this process of emergence by using a suitable *cognitive model*. This is the case of electronic devices considered to be *working* if the observer has a suitable cognitive model and expectancies. A formal definition of emergent properties was introduced by Baas (Baas, 1994), as reported in Chapter 3. If we denote with:

- S^1, a set of interacting elements having observable properties at the level of single elements Obs1 (S^1), and
- S^2, a second order structure, which is the result R of applying interactions Int^1 to the elements of S^1, whose observable properties are Obs^1 (S^1) : S^2 = R (S^1, Obs^1 (S^1) , Int^1),

then a property P of S^2 is *emergent* if and only if it is observable at the S^2 level but not at a lower level, i.e. at the S^1 level.

It is possible to identify **subsystems**. These are intended as systems having roles and functions within an overall system (e.g., a company).

The observer identifies subsystems. In any case, it is possible to consider them with their identity and boundaries independently (*autopoietic* systems are an example, see Appendix 1). Different subsystems may be superimposed with respect to some variables giving rise to structures showing interdependence, e.g., among economy, education and government, intended as three subsystems of society.

Some examples are introduced in Table 1.1. See Chapter 3 for more theoretical discussions related to *emergence*.

Sets	Structured Sets	Systems	Subsystems
Electronic components	Electronic components structured on a board, following a circuit design	An electronic circuit board with components which interact when power is supplied.	Group of components classed by function such as amplifiers, regulators, decoders.
Students	Students grouped by sex or in alphabetical order	School	Classrooms
Cells	Cells by type	Living being	Organ
Words	Words in alphabetical order or in a syntactical structure	A book, a story, a poem	Verses, Stanzas, Chapters
Musicians	Musicians grouped by their language, sex or age	Orchestra	Musicians playing same kind of musical instrument
Soldiers	Ordered by age	Army	Division
Workers	Ordered by activity	Corporation	Department
Animals	Animals by type, ordered by age, grouped by color, staple food, etc.	Herds, swarms, flocks, packs	Single animals considered *while* part of a system, such as parents

Table 1-1. Examples of sets, structured sets, systems, subsystems.

Systems, resulting from processes of emergence (Chapter 3), are not just extensions and amplifications of the characteristics of their elements. Systems have their own identity and peculiarities, and have to be specifically understood and managed. When a device is said to be **working** it shows this kind of transformation: a working device is no longer the same as an inactive one. Examples are: radio, TV, HI-FI systems, cars and computers.

We have moved from the concept of *structured set* to the concept of *system*.

This transformation presents many interesting aspects which must be taken into account when managing systems. Depending on the level of description it may be:

a)   predictable or unpredictable;

b)   reproducible or irreproducible;

c)   reversible or irreversible;

d)   stable or unstable.

The transformation *from a set to a system* is not *equivalent* to the process of *only* setting new configurations of elements or new relationships between them. As described above, it is a process of *emergence*. When considering the patterns in which atoms are arranged in a solid we see that they depend on parameters such as chemical composition, temperature and magnetic field. A **phase transition** is a change in the arrangement of atoms. In Physics a *phase transition* is the transformation of a system from one *phase* to another. A phase is a set of states having uniform physical properties such as liquids, solids and gases. In Systemics phase transitions are considered as occurring not only in physical systems, but also in other contexts such as learning or economic processes.

It is also possible to consider processes of **degeneration:**
- from systems to sets, when for any reason the interactions between elements are not active anymore;
- from structured sets to sets, when for any reason the elements no longer interact by following a *suitable* (for an observer) structure;
- from systems to structured sets, when the interactions, making *emergent* a system because of *cooperative, synergic* effects between components, cease to be of cooperative nature.

In Chapter 3, dedicated to *emergence,* emphasis will be placed upon different conceptual levels of occurrence of systemic properties: from systems to multiple-systems, to Collective beings.

1.6   Other approaches

Different disciplinary approaches to Systemics have already been introduced in Section 1.3. Several others (ordered by year of the cited publication), are listed below.

- **Max Wertheimer**, **Wolfgang Köhler** and **Kurt Koffka**, who founded Gestalt Psychology at the beginning of the Twentieth century. **Kurt Koffka** wrote a celebrated book about the principles of Gestalt theory (Koffka, 1935);
- **Kenneth E. Boulding** in *economics* and *management* (Boulding, 1956);
- **Magoroh Maruyama** for the *second cybernetics* (Maruyama, 1963);
- **C. West Churchman** founding father of the *systems approach* (Churchman, 1971; 1979) as well as of the fields of *operations research* (Churchman, 1961) and *management science* (Churchman and Verhulst, 1960);
- **Gregory Bateson** in *anthropology* (Bateson, 1972);

- **Aurelio Peccei and Alexander King** founders in the late 1960s of the *Club of Rome* introducing the need to study possible alternative to growth for global evolution of the world (Meadows *et al.*, 1972; 1993);
- **Russell L. Ackoff** in *social systems* (Russell, 1974);
- **Humberto Maturana** (1929-) and **Francisco Varela** (1946-2001) introduced in the 1970s the concept of autopoiesis (the process whereby an organization produces itself) later discussed in the book (Varela *et al.*,1974; Maturana and Varela, 1980);
- **John P. Van Gigch**, in *systems applications* (Van Gigch, 1978), *epistemology* (Van Gigch, 2003);
- **Peter Checkland** in *systems practice* (Checkland, 1981);
- **Stafford Beer** in management (Beer, 1972);
- **Peter M. Senge** in *system dynamics* (Senge, 1990) and *learning organizations* (Senge *et al.*, 1994);
- **B. H. Banathy** in *education* (Banathy, 1991) and *social systems design* (Banathy 1996; 2000);
- **Ian I. Mitroff** in *management* (Mitroff and Linstone, 1993);
- **Michael C. Jackson** in *management science* (Jackson, 2000).

References

Ashby, R., 1956, *An Introduction to Cybernetics.* Wiley, New York

Baas, N. A., 1994, Emergence, hierarchies and hyperstructures. In: *Alife III, Santa Fe Studies in the Science of Complexity*, Proc. Volume XVII, (C. G. Langton, ed.), Addison-Wesley, Redwood City, pp. 515-537.

Babloyants, A., 1986, *Molecules, Dynamics & Life: An Introduction to Self-Organization of Matter.* Wiley, New York.

Banathy, B. H., 1991, *Systems Design of Education. A Journey to Create the Future.* Educational Technology Publications, Englewood Cliffs, NJ.

Banathy, B. H., 1996, *Designing Social Systems in a Changing World.* Kluwer, New York.

Banathy, B. H., 2000, *Guided Evolution in Society: A Systems View.* Kluwer, Dordrecht.

Barrow-Green, J., 1997, *Poincaré and the Three Body Problem.* American Mathematical Society,. Providence, RI..

Bateson, G., 1972, *Steps to an Ecology of Mind.* Chandler Publishing Company, San Francisco.

Beer, S., 1972, *Brain for a firm.* Wiley, Chichester, UK.

Beer, S., 1994, *Beyond Dispute.* Wiley, New York.

Bohm, D., 1992, *Thought as a System.* Routledge, London.

Boulding, K., 1956, General Systems Theory: The skeleton of science, *Management Science* 2:197-208.

Boulding, K., 1985, *The World as a Total System.* Sage Publications, Thousand Oaks, CA.

Briggs, J. P., and Peat, F., D., 1984, *Looking Glass Universe: The emerging science of wholeness.* Simon & Schuster, New York.

Cantor, G., 1884, *Über unendliche, lineare Punktmannigfaltigkeiten, Arbeiten zur Mengenlehre aus dem Jahren 1872-1884.* Teubner, Leipzig, Germany.

Capra, F., 1996, *The Web of life.* Anchor/Doubleday, New York.

Checkland, P., 1981, *Systems Thinking, Systems Practice.* Wiley, New York.

Checkland, P., and Scholes, J., 1990, *Soft Systems Methodology in Action.* Wiley, New York.

Churchman, C. W., 1961, Decision and value theory. In: *Progress in Operations Research* (R. L. Ackoff, ed.), Wiley, New York, Vol. 1, 34-64.

Churchman, C. W., 1968, *The Systems Approach.* Delacorte Press, New York.

Churchman, C. W., 1971, *The Design of Inquiring Systems: Basic Concepts of Systems and Organization.* Basic Books, New York.

Churchman, C. W., 1979, *The Systems Approach and Its Enemies.* Basic Books, New York.

Churchman, C. W., and Verhulst, M., (eds.), 1960, *Management Sciences.* Pergamon, New York.

Crick, F., 1994, *The astonishing hypothesis: the scientific search for the soul.* Scribner, New York.

Eilenberg, S., and Mac Lane, S., 1945, General theory of natural equivalences, *Transactions of the American Mathematical Society* **58**:231-294.

Emery, F. E., (ed.), 1969, *Systems Thinking: Selected Readings.* Penguin, New York.

Flood, R. L., and Carson E., 1988, *Dealing with Complexity. An Introduction to the Theory and Application of Systems Science.* Kluwer, New York.

Flood, R. L., and Jackson, M. C., (eds.), 1991, *Critical Systems Thinking: Directed readings.* Wiley, Chichester, UK.

Forrester, J. W., 1961, *Industrial Dynamics.* MIT Press, Cambridge, MA.

Forrester, J. W., 1968, *Principles of Systems.* Wright-Allen Press, Cambridge, MA.

Gibbons, M., Limoges, C., Nowotny, H., Schwartzman, S., Scott, P., and Trow, M.,1994, *The New Production of Knowledge - The Dynamics of Science and Research in Contemporary Societies.* Sage, London.

Graham, R., and Haken, H., 1969, Analysis of quantum field statistics in laser media by means of functional stochastic equations, *Physics Letters* A **29**:530-531.

Haken, H., 1981, *Erfolgsgeheimnisse der Natur.* Deutsche Verlags, Stuttgart.

Haken, H., 1983a, *Synergetics, an Introduction: Nonequilibrium Phase Transitions and Self-Organization in Physics, Chemistry, and Biology.* Springer, New York.

Haken, H., 1983b, *Advanced Synergetics.* Springer, Berlin-Heidelberg-New York.

Haken, H., 1987, Synergetics: An Approach to Self-Organization. In: *Self-organizing systems: The emergence of order* (F. E. Yates, ed.), Plenum, New York, pp. 417-434.

Haken, H., 1988, *Information and Self-Organization. A macroscopic approach to complex systems.* Springer, Berlin

Hall, A. D., and Fagen, R., E., 1956, Definition of a system, *General Systems Yearbook* **1**:18-28.

Heims, S. J., 1991, *The Cybernetics Group.* MIT Press, Cambridge, MA.

Heisenberg, W., 1971, *Physics and Beyond.* Harper & Row, New York.

Hernandez-Bermejo, B., and Fairén, V., 1997, Lotka-Volterra representation of general nonlinear systems, *Mathematics Biosciences* **140**:1-32.

Jackson, M. C., 2000, *Systems approaches to management.* Kluwer, NewYork.

Klir, G. J., (ed.), 1972, *Trends in General Systems Theory.* Wiley, New York.

Klir, G. J., (ed.), 1991, *Facets of Systems Science.* Kluwer, New York.

Klir, G. J., 1969, *An approach to General systems theory.* Van Nostrand, New York.

Klir, G. J., 2001, *Facets of Systems Science. Second Edition.* Plenum, New York.

Klir, G. J., and Yuan, B., 1995, *Fuzzy sets and Fuzzy Logic: Theory and applications.* Prentice Hall, Englewood Cliffs, NJ.

Koffka, K., 1935, *Principles of Gestalt Psychology.* Lund Humphries, London.

Kuhn, T. S., 1962, *The Structure of Scientific Revolutions.* University of Chicago Press, Chicago, IL.

Lorenz, E., 1963, Deterministic Non Period Flow, *Journal Of The Atmospheric Sciences* **20**:130-141.

Lotka, A. J., 1920, Undamped Oscillations derived from the law of mass action, *Journal of the American Chemical Society* **42**:1595-1599.

Maruyama, M., 1963, The second cybernetics: Deviation-amplifying mutual causal processes, *American Scientist* **51**:164-79.

Maturana, H. R., and Varela, F., 1980, *Autopoiesis and Cognition: The Realization of the Living.* Reidel, Dordrecht.

Meadows, D. H., Meadows, D. L., and Randers, J., 1993, *Beyond the Limits: Confronting Global Collapse, Envisioning a Sustainable Future.* Chelsea Green Publishing Company, White River Junction, VT.

Meadows, D. H., Meadows, D. L., Randers, J., and Behrens III, W. W., 1972, *The limits to growth: a report for The Club of Rome's project on the predicament of mankind.* Universe Books, New York.

Mesarovic, M. D., 1968, On some mathematical results as properties of general systems, *Mathematical systems Theory* **2**:357-361.

Mesarovic, M. D., 1972, A Mathematical Theory of General System Theory, In: *Trends in General System Theory* (G. J. Klir, ed.), Wiley, New York, pp. 251-269.

Mesarovic, M. D., and Takahara, Y., 1989, *Abstract Systems Theory.* Springer, Berlin

Miller, J. G., 1978, *Living Systems.* McGraw-Hill, New York.

Minati, G., 2001, *Esseri Collettivi.* Apogeo, Milano, Italy.

Minati, G., and Pessa, E., (eds.), 2002, *Emergence in Complex Cognitive, Social and Biological Systems.* Kluwer, New York.

Minorsky, N., 1962, *Non linear oscillations.* Van Nostrand, Princeton, NJ.

Mitroff, I., and Linstone, H. A., 1993, *The Unbounded Mind: Breaking the Chains of Traditional Business Thinking.* Oxford University Press, New York.

Nicolescu, B., 1996, *La Transdisciplinarité,* Rocher, Paris.

Nicolis, G., and Prigogine, I., 1977, *Self-Organization in Nonequilibrium Systems: From Dissipative Structures to Order through Fluctuations.* Wiley, New York.

Noonburg, V. W., 1989, A neural network modeled by an adaptive Lotka-Volterra system, *SIAM Journal on Applied Mathematics* **49**:1779-1793.

Noszticzius, Z., Noszticzius, E., and Schelly, Z. A., 1983, A Lotka-Volterra model for the Halate driven oscillators, *The Journal of Physical Chemistry* **87**: 510-524.

Porter, W. A., 1965, *Modern Foundations of Systems Engineering.* Mac Millan, New York.

Prigogine, I., 1978, Zeit, Struktur und Fluktuationen (Nobel-Vortrag), *Angewandte Chemie. Sonderdruck,* Heft 9, Verlag Chemie, **90**:704-715.

Prigogine, I., 1981, *From Being to Becoming: Time and Complexity in the Physical Sciences.* W. H. Freeman & Co., New York.

Rapoport, A., 1968, General System Theory. In: *The International Encyclopedia of Social Sciences* (D. L. Sills, ed.), Macmillan & The Free Press, New York, Vol. 15, pp. 452-458.

Rosen, R., 1978, *Fundamentals of Measurement and representation of Natural Systems.* North-Holland, New York.

Rosen, R., 1985, *Anticipatory systems.* Pergamon Press, New York

Roussel, M. R., 1997, An analytic center manifold for a simple epidemiological model, *SIAM review* **39**:106-109.

Russell, L. A., 1974, *Redesigning the Future: A Systems Approach to Societal Problems.* Wiley, New York.

Senge, P. M., 1990, *The Fifth Discipline: The Art & Practice of the Learning Organization.* Doubleday Currency, New York.

Senge, P. M., Ross, R., Smith, B., Roberts, C., and Kleiner, A., 1994, *The Fifth Discipline Fieldbook: Strategies and Tools for Building a Learning Organization.* Doubleday Currency, New York.

Skinner, B. F., 1938, *The behavior of organisms: an experimental analysis.* Appleton-Century-Crofts, New York.

Skinner, B. F., 1953, *Science and Human Behavior.* Macmillan, New York.

Smuts, J. C., 1926, *Holism & Evolution.* Macmillan, New York.

Sparrow, C., 1982, *The Lorenz Equations: Bifurcations, Chaos, and Strange Attractors.* Springer, New York.

Sutherland, J., 1973, *A General Systems Philosophy for the Social and Behavioral Sciences.* George Braziller, New York.

Tainaka, K., 1989, Stationary Pattern of Vortices or strings in Biological Systems: lattice version of the Lotka-Volterra Model, *Physical Review Letters* 63:2688-2691.

Umpleby, S. A., and Dent, E. B., 1999, The Origins and Purposes of Several Traditions in Systems Theory and Cybernetics, *Cybernetics and Systems* 30:79-103.

Van der Pol, R., and Van der Mark, J., 1927, Frequency demultiplication, *Nature* 120:363-364.

Van Gigch, J. P., 1978, *Applied General Systems Theory.* Harper & Row, New York.

Van Gigch, J. P., 2003, *Metadecisions: Rehabilitating Epistemology.* Kluwer, New York.

Varela, F., Maturana, H. R., and Uribe, R., 1974, Autopoiesis: The organization of living systems, its characterization and a model, *BioSystems* 5:187-196.

Verhulst, F., 1990, *Non linear differential equations and dynamical systems*, Springer, Berlin.

Vitiello, G., 2001, *My double Unveiled: the dissipative quantum model of brain.* Benjamins, Amsterdam.

Volterra. V., 1926. Variations and fluctuations of the number of individuals in animal species living together. In: *Animal ecology*, (R. N. Chapman, ed.), McGraw-Hill, New York, pp. 409-448.

Von Bertalanffy, L., 1950, The Theory of Open Systems in Physics and Biology, *Science* 111:23-29.

Von Bertalanffy, L., 1952, *Problems of Life. An Evaluation of Modern Biological and Scientific Thought.* Harper & Brothers, New York.

Von Bertalanffy, L., 1956, *Les problèmes de la vie.* Gallimard, Paris.

Von Bertalanffy, L., 1968, *General System Theory. Development, Applications.* George Braziller, New York

Von Bertalanffy, L., 1975, *Perspectives on General System Theory: Scientific-Philosophical Studies.* George Braziller, New York.

Von Foerster, H, 1979, Cybernetics of Cybernetics. In: *Communication and Control in Society* (K. Krippendorff, ed.), Gordon and Breach, New York, pp. 5-8.

Whitehead, A. N., 1929, *Process and Reality.* Macmillan, New York.

Wiener, N., 1948, *Cybernetics or Control and Communication in the Animal and the Machine.* MIT Press, Cambridge, MA.

Wiener, N., 1961, *Cybernetics: Or control and communication in the animal and the machine.* Second edition. MIT Press, Cambridge, MA.

Zadeh, L. A., Klir, G. J., (ed.), Yuan, B., (ed.), 1996, *Fuzzy Sets, Fuzzy Logic, and Fuzzy Systems: Selected Papers by Lotfi A. Zadeh.* World Scientific, Singapore.

Zwillinger, D., 1997, *Handbook of Differential equations* (3rd edition). Academic Press, Boston, MA.

Chapter 2

GENERALIZING SYSTEMICS AND THE ROLE OF THE OBSERVER

This chapter is dedicated to *generalizing Systemics*, through the identification of systemic properties, of invariants in the systemic cultural framework, of general assumptions and approaches.

As introduced later in this book, learning to *identify* the systemic aspects of problems (Flood and Jackson, 1991; Watzlawick *et al.*, 1967; 1974) also helps one to realize just how suitable and appropriate systemic approaches are. Regarding this aspect, a list containing a more detailed description of the most important kinds of systemic features is given in Appendix 1.

When we deal with problems entailing such features, then a systemic approach is required, using suitable theoretical tools, such as those introduced in Chapter 1.

To provide some idea of the conceptual extent of attributes and specifications relating to the concept of system, a few of them will be mentioned here: Adaptive, Autonomous, Autopoietic, Chaotic, Complex, Dissipative, Hierarchical, Open and Closed, Self-Organized. Some systemic features of problems can, however, be dealt with by using *System Archetypes*, introduced in Chapter 7.

The crucial aspect common to systems thinking and to the recent evolution of scientific thinking regards the **role of the observer**. The theoretical role of the observer in Systemics has been mentioned above, with particular reference to emergence. This aspect will be discussed below in Chapters 3 and 4. Focus will now be placed upon the theoretical role of the observer in modern science by introducing, first of all, the basic ideas and conceptions of one of the founding fathers of Systemics: Heinz von Foerster.

2.1 The contribution of Von Foerster

Heinz von Foerster (1911-2002), together with Warren McCulloch, Norbert Wiener, John von Neumann and others, was the architect of *cybernetics*. He developed the so-called *second-order cybernetics*, or *cybernetics of cybernetics* (Von Foerster, 1979; 1981), focusing on self-referential systems and eigenbehaviors for the explanation of complex phenomena. Von Foerster defined it as the "cybernetics of observing systems".

He introduced an important distinction between *trivial* and *non-trivial machines*, very important forrecognizing the complexity of cognitive behavior.

A *trivial machine* is a machine whose *operations* are not influenced by the outcomes of previous operations performed by the machine itself. The behavior of such a machine can be analytically described and predicted.

A *non-trivial machine* is a machine whose *operations* are influenced by the outcomes of its previous operations, so that the problem of deducing the machine structure from only a knowledge of its behavior becomes unsolvable.

He formulated the shift from first to second order cybernetics (Von Foerster, 1981) as follows:

> *While in the first quarter of this century physicists and cosmologists were forced to revise the basic notions that govern the natural sciences, in the last quarter of this century biologists will force a revision of the basic notions that govern science itself. After that "first revolution" it was clear that the classical concept of an "ultimate science" that is an objective description of the world in which there are no subjects (a "subjectless universe"), contains contradictions.*

This view had a determining influence on many cognitive scientists (Varela *et al.*, 1991) and (radical) constructivists (Butts and Brown, 1989; Von Glasersfeld, 1991; 1995), and was summarized in his book "Understanding Understanding: Essays on Cybernetics and Cognition" (Von Foerster, 2003).

The outstanding importance of Von Foerster's contribution lies in his making evident not only that (1) the process of knowing itself should be studied, but that (2) a subject considered only in a very limited way by classical physics should be the focus of this new approach: *the observer.*

With reference to the first aspect, it would be possible to view Second Order Cybernetics as offering theoretical ground (e. g, Von Neumann, 1996) when considering *cognitive neuroscience* in the context of *cognitive science*, that is, the interdisciplinary study of mind and intelligence, embracing philosophy, psychology, artificial intelligence, neuroscience, linguistics, and anthropology.

As is well known, Cognitive science had its intellectual origins in the mid-1950s when researchers in several fields began to develop theories of mind based on the concepts of Representation and Computation. The main hypothesis of most current exponents of Cognitive Science is that it is possible to understand mental processes in terms of representational structures embodied within the mind and of computational procedures which operate upon those structures. This assumption includes connectionist theories based on artificial neural networks.

With reference to the second point, the crucial aspect, common to many scientific as well more general cultural domains, is the emphasis on *the role of the observer, no longer considered as passive, , but active.* For a long time (and often even today) the most popular philosophical framework adopted by scientists was that offered by *realism,* based on the assumption of an *external* reality, whose existence and features are independent from the observer (Churchland, 1979; 1995). However, after the Second World War most researchers working within the domains of social and human sciences began to underline that basic scientific assumptions as well as scientific results were essentially determined by the attitudes, goals, knowledge, and mental schemata of the observer (Berger and Luckmann, 1966; Watzlawick, 1983). In other words, it seemed that it was impossible to do science without taking into account the *subjectivity* of the observer.

However, a similar approach had already been introduced at the beginning of the Twentieth Century by theoretical physicists. The first time the role of the observer was taken into account was within the *Theory of Relativity,* which showed how a change in very elementary observational features, such as the place and time of observation, or the state of motion of the observer, could bring about profound transformations of the whole picture, as a result of the observations themselves. However, greater acknowledgement of the fundamental role of the observer came only with the advent of Quantum Theory. Here, the act of observation itself introduces a perturbation of the state of the system such that it becomes impossible, in principle, to ascertain the state before the observation was made. Experimental outcomes can be nothing but a by-product of the interactions taking place within a complex system, made up of the observer and the world in which he/she is embedded, a system which, however, is *entangled*, as these entities cannot be separated from one another. This implies that, when an observer decides to design an experiment to measure a given physical quantity with unlimited precision, this same decision automatically excludes the possibility of measuring another quantity with unlimited precision. This circumstance is expressed by the celebrated *Uncertainty Principle*, also called the *Heisenberg Uncertainty Principle*, proposed in 1927 by the German physicist Werner Heisenberg (1901-1976). Referring to atomic or subatomic particles, such as an electron, it states that "The more precisely the position is determined, the less precisely the momentum (mass times velocity) is known in this instant, and *vice versa.*" *This new approach made unsustainable the assumption of an existing general deterministic order which we are unable to detect* This *Principle*, while having little to do with the role of the observer as introduced in *second order cybernetics*, nevertheless stimulated very important reflections about the impossibility of reaching *objective* knowledge (Heisenberg, 1971).

As mentioned above, a very important theoretical shift took place with the introduction of *second-order cybernetics.* The observer is considered here as a cybernetic system, trying to develop carry out a model of other cybernetic systems. The process may be viewed as "cybernetics of cybernetics", i.e., a "meta" or "second-order" cybernetics.

A "first-order" cyberneticist will study a system as if it was a passive, objectively given "object", which can be observed, manipulated, from an objectivistic point of view. A second-order cybernetician dealing with an organism or a social system, on the other hand, identifies a system as an agent interacting with another agent, the observer. The observer and the observed system cannot be separated, and the results of observations will depend upon their mutual interaction.

2.2 The role of Observer in scientific theories

Modern science now accepts that objectivism, *as the only cognitive strategy*, is a gross simplification that might just work at certain simple levels of description. The level of description, however, has to be much higher when dealing with complex systems. In this case, as already underlined, one must resort to conceptual approaches, methodologies and conceptions allowing simultaneous multiple perspectives. Systemics is just a suitable *conceptual paradigm* for complexity, allowing one to deal with interactions, multiple representations, multi-agent modeling, and non-linearity. In this context objectivistic assumptions must be reconsidered. As shown in the previous section, within modern science the **observer is considered as an integral part** of the phenomena. It means that, as mentioned for emergence phenomena, the cognitive model of the observer, the establishment of meta-levels (*models of models*), as introduced in *logical openness* (see Appendix 1), and the processes of interaction, are part of the description of the phenomena. To think that it is at least inadequate and ineffective to believe that what is intended as **real** could not exist independently from the observer, viewed as an agent detecting and creating reality (Berger and Luckmann, 1966), may be understood as equivalent to adopting the anthropocentric and ethnocentric views typical of classical humanistic philosophical approaches. Science now accepts that the observer, *sees* through his/her visual system and his/her cognitive models. That is, detecting what they are ready to see, expecting to see and able to be seen. This process is well known in the animal world, both at a physiological level (consider the well known studies on the frog's visual system able to *see* only certain kinds of movements, those of its prey, while stationary objects actually *disappear*) and at a cognitive level. These considerations will be crucial for explaining, as introduced below, the emergence of **Collective Behaviors** (as in the process of emergence of flocks). To think that what is visible is what *really* exists (i.e. independently from the observer, not only detecting but even creating reality through its cognitive system) is equivalent to assuming that it is possible to use *empty* models, in the tradition of the classical *tabula rasa* approach introduced by the exponents of *pre-cognitive science* such as the empiricist John Locke (1632 - 1704) in his book *An Essay Concerning Human Understanding* (1690).

This new approach can be illustrated by the distinction between

- Trying to understand how something really is, and
- How appropriate it is to consider something (which model to adopt) to be more effective.

For example, if while walking we are struck by something from behind, there are two possible reactions:
a) *Looking* for what struck us, or
b) *Realizing* what struck us.

To adopt the hypothesis of the existence of *empty models*, that is to believe that what is seen is that which *linearly* objectively exists, without any information processing by the observer using a cognitive model, is, at the least, naive and not very useful. Activities such as cognitive processing, modeling, deciding, communicating, and possessing a certain behavior are based on using *languages*. Languages do not objectively exist, as objects, but are strictly related to **usage**: it is not even possible to decide what *a language is and what it is not*. It depends on how it is used. Besides one must take into account that languages have various representational powers.

When *using* certain concepts it may not be necessary to *represent* them, contrary to Whorf's hypothesis (Von Bertalanffy, 1968; Caroll, 1956). Namely, some primitive natural languages lack the representation of concepts such as numbers and arithmetical operations. Users of these languages have no possibility of representing numerical concepts, but *a loss of a sheep from the herd is immediately perceived by the users of such a language*. This perception and usage may occur through the correspondence with certain elements (e. g.,   rocks) used to *check quantities without counting*. This brings to mind what Zadeh stated, when considering the Computational Theory of Perceptions (CTP) and of *Soft Computing* as an "an association of computing methodologies which includes, as its principal members, fuzzy logic (FL), neuro-computing (NC), evolutionary computing (EC) and probabilistic computing (PC)".

An analogous situation occurs when there is the emergence of a *collective representation by individuals unable to generate a representation alone* (see Section 3.3), as happens with ants (Deneubourg *et al.*, 1989; Franks *et al.*, 1991).

In modern science the **complementary and not alternative** use of the two, objectivistic and non-objectivistic approaches mentioned above, emerges as the most effective strategy, in turn based upon the *level of description assumed*. In some cases it may be more effective to use the former approach and in others the latter, or even both simultaneously.

These two approaches correspond intuitively to the concepts of *discovery* and *invention*.

With regard to the subject of memory, the objectivistic approach implies that memory is viewed as a *store;* therefore the most effective use of it is to *find* what has to be recalled. The most important steps of the memorization process are supposed to be coding, ways of memorizing and linking, very important in determining the retrieval efficiency.

A conceptually different, non-objectivistic, approach is to identify the retrieval process with the *reconstruction* of what is expected to be remembered from the information and the details searched for and found. Memory is in this case the name of a process integrated within the processing of the whole cognitive system.

In these introductory examples it is easy to see how the two approaches are complementary or, even better, how effectively they can be used.

It should not be assumed that one approach is better than the other or that one is right and the other wrong. The conceptual approach is that of greater suitability and effectiveness by using the two approaches not alternatively, but dynamically, in a complementary way.

An example of the **usage** of different models is given by the history of the conceptions adopted by human beings to interpret, describe and understand planetary motion. Starting from a geocentric conception, a heliocentric one was then adopted, being invented/discovered on the basis of observations and experimental data. We cannot, however, assert that one is right and the other wrong: the second is only much simpler, fitting with experimental data and more suitable for practical applications. To be consistent and to explain experimental data adopting the former would be very complicated if not impossible. It is possible to state that the latter is *waiting* to be superseded and to become a particular case of an extended new conception, even more suitable and effective. *Simplicio*, the Aristotelian philosopher and supporter of the Ptolemaic system in Galileo's "Dialogo sopra i due massimi sistemi del mondo" (1632), a very problematic question arises: what if Simplicio had had access to computers? That is, if he had not needed sophisticated models because of the availability of very high computing power? Does this not correspond to our own situation?

The **Usability** of conceptual devices and their suitability for the context, for the problems to be faced, is the crucial point. A map based on the assumption that the earth is flat is very useful to plan and check short journeys. On the other hand, if very long journeys have to be undertaken, ffrom one continent to another, it is more suitable to adopt the idea that the earth is spherical.

We must always recall that observers act by processing information through models available within their own *Cognitive Systems* (see Section 3.4), by using languages and logic processes of their inferential systems. Among the latter the main ones are:

Deduction

Deduction is a kind of *inference* (the process of *making inferences* may be understood as *generating conclusions* from premises), starting from the necessary premises: the latter contain everything necessary to reach the conclusion. Therefore, in a valid deduction, the conclusion cannot be false if all premises are true.

In the case of deduction the most widely used rule is the so-called *Modus Ponens*. In it one starts from the application of a general rule (R): x--->y, which is expected to be true if premises are true. When a particular case (C) holds, the resulting conclusion (Res) is obtained because the rule (R) and (C)---->Res is applied.

Here is an example:

- All the pieces in this box are black, - rule (R).
- Those pieces come from this box, - case (C).
- Therefore those pieces are black, - result (Res).

Induction

Induction (Holland *et al.*, 1986) is an inference, which from a finite number of particular cases leads to another case or to a general conclusion. For instance, if from a bird watch the passage of only black ravens has been observed, then it is possible to induce that the next raven detected will be black or that all ravens are black.

In the case of the *induction* of a rule or of a result (Res) from a set of configurations (Cn) of elements, we start from the observations: C--->Res, C'--->Res, C''--->Res, ..., and then we assume valid the general rule Cany--->Res.

An example is:

- Those pieces come from this box, - case (C).
- Those pieces are red, - (Res).
- All the pieces in this box are red, - (R).

Abduction

In the case of abduction a reasoning of this kind is adopted:

- The starting point is a collection of data D;
- The hypothesis H, if true, could explain D;
- No other hypothesis can explain D better than H;
- Then H is probably true.

There is a *hypothesis inventing process* that may be even viewed as a *selection* among the most suitable ones for explaining D.

With abduction a process of *clustering* is carried out, grouping together variables that are most probably related (or, more precisely, that it is suitable to think they are): "*Because B is true probably A is also true, since if A were true the truth of B would be obvious*".

Charles S. Peirce defines his concept of *abduction* in the following way: "Abduction is the process of forming an explanatory hypothesis. It is the only logical operation which introduces any new idea" (Peirce, 1998).

Paraphrasing Foerster (Von Foerster *et al.*, 1974), there is *no information, or anomalies* in the environment. If a given phenomenon looks strange, this means that the theoretical framework used to interpret this phenomenon is inappropriate. This cognitive process of reformulation of the model is labeled *abduction*, and its aim is to "normalize" anomalies (Andreewsky and Bourcier, 2000).

2.3 Uncertainty Principles in Science

The quest for Uncertainty Principles in science (and their discovery in specific domains) arose as a result of the need for dealing with systems in which the act itself of observing or of monitoring was interfering with the activity of the systems themselves. This interference, as is well known, was observed very early on in the history of physics, when researchers began investigating phenomena on an atomic scale. However, more recently its occurrence was also observed by engineers working with complex systems, such as parallel computing arrays (*cfr.* Cheng and Ushijima, 1991).

In order to set a framework for understanding the role played by the different Uncertainty Principles so far introduced, it is convenient to start again from the pioneering intuition of H. von Foerster (Von Foerster, 1981) while trying to build a general theory of the behavior of a complex system whose subsystems are an *observed system* (a sort of environment) and an *observing system* (the observer, or, in other contexts, the scientist). Such a system can be depicted as in Figure 2.1.

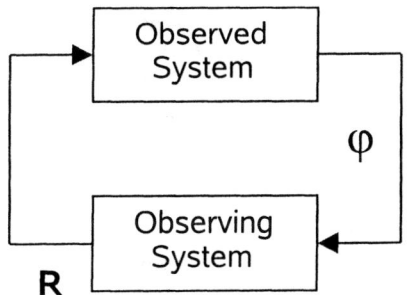

Figure 2-1. Observed and observing systems.

As we can see from the figure, the Observed System initially inputs its state φ to the Observing System (here the mathematical nature of φ does not matter: it may be a number, a function, a functional, etc., according to the theoretical framework adopted). The latter, in turn, perturbs the Observed System, owing to the act of observation itself, and the perturbation can be described as the action of an operator R which, acting on φ, gives rise to a new state of the Observed System $R\varphi$. Such a change implies that, as a consequence of the observation, the input to the Observing System is no longer φ, but $R\varphi$. As the latter continues to make observations, it will in turn transform this input to $RR\varphi$. If we continue further along this road, after n environment-observer interactions, the state of the environment itself will be

R^n φ, where the symbol R^n denotes the n-th iteration of applying the operator R . Generally speaking, R^n φ will be very different from φ, so that this argument seems to leave no hope, for the Observing System, of detecting some stable feature of its environment. However, there is still a possibility: let us introduce an environmental state φ^* fulfilling the relationship:

$$R \varphi^* = \varphi^*$$

In mathematical terms, φ^* is called a *fixed point* of the operator R . It is now easy to see that, if the environment falls into the state φ^*, then it will remain indefinitely in this state, despite the perturbations induced by the observer. The latter, in turn, will detect that the environment is in the invariant state φ^*.

The proposal of Von Foerster gives rise to a number of difficult conceptual problems, listed as follows:

a.   Can such an argument always be translated into a mathematical form? In other words, how is it possible to define in a formalized way the operator R and the state φ? Or, conversely, when is it not possible to introduce such a definition?

b.   The argument of Von Foerster is useful if we assume that the form of the operator R is such that, when starting the process of mutual interaction between observer and environment from an initial environmental state φ_0, different from φ^*, the iterated application of R gives rise (eventually, after an infinite number of steps) to the state φ^*. In mathematical terms, an operator endowed with such a property is called a *contractive operator*. But what are the necessary conditions for R to be contractive? And what happens if R has a multiplicity of different fixed points?

c.   The operator R could have a form such that its fixed point is trivial. In this case, even if R were to be contractive, the behavior of the whole observer-environment system would fall into triviality and the whole theory would become useless. A simple example of such a situation is given by the case in which the environmental state is described by a real number x and the action of the operator R consists simply in the multiplication of this number by a given constant k , assumed to be different from zero. In this case the fixed point x^* is defined by the condition:

$$k x^* = x^*$$

whose only solution is the trivial one, that is $x^* = 0$!

Although problems 1) and 2) are still waiting for a solution, problem 3) can be circumvented by relaxing the requirement of the fixed point and by substituting it with a wider one:

$$R \varphi^* = \lambda \varphi^*$$

with the further condition that φ^* be nontrivial. In mathematical terms, the number λ is called an *eigenvalue* of the operator R, whereas the terminology adopted for φ^* depends on the mathematical context used to describe the latter. For instance, if φ^* is a vector, one speaks of *eigenvector*; on the contrary, if φ^* is a function, it is called an *eigenfunction*, and so on. In general, we will refer to φ^* as an *eigenstate*.

A suggestive interpretation of this relationship associates φ^* with an *observable quantity* of the environment, λ the *result of an observation*, and to R, the *observing action*. Starting from this interpretation, we can search for a framework for discussing the roles played by the various Uncertainty Principles used in Science. The discussion starts from the use of some interesting arguments (and related definitions) put forward by Diettrich within the context of his *Theory of Cognitive Operators* (for a summary, see Diettrich, 2001). The main hypothesis adopted by Diettrich is the existence of phylogenetically acquired mental operators (a subclass of possible Rs), which are responsible for our perceptual capacities (as well as for the related motorial ones) in everyday experience. We refer to these operators as "primitive operators". Problems arise, according to Diettrich, when designing a procedure for performing a scientific experiment (in physics, for example) or when we design a procedure for performing a "mental experiment" on abstract entities, defined within an abstract theory. In this case, we are faced with the possibility of a contradiction between our traditional world picture, in turn based on the eigenvalues of our inborn primitive mental operators (we will denote such a world picture as "classical world picture"), and the world picture resulting from the new operators associated with our physical or mental experiments. In order to discuss what could occur, we will introduce a suitable notation, based on the one already used above. More precisely, let us denote with P a generic primitive operator and with φ a generic eigenstate of P. A classical world picture will then be associated with the condition:

$$P \varphi = \lambda \varphi$$

Let us now introduce a generic operator, associated with our physical experiment (or our "mental experiment") denoted with O. By applying this operator to both members of the above relationship we obtain:

$$O P \varphi = \lambda O \varphi$$

Two possible cases can then arise: 1) operators O and P are mutually commuting, that is they satisfy the condition $O P = P O$; 2) operators O and P are not commutative, i.e. $O P \neq P O$. In the former case we can rewrite the above relationship in the form:

$$P O \varphi = \lambda O \varphi$$

This new relationship tells us that $O\varphi$ is still an eigenstate of our primitive operator P. This observation allows an interesting interpretation: $O\varphi$ denotes the result of the application of our experimental scientific device to the environment, in turn related to a particular world picture arising from the action of the device itself, and the fact that it continues to be an eigenstate of P means that the world picture arising from our experimental apparatus does not contradict our classical world picture, as represented by P. In other words, if the operators associated to our scientific measurements commute with our primitive cognitive operators, we will obtain a world picture which can still be expressed in terms of our everyday experiences. Such a new picture will, of course, in a sense, be an extension of our classical world view (in this case Diettrich speaks of *quantitative extension*) but it will be fully compatible with it, and understandable by resorting to the classical view. A typical case of such a situation is offered by classical mechanics which, as a matter of fact, represents nothing but an extension and a formalization of our everyday experience of the motion of bodies.

An entirely different situation occurs when the operators O and P do not commute. In this case, which Diettrich refers to as *qualitative extension*, the world picture arising from our scientific measures no longer agree with our classical world picture, and cannot be explained within the classical context. A typical (and celebrated) case is given by Quantum Mechanics. This case is of a particular historical and conceptual importance as, within the formalism of this theory, it is possible, probably for the first time, to connect the lack of commutativity with the appearance of Uncertainty Principles dealing with physical quantities which, at first sight, could seem entirely understandable in classical terms. It will be interesting to present here the reason for the occurrence of Uncertainty Principles in the absence of commutativity. It must first be mentioned that the set of possible experimental devices, or measuring apparatuses, giving rise to world pictures differing from the classical one is, in principle, unlimited. In mathematical terms, this amounts to saying that the set of the operators non commuting with P is infinite. Such a circumstance entails that not only each operator, not commuting with P , will be associated with a non-classical world view, but that there will be an infinity of non-classical world views, each differing from the others. This is due to the fact that different non-classical operators can also be mutually non-commutative. Now, if a given observable quantity α is an eigenstate of a given non-classical operator A and another observable quantity β is an eigenstate of another non-classical operator B, with A and B not commuting, automatically there will be an Uncertainty Principle connecting our uncertainty about the measurement of α with our uncertainty about the measurement of β. Further, the mathematical machinery of quantum theory leads us to the conclusion that such an Uncertainty Principle will have, in the simplest cases, the form:

$$\Delta\alpha \ \Delta\beta \geq k$$

where k is a suitable constant.

An entirely analogous argument could be made for the "mental experiments" which, in turn, are connected to mathematical activity. For this, we must first take into account that there are currently two main views about the nature of mathematics: the *formalist* one (exemplified by Hilbert's program) and the *constructivist* one. According to the former, mathematical entities live within an abstract world, where they are endowed with a sort of "objective existence", in turn made possible by the absence of contradictions. Within such a Platonic world mathematicians can only 'discover' entities which exist, beyond space and time, independently from their searching. In other words, these entities are nothing but expressions of an (eternal) logical necessity. On the contrary, the constructivists assert that mathematics, like physics and other sciences, is nothing but a human construction, produced in certain ways as a consequence of given particular goals, natural and artificial tools, and environmental features. In short, mathematics is built for human needs and, as such, it is grounded on our primary perceptual, motorial, and cognitive capacities, the same which are associated with the primitive operators previously introduced and to the classical world view.

As is well known from the history of mathematics, the formalist program, aiming at a complete axiomatization of all mathematics, and to a general control over all mathematical formalisms, through a super-discipline named Metamathematics, ran into failure, owing to theorems such as the celebrated Gödel incompleteness theorems. Despite this, however, it is important to discuss here whether a systemic view is or is not compatible with either of the two views of mathematics introduced above. On this point, it should be recalled that, as already stated in Section 1.1 of this book, a systemic view implies, in principle, the exclusion of the existence of a unique view of a system, or of a unique theory (the only true one!) for describing and explaining its phenomenology. On the contrary, the formalist view assumes the objective existence of one, and only one, mathematical world, which, *per se*, should be described by only one omnicomprehensive theory, the only one true. For the formalist, the fact that such a theory is still lacking is due only to the insufficient development of mathematics; in any case, such a theory *already exists*, and it is only waiting to be discovered. It is thus obvious that the formalist program is in total disagreement with the systemic view. The latter, therefore, is more compatible with the constructivist approach, and it is useful to recall that most researchers holding a systemic view are strongly influenced by the philosophical aspects of constructivism, known as *radical constructivism* (see Von Glasersfeld, 1995).

If we adopt a constructivist view, then this implies that even basic mathematical constructs are based on primitive cognitive operators, as for physical ones . We can, thus, introduce the notion of a *classical* mathematics, that is a mathematics whose cognitive operators commute with the primitive cognitive operators. Classical mathematics gives rise to a quantitative extension of the classical view related to operations of everyday experience, such as counting, adding, manipulating objects, measuring lengths, and so on. As a matter of fact, the constructs of classical mathematics can always be understood in terms of intuitive, and elementary, operations such as those quoted above. Perhaps this circumstance could explain what Wigner calls "the unreasonable effectiveness of mathematics" in describing the natural world (Wigner, 1960). Going beyond classical mathematics, it is, however, possible to construct even *non-classical* mathematics (which, as a matter of fact, has been built). The latter, based on cognitive operators which do not commute with the primitive ones, gives rise to constructs which cannot be fully understood in terms of elementary operations in everyday life. In any case, owing to the unlimited number of different possible operators which do not fulfill the commutative property, we should expect the appearance, even within non-classical mathematics, of Uncertainty Principles. And, in fact, one such principle has effectively been introduced: the Gödel incompleteness theorems already cited above! In short, we can interpret the content of these theorems as asserting that no axiomatic system (beyond a given degree of complexity) can exhaust the possibilities offered by a non-classical mathematical world, owing to the presence within it of formal expressions which, in a sense, are unreachable by starting only from axioms, and equipped only with logical inference rules. Thus, we could assert that, to a certain degree, the Uncertainty Principles introduced in Physics (such as the celebrated Heisenberg one) and the Gödel theorems in mathematics are but different sides of the same coin, related to different domains of application of the cognitive operators, or to different, but strongly related, cognitive activities of the same observer.

Now that a framework has been set for understanding the role played by Uncertainty Principles, we can now mention the practical work carried out within the disciplinary domains in which Uncertainty Principles have been stated in an explicit form. As often mentioned, quantum mechanics offered in the past the best context for this. Even non-specialists know that within this theory the uncertainty associated with the location of a particle, denoted by Δx, and that associated with its momentum, denoted by Δp, are connected by the Heisenberg relationship

$$\Delta x \ \Delta p \ \geq h/4\pi$$

where h is Planck's constant. Such a principle, however, can also be expressed in another form, related to the uncertainty in the values of energy, ΔE, and time, Δt, giving:

$$\Delta E \ \Delta t \ \geq h/4\pi$$

Despite the fact that some physicists provide different interpretations to the two forms of the Uncertainty Principle, it can be seen that both are nothing but particular cases of a more general Uncertainty relationship. To introduce the latter, it is useful to recall that within quantum mechanics the physical quantities, in conformity with the views contained in this Section and described above, are associated with suitable *operators*, acting on functions describing the state of a system. The actual values taken by a particular physical quantity, and obtained as a result of a given measuring procedure, are nothing but the eigenvalues of the operator associated with the quantity itself. We must further recall that, for reasons of mathematical economy, the structure of quantum mechanics has been shaped on the basis of that of classical mechanics. The latter, as is well known, relies upon two main dependent variables, spatial location and momentum, which, technically, are said to be each other's *conjugate*. However, within the most sophisticated development of classical mechanics, so-called Analytical Mechanics, introduced by Euler, Lagrange, Hamilton, Jacobi and many others, it can be shown that it is possible to introduce other pairs of dependent variables, mutually conjugate because they play the same roles played by location and momentum in elementary classical mechanics. These new variables can be, of course, obtained from location and momentum through suitable transformations, and they are introduced mainly to simplify the structure of problems in classical mechanics which, if expressed in terms of location and momentum variables, would be very difficult to solve.

It is now possible to show that, if two physical quantities are conjugate, in quantum mechanics the associated operators do not commute; as a consequence their uncertainties are connected by a suitable uncertainty principle. More precisely, if we deal with two conjugate physical quantities *a* and *b,* whose associated quantum-mechanical operators are *A* and *B*, then these operators will fulfill a relationship of the kind:

$$A B - B A \neq 0$$

and the uncertainties Δa and Δb will be connected by the general form of the Uncertainty Principle:

$$\Delta a \; \Delta b \geq h/4\pi$$

The Heisenberg form of Uncertainty relations is only a particular case of the more general formula shown above.

Despite the fact that Werner Heisenberg is universally considered as the first scientist to introduce the Uncertainty Principle in the 1920s, it should be recalled that such a principle was already known, although not recognized as such, within the domain of Fourier Analysis. Such a circumstance was acknowledged only much later and only more recently have engineers began to make common use of the Uncertainty Principle in Signal Analysis. It is instructive to look at the form of this principle in this domain, because it suggests other interpretations of the results already obtained within quantum mechanics. In this case, we are dealing with signals, defined by a suitable quantity (signal amplitude), varying as a function of time. Since the nineteenth century, one of the most widely used methods for extracting the available information from a signal was to express it as a weighted sum of a number (theoretically infinite) of perfectly periodical signals (e.g., sinusoids), having frequencies which are multiple integers of a fundamental base frequency. Such a decomposition is known as *Fourier Analysis*. The latter, based on a sound mathematical theory (to which even N. Wiener, the founder of Cybernetics, contributed), and today implemented through very efficient computer programs, allows one to find which frequency components predominate in the signal under study. This knowledge then allows further inferences to be made about the nature of the phenomena producing the signal.

However, despite its effectiveness, Fourier Analysis entails difficulties when attempting to answer to one of the fundamental questions of interest for the researcher: *when* does a given frequency component enter into play, contributing to the observed form of the signal under study? One sophisticated mathematical theory showed that an Uncertainty Principle exists: as the precision in knowing the instant in time at which a given frequency component enters into play increases, the attainable precision in the determination of *which frequency* enters into play decreases. In other words, if you want to know exactly when a given component contributes to the signal, you must accept a lower level of precision in the frequency of that component. The converse is also true: if you want to know precisely the frequency of a component, you cannot determine exactly when it will contribute to the overall signal. In technical terms such an Uncertainty Principle can be expressed as :

$$\Delta \omega \, \Delta t \geq 1/2$$

where $\Delta \omega$ is the so-called spectral width of the signal (the uncertainty in the frequency of its components), whereas Δt is the so-called temporal width (the uncertainty in the time location of its components). An elementary exercise of physics shows that this principle is nothing more than the Heisenberg Uncertainty Principle cited above. Namely, for a single quantum of energy we have the elementary Planck formula:

$$E = h \, \omega /2 \, \pi$$

And, substituting the latter in the second form of the Heisenberg Uncertainty Principle, we obtain precisely the Uncertainty Principle holding within Fourier Analysis.

Why this immediate relationship between these two Uncertainty Principles? The physical rationale is that within quantum mechanics the particles can be also be considered as waves, and thus as signals, and the Uncertainty Principle simply expresses the fact that we cannot determine simultaneously both the frequency and the time location of the components of these waves with arbitrary precision.

More recently the Heisenberg Uncertainty Principle has been related to the theory of statistical estimates. Within this theory, there is a relationship, known as the Cramer-Rao inequality, or Cramér-Rao bound, giving the ultimate limit for the resolution of an unbiased estimator (see textbooks such as Bury, 1976; Kay, 1993). If one considers the physical measuring devices as estimators of unknown quantities, it becomes possible to apply the theory of statistical estimates and, within this context, it has been shown that from the Cramér-Rao inequality we can re-derive the Heisenberg Uncertainty Principle (*cfr* Raymer, 1994; Roy, 1998; Hall and Reginatto, 2002). This circumstance opened a new, wide and unexplored, domain of research. Amongst the more recent results obtained in this field we will cite only the introduction of suitable generalizations and corrections of the old Heisenberg Uncertainty Principles (Brody and Hughston, 1997).

The need for suitable generalizations of the Uncertainty Principle currently known arises within the context of studying and managing highly complex systems, such as socio-economic and cognitive ones. Namely, the latter are characterized by the fact that every model adopted to describe them is, in principle, partial and can capture only certain particular features of the system under study, neglecting other features which are no less important. Thus, owing to the practical impossibility of using a single, unique model of the system (the *true* model), in these cases, we have to resort to a multiplicity of different models. As each of them allows only a partial description, suitable, generalized Uncertainty Principles are necesssary, stating, for each partial model, what we gain and what we lose by using it or, in other terms, the increase in uncertainty with respect to the knowledge of given system features when we decrease the uncertainty of others. The theory of these generalized Uncertainty Principles is still under development and will undoubtedly constitute one of the fundamental tools of future Systemics.

2.4 The DYnamic uSAge of Models (DYSAM)

2.4.1 Background to DYSAM

As shown in the previous sections, the study of systems whose complexity exceeds a given threshold cannot be based on the use of a single model. It is necessary to resort to the simultaneous use of several different models of the same system. Here we will introduce a systemic interpretation and generalization of many approaches, already present in the literature relating to this subject. Their common feature is that they do not search for *the* single optimum, *the* right or best solution.

As entry point in this list of approaches we have chosen the well known **Bayesian method**. It is a statistical treatment using an approach based on a "continuous exploration" of the events occurring within the environment. It can even include interpretations of the probabilistic structure of these events overcoming the conventional framework for data treatment based on the dichotomy "match/non-match".

This approach is based on Bayes' theorem (Bayes, 1763), named after the reverend Thomas Bayes (1702-1761), which states that the probability of A given B is equal to the probability of A times the probability of B given A (the probability of B) divided by the probability of B.

In more formal terms we may think of all possible, mutually exclusive, hypotheses which could condition a specific event. The problem solved by

Bayes' theorem is the inverse of that of classical *conditional probability*: what is the probability of a hypothesis given the occurrence of an event?

The Bayes theorem, as generalized by Laplace (Laplace, 1814/1951), is the basic starting point for inference problems using probability theory as the main logical tool (Bretthorst, 1994).

It is interesting to note the philosophical continuity between the approach expressed by the Bayes theorem and the **Pierce abduction** introduced in the previous Section.

The next element of our list refers to a subject studied in Artificial Intelligence, so-called **Machine Learning.**

The research area of machine learning has a long history with particular reference to topics such as decision-making, knowledge acquisition or extraction, and knowledge revision or maintenance. Machine learning provides a large number of techniques based, for instance, upon Neural Networks and Genetic Algorithms. Since one of the purposes of *machine learning* is to make decisions, Bayesian statistics are often used.

We refer here to a specific way of using machine learning techniques and algorithms, so-called **Ensemble Learning**.

Ensemble learning is a relatively new concept suggested by Hinton and van Camp in 1993 (Hinton and Van Camp, 1993).

The basic idea of *ensemble learning* is to combine an uncorrelated collection of learning systems all trained in the same task. In general, this approach is based upon controlling the ensemble performance by stabilizing the solution through the reduction of dependence on the training set and the optimization algorithms used by the members of the ensemble.

The underlying idea is that an ensemble of, for instance, neural networks will perform better than any individual neural network.

Ensemble learning algorithms have become very popular over the past few years because these algorithms, which generate multiple basic models using traditional machine learning algorithms and combine them into an ensemble model, often showed significantly better performance than single models. To this regard, two of the most popular algorithms are *bagging and boosting* because of their good empirical results and theoretical support (for details, see Bauer and Kohavi, 1999; Breiman, 1996; Schapire, 1990; Schapire *et al.*, 1997; Schwenk and Bengio, 2000).

Regarding *classifiers*, the machine learning community has shown considerable interest in the topic of ensemble algorithms which train multiple classifiers with known algorithms, and combine their answers to reach a more accurate final answer. In several problem domains this approach is very effective as, at least partially, explained by Dietterich (Dietterich, 1997). More details can be found in other works (Butz, 2002; Feng *et al.*, 1994; Herbrich, 2001; Lanzi *et al.*, 2002).

Using this approach, many different strategies have been employed, the majority of which fall into four categories – the statistical, rule-based, memory-based and combined strategies.

A comparison of the results indicates that combined systems perform better than any of the other approaches. Ensemble methods generally outperform any single technique due to the fact that they overcome three fundamental machine learning problems, i.e., those related to insufficient training data, inbuilt classifier bias, and searching only a finite subspace of hypotheses (Dietterich, 2000).

One version of this approach is the so-called *Online Ensemble Learning*.

Usually *ensemble models* have largely been learnt only in batch mode: the learning process is subdivided into *epochs*, so that all training examples are presented within a single epoch; the updating takes place only at the end of each epoch. Online learning is an approach focused upon updating models in correspondence to the presentation of each training example. This kind of learning is especially useful when data are continuously arriving, making storing data impractical , or when the dataset is very large. In the latter case the time necessary for multiple runs would become a problem.

Another approach worth mentioning is so-called **Evolutionary Game Theory** (Maynard-Smith, 1982; Weibull, 1995). A few words need to be spent on classical *game theory*. This theory is based on the **von Neumann** "minimax theorem," stated in 1928. The consequences and the applications of this theorem have been studied in the book written jointly by von Neumann and Oskar Morgenstern in 1944, *Theory of Games and Economic Behavior*. The minimax theorem is based on the assumption that for a large class of two-person games, there is no point in playing (see, for instance, the so-called *Prisoner's dilemma* introduced below). This occurs if both players are assumed to choose, as the "optimal" playing strategy, the one expected to minimize the maximum loss. By playing in this way a player is *statistically* sure of not losing more than this minimum, usually called the minimax value. The theorem states that, since that minimax value considered by a player is the *negative* of that, similarly defined, of the opponent, the game outcome in the long run is completely determined only by the game rules.

A typical example used to introduce the general ideas underlying the theory of games is the *Prisoner's Dilemma*, originally formulated in 1950 by the mathematician Albert W. Tucker (on this subject, see Axelrod, 1984; 1997). The two players can choose between two moves, either "cooperate" or "defect". The characteristic of the game is that each player gains when both cooperate, but if only one of them cooperates, the other one, defecting, will gain more. In the case both defect, both lose (or gain very little). The point is that, when each player must decide his/her move, he/she is unaware of the move made by the other player.

The game was named in this way because of the reference to the following hypothetical situation: two criminals are arrested under the suspicion of having committed a crime together.

The police do not have sufficient proof against them. The two prisoners are kept isolated from each other. The police visit each of them and offer a deal: the one who offers evidence against the other one will be freed.

Three situations are possible:

- If neither of them accepts the offer, they are in fact cooperating against the police, and neither of them will be sentenced to imprisonment because of lack of proof. In this case they both gain.

- If one of them betrays the other by confessing to the police, the defector will gain more, since he is freed. The one who remained silent, the betrayed, will receive the full punishment, since he did not help the police, and there is sufficient proof.

- If both betray, both will be punished, but less severely than if they had refused to talk. The dilemma resides in the fact that each prisoner has a choice between only two options, but cannot make a sound decision as he does not know what the other one will do.

Such a situation of losses and gains applies to many real life situations, in which the cooperating agent whose action is not returned will lose resources to the defector, without either of them being able to collect the advantages from the "synergy" of their cooperation.

The theoretical and experimental analysis of this game was devoted to the eventual occurrence of effective strategies, especially when the game has *more than two players*.

The situation becomes particularly interesting when the game is played in an *iterated* way within the same group of players, thereby allowing partial time histories of game behavior to influence future decisions. This version, the so-called *iterated prisoner dilemma game*, has been of interest for game theorists (Pessa *et al.*, 1998).

We cannot avoid mentioning here the research carried out by Robert Axelrod while searching for a winning strategy for *iterated prisoner dilemma* games. In 1981 he organized a computer tournament between 200 different strategies proposed by 14 game theorists. The game has been played in computer tournaments, using various computer procedures against one another. A very simple strategy, called "tit for tat" by its submitter Anatol Rapoport, was the overall winner. The strategy was based on cooperating on the first move and then doing whatever the opponent did in the previous move. It is a strategy based on *reciprocity*. This strategy had great influence on understanding and modeling the evolution of cooperative behavior in societies.

However, this result leaves unanswered the general question about the existence of a unique optimal strategy for games with incomplete information such as the iterated prisoner dilemma games. On this point it should be stressed that the establishment of a strategy in such games can be seen as entirely equivalent to reaching an equilibrium point in complex dynamic system behaviors, so that we can identify the strategies with the equilibrium points of evolutionary histories of game behaviors. The question of the existence of a unique optimal strategy thus translates into the question of whether a given game with incomplete information allows for a unique equilibrium point. The most important result on this was obtained by J. Nash, who showed how, in a game with incomplete information whose complexity is great enough (such as in the iterated prisoner dilemma game) we never have a single equilibrium point, but a multiplicity of different equilibrium points. Such a result destroys the hopes of those looking for the *best* strategy in such games, because a strategy, as such, simply does not exist in principle. What we can have is only a momentary choice of a given strategy (equivalent to reaching one of the various equilibrium points) dictated by the particular history of interactions between the players which occurred during the game. Of course, the effect of such a history is strongly dependent on the cognitive features of the players themselves, that is on their memory span (how many past moves can be remembered?) and on their ability to infer rules on the basis of observed previous behaviors. The existence of such a complicated situation definitely rules out the possibility of applying traditional economic theories based upon rational behavior to the study of games such as those mentioned above and, perhaps, to all real economic behaviors. That is, as the concept itself of unique equilibrium point and of optimal strategy or solution is lacking, it is impossible to resort to the latter for explaining real behaviors within complex environments.

Technically, the reference is to the so-called *folk theorem* of evolutionary game theory (Hofbauer and Sigmund, 1998) which has been proven by several authors simultaneously in the early 1950s. These results originated from John Nash's work on equilibria in game theory (Nash, 1950a; 1950b; 1951). Nash's 27-page dissertation, "Non-Cooperative Games," written in 1950 when he was 21, was later honored with the Nobel Prize in Economics in 1994.

In the historical survey on game theory published by Robert J. Aumann (Aumann, 1987) it is possible to find references and information about the non-cooperative equilibrium concept and bargaining model introduced by Nash in 1950-1951. Applications include those taking an explicitly non-cooperative game theoretical approach for modeling oligopoly, as in Shubik (Shubik, 1959).

As a consequence of this new perspective, even Artificial Intelligence has begun to investigate the usage of evolutionary processes to produce systems with greater abilities than those of the originally designed systems, abilities emerging when the systems themselves are placed in competition/cooperation with other systems. It is to be expected that cooperative/competitive strategies emerge from these evolutionary processes, such as the so-called *Evolutionary Stable Strategies (ESS)*. This approach has been applied to model ecosystems, (Huberman and Hogg, 1988; 1993), biological systems (Hines, 1987; Schuster, 1998), and markets (Gints, 2000). According to this approach a theory of human behavior could be based upon two simple ideas:

- people (or intelligent agents) form rules and habits of action regulating their decisions in complex social environments (as opposed to precisely calculating the right decision in every situation);
- the rules which people (or intelligent agents) use are those which survive a process of evolutionary selection, where the suitability of a rule depends on how well it works.

These two ideas constitute the core of the evolutionary game theory.

In the field of economics *Evolutionary Game Theory* is one of the most active and rapidly growing areas of research. Traditional game theory models assume that all players are fully rational and have a complete, detailed knowledge of the game. Evolutionary models assume that people (agents) choose their strategies through a trial-and-error learning process. During this process they discover that some strategies work better than others. In games the process is repeated many times: strategies with little effectiveness tend to be weeded out, and an equilibrium may emerge. Larry Samuelson has been one of the main contributors to the evolutionary game theory literature. He examined the interplay between evolutionary game theory and the equilibrium selection problem, with special reference to non-cooperative games (Samuelson, 1997). Topics in this field of research include evolutionary stability, backward and forward induction, the dynamics of sample paths, the ultimatum game, drift and noise.

These approaches are particularly suitable for implementation using sub-symbolic techniques, such as Neural Networks (Bishop, 1995; Pessa, 1994; Rojas, 1996), where the system may learn from a training set. In this case the training set is organized on the basis of a usage of multiple approaches, allowing the system to use mistakes of one model combined with the mistakes of others to point towards the correct solution.

To end this section, we mention that other approaches present in the literature on multiple-model usage are related to dynamic user modeling for software (Runeson and Wohlin, 1982) and to mixing and matching methodologies (Mingers and Gill, 1997).

2.4.2 Implementing DYSAM

The notion of the DYnamic uSAge of Models (DYSAM) (Minati, 2001; Minati and Brahms, 2002) is related to situations in which the system to be studied is so complex that we cannot, *in principle*, fully describe it using a single model nor a sequence of models, each being a refinement of the preceding one. **This happens when the process of emergence gives rise to the dynamic establishment of different systems, multi-systems, as with Collective Beings.** Within classical scientific inquiry such a case was never taken into consideration, as the systems investigated were so simple (a planet revolving around the sun, a steam engine, a moving electric charge) that a single model was largely sufficient to describe their behavior. The problem was rather the choice of the right model (or the most suitable one), taking for granted that the problem had one and only one solution. However, it is clear that such a fortunate situation does not occur with biological, economic, social and cognitive systems. It should also be noted that even at the beginning of the Twentieth Century it was shown by the Founding Fathers of Quantum Theory that even for most physical systems we could not, in principle, describe all possible behaviors by resorting to a single kind of model.

An account of DYSAM can be given by answering three fundamental questions:

- What is it?
- When should it be introduced?
- How does it work?

Regarding the first question, a first rough answer is that DYSAM refers to the ability to select, among those available, a number of different models of a given system, in order to use them in a systemic way, when faced with the problem of understanding the behavior of that system. The expression "in a systemic way" means that the investigator, while being aware of and notwithstanding the intrinsic limitations of each different model of the same system, makes use of their results in an "intercrossed way", taking into account even the mistakes and not simply avoiding them. Such a process, in turn, depends upon:

a) past, current or expected contextual information;
b) the global context, without reference to the specific decision, or selection to be made;
c) the behavioral strategies adopted;
d) any kind of recorded information used to reach a decision;
e) processes occurring within the investigator cognitive system and regarding emotional activity, affection, attention, perception, inference and language production or understanding.

More precisely, we can say that DYSAM occurs when, in concomitance with specific circumstances described in points a) - e) above, the investigator, equipped with a *general optimality criterion* deriving from the goals pursued in studying a given system, finds that a number of different models of this system fulfill such a criterion to exactly the same degree. Thus, DYSAM consists of using the set of these models (*equivalent* with respect to their optimality) to:

- compute an *evaluation function*, which associates a *strategic value* to each operation related to the use of a particular subset (even down to a single element) of the whole model set;
- derive a *strategy* for the use of these models based on a *local optimality criterion*.

Thus, DYSAM can be viewed as equivalent to the behavior of a player in a chess game where, in each situation, a number of alternative moves are possible, all equivalent with respect to the final goal (to win the game), and where a strategy, based on evaluations of possible game situations and on local heuristics, is required to select the next move to be made.

This definition of DYSAM helps us to answer the second of the three fundamental questions presented above, which is to understand in which situations a recourse to DYSAM is mandatory. These include the following cases:

- a system which can be described only through a number of *different partial representations* (here the concept of "representation" is assumed to be defined in a well-specified mathematical way). Such a case, for instance, occurs when speaking of corpuscular and wave representations of atomic phenomena, or when speaking of the different, not unitarily equivalent, representations of a quantum matter field (which can be thought of as different matter "phases"); contrary to common belief, even the collective and the individual description of system behaviors are nothing but two different partial representations of them and, as such, neither can pretend to fully describe that system;

- a system allowing for a number of different equilibrium behaviors, which have the same probability.; Such a case occurs in spontaneous symmetry breaking phenomena well known in physics and used in models of intrinsic emergence (Umezawa, 1993; Crutchfield, 1994; Pessa, 1998). A similar circumstance, however, also characterizes the behavior of neural networks, cellular automata and artificial life systems;
- a system whose models must allow for the introduction of noise or fuzziness, related to individual and unforeseeable phenomena, as is the case for biological, socio-economic or cognitive systems.
- a system whose models must necessarily incorporate a model of the observer of that system, or a model of the model builder.

The various cases appearing in this list may partially overlap and theoretical physics offers a number of such examples. What should be stressed is that a DYSAM-based approach may be suitable for modeling emergent, dynamic phenomena with particular reference to those associated with *intrinsic emergence*. In this case it is not only impossible to foresee a specific behavior (even if compatible with model assumptions), but its settling may give rise to profound changes in system structure leading to the need for formulating a new model of the system itself. Such a conceptual schema may allow for a better approach to the problem of modeling emergence while taking into account the crucial role of the observer, using the latter's own models. It is not a matter of recognizing in the emergence process what is expected by the observer but to *reconstruct* the emergence process in a plausible way using available models.

Turning now to the third fundamental question, the DYSAM approach may be considered as being articulated in five components (for a graphical description of the conceptual scheme of DYSAM, see Figure 2.2):

A. Identification of a database of symbolic and non-symbolic, deterministic and non-deterministic, dynamic and non-dynamic, context- and non-context-sensitive, etc., models;
B. Identification of information about the context of the system/phenomenon to be modeled;
C. Selection of models considered by the user to be suitable for the decision process later described in E;

And, in real time or not, in parallel, synchronously or sequentially, depending on the kind of process to be dealt with by DYSAM:
A. Dynamic identification of levels of representation of the case to be modeled which allow *multi-model*-based processing;
B. *Dynamic usage of models* taking place when the **decider** is:

- an artificial system, such as a computer program, using (1) inter-crossed results from different models (i.e., decisions are taken by considering results of different models reciprocally *linked*), (2) what are assumed to be *errors* in single models, in order to evaluate and decide, (3) results with reference to the *recorded usage* of models in different contexts, *learning* processes and contextual information;
- a human being adopting strategies which can be modeled by using even **combined** different approaches such as: optimizing, value-based acting, modeled by (evolutionary) game theories, etc.

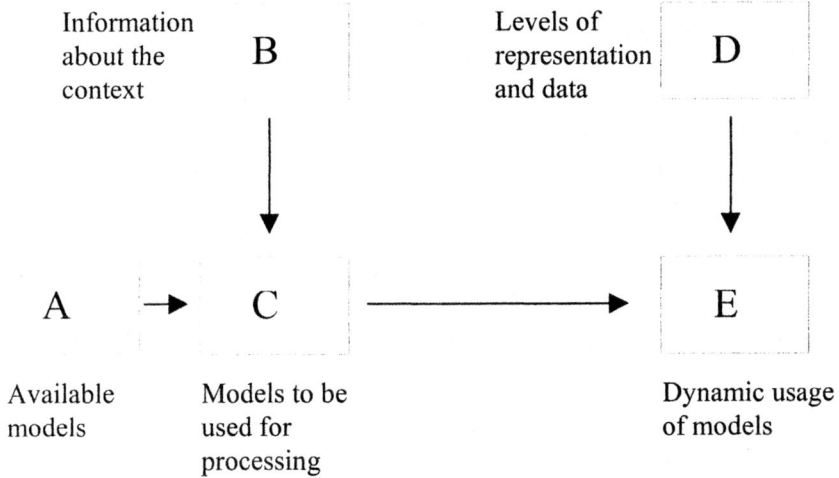

Figure 2-2. Conceptual scheme of a **DYnamic uSAge of Models (DYSAM).**

The core of the DYSAM approach is well represented by component E. In short, the concept of DYnamic uSAge of Models refers to multiple modeling, to multiple cognitive models used by the agent (assumed to make decisions or strategies for using resources) and to their simultaneous processing, at all the various possible levels of representation, and to account for the different strategies considered within the same level of representation.

It should be noted that DYSAM exploits the parallel processing (even though it is simulated in a sequential way) of different models related to the same level of description, so as to take into account different results *simultaneously*. The goal is not restricted to compare *only* effectiveness and suitability, but to carry out a learning, evolutionary, emergent system of approaches which are able to use in the *locally best* way the resources available.

Thus, DYSAM is not a single, procedural, rule-based methodology, but a systemic general model, a *meta-model* (i.e., a model of models), used to carry out single, contextual methodologies.

Realizing how much a methodology may be related to DYSAM is equivalent to enquiring about its robustness, i.e., how systemic it is, because all possible local aspects have been considered.

DYSAM should be applied as a methodology for decision-making, for dealing with complex systems and for designing the usage of various resources to reach a goal.

Multiple representations and multiple strategies must be taken into consideration when dealing with dynamical systems whose complexity calls for approaches beyond traditional ones. DYSAM is particularly useful when considering Collective Beings (described below in Chapter 3) where elements dynamically use different cognitive models.

A very good conceptual framework for the application of DYSAM is that introduced in Economic Theories and then implemented using game theory, already considered in the previous section. Classical Economic theories proposed the utility function as an objective datum, independent from the economic agent itself. The problem of decision has been classically intended as a problem which is conceptually equivalent to solving optimization processes between alternative actions. The evolutionist approach, on the contrary, adopts a very different point of view. In evolutionary theory the utility function is considered momentarily as a result of organism-environment interactions. In the previously mentioned Evolutionary Game Theory (Taylor and Jounker, 1978) optimization is assumed to take place, to **emerge**, through a selection mechanism (Ackley and Littman, 1990), based upon the replication of successful strategies.

This approach works especially in situations like that in the iterated *prisoner dilemma game* (see above), now considered within a *systemic* framework (Cho, 1995), that is, by considering the human subject taking decisions (the player) and its environment (in this case, the rival player) as subsystems of a more general subject-environment system. From this point of view each of them has the goal of reaching an *equilibrium* state of the *whole system*, instead of each subsystem taken individually. Contrary to classical theories, this implies that the behavior of the individual player cannot be described as aiming at the maximization of a suitable individual utility function. The strategic choices of each player change with time as a function of accumulated experience during the game and, chiefly, of the results of the interaction with the other player. Strategies are thus viewed as dynamical equilibrium configurations of the whole system constituted by the two players. And, as we have already stressed, Nash's results gave the mathematical foundation to this point of view.

In classical Decision Theory the focus is on finding a fixed, objective, utility function. In the evolutionary approach what is fixed is only the form of the law through which a subject's decisions vary as a function of the results obtained in the previous move, and of their initial motivational factors. DYSAM is a suitable approach when in the decider(s)-environment system there is more than one equilibrium state, which means more than one utility function, more than one model.

We have already mentioned the suitability of the DYSAM approach for phenomena where, *in principle,* it is impossible to use one single model.

This refers particularly to emergent phenomena (introduced below in Chapter 3) for which we do not have, by definition, a model for what is emerging.

Examples of the application of DYSAM include deciding corporate strategies; usage of remaining resources in a damaged system (i.e., disabled); use of various single resources in collective ways (such as public facilities and information); learning the usage of the five sensory modalities in the evolutionary age (the purpose is not to select the best one but to use all of them together).

When is DYSAM assumed to be inappropriate? When there is not sufficient time for parallel processing of multi-models; when the complexity of the considered system is not known; when too many models are available without distinctions being made; when it is not possible to represent different levels, and only a single model is available.

DYSAM is conceptually related to *logical openness* (Minati *et al.*, 1998) introduced in Appendix 1 and related to the ability of the system to *decide* which level of openness to use depending on the context. This contrasts with the classical view which states that a system may be open *or* closed. In this modern view openness is a property which is dynamically managed at different levels by the cognitive system. The logical approach to systemic openness is applied in disciplinary fields including the *Human Machine Interface* (Carroll, 1989) and *User Modeling* (Allen, 1990; Kobsa, 1993; McTcar, 1993; Stewart and Davies, 1997).

2.4.3 A model of DYSAM

The previous discussion about DYSAM, while stressing the fact that it must be viewed as a conceptual framework for dealing with complex systems rather than a precisely defined procedure, neglected in some way the fact that DYSAM can be concretely implemented, and in a virtually unlimited number of different ways. Here, contrary to the approach taken in the Sections above, we will focus on a particular implementation of DYSAM to illustrate, on one hand, the problems connected with the implementation itself and, on the other, peculiarities in the behavior of the observer/researcher adopting a particular DYSAM implementation.

In order to avoid high computational costs while allowing, at the same time, a computer simulation of the time evolution of the observer/environment system, we will assume that the models available to the observer can be represented through particular kinds of *neural networks*. As is well known, a neural network is a generic system consisting of *units*, susceptible of being activated as a function of suitable input signals, and by *weighted interconnections*, allowing for transmission of the activation signals from the output of given units to the input of others. Each model available to the observer will be represented by a neural network with two layers of units and feedforward connections, as depicted in Figure 2.3 below.

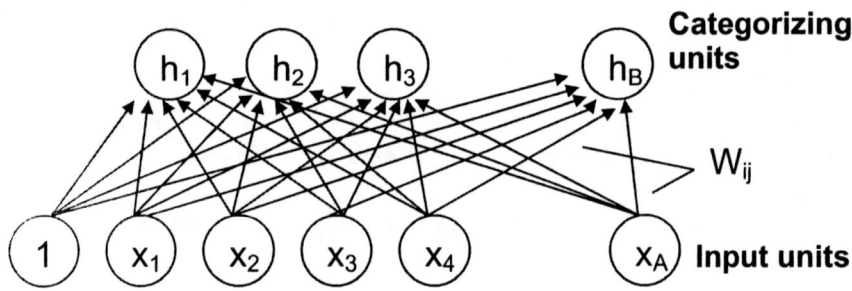

Figure 2-3. Representation of a neural network with two layers of units and feedforward connections

The drawing represents a *categorizer* neural network, where each categorizing unit is associated with a different category and receives inputs, through the feedforward connections with connection weights w_{ij}, from the input units, which are in turn activated, at each instant in time t, by the signal coming from the external environment at that same instant. We will assume that this kind of network, within our model, will categorize the input vector $x_i(t)$, occurring at time t, according to the following procedure:

- computation of the various Euclidean distances between the vector $x_i(t)$ and the vectors w_{ij}, where the symbol denotes the vector of weights

associated with the connection lines from the input units to the input of the i-th categorizing unit;
- insertion of the input vector into the category whose weight vector has the minimum Euclidean distance from it.

The Euclidean distance d between two vectors x_i and y_i is given by the well-known formula:

$$d = [\Sigma_i (x_i - y_i)^2]^{1/2}$$

The logical scheme underlying our model can now be described as follows:

a. at each time instant t the environment sends a signal to the organism; this signal is represented by an input vector; the latter, however, is associated with the index of the *correct category* into which that signal has to be placed by the organism; the latter, on the other hand, has no knowledge of the correct category;

b. at the same instant the organism can rely upon a repertoire of *active networks* (corresponding to the possible models which could be used simultaneously within a DYSAM procedure), that is, of networks whose *fitness* is greater than a given minimum threshold *fth*; if, at this instant in time, this repertoire is empty, then a new network is created, in which the weight vector associated with the correct category coincides with the input vector, whereas the weight vectors of the other categories are chosen randomly; an initial fitness slightly greater than the minimum fitness threshold is then associated with this new network; after this creation the system goes to the new input vector present at time $t + 1$;

c. in the case where the repertoire of active networks is not empty, first a control is made on the whole set of networks, to detect those whose fitness is lesser than a given minimum admissible value *fmin*; the latter are immediately eliminated so as to disappear from the available network set; then the set of remaining networks is checked to ascertain whether there is a network with at least one of its weight vectors whose Euclidean distance from the input vector is less than a given critical distance *dmax*; the networks endowed with such a property are counted as active networks, even if their fitness is less than the minimum threshold;

d. for every active network we determine the index c of the category into which it puts the input signal, according to the procedure based on the method of the minimum Euclidean distance described above; then the network considered receives a *reinforcement signal*, which depends upon the correctness of the categorization it carried out, that is upon the comparison among the index of the category it chose and the index cd of the correct category associated with the input vector; the reinforcement signal r is computed using the formula:

$$r = [1/(1 + /c - cd /)] (fmax - f)$$

where f denotes the current fitness of the considered network, and where $fmax$ is the maximum admissible value for the fitness of every network in the repertoire;

e.   if a network receives a reinforcement signal exceeding a given reinforcement threshold rth, then its fitness is increased by a suitable amount df; moreover, the network itself undergoes an unsupervised training on the particular input pattern taken into consideration for a number $tapp$ of time steps, following the so-called Kohonen's rule (Kohonen, 1984), in which the weight vector closest (lowest Euclidean distance) to the input vector is updated according to the formula:

$$\Delta w_{ij} = \alpha(t)(x_j - w_{ij})$$

where the *learning parameter* $\alpha(t)$ changes with time according to:

$$\alpha(t) = \eta_0 \, exp(- \beta_0 \, t)$$

in the opposite case its fitness is decreased to a lower value f' given by:

$$f' = f(1 - p)$$

where p is a suitable parameter; after updating the fitness of all active networks, the system goes to the input presented by the environment at time $t + 1$ and the whole cycle restarts.

From this description of the model operation, it is clear that its evolutionary features depend crucially upon the characteristics of the sequence of inputs coming from the environment. The latter, in turn, are specified by:

• the number of different possible input patterns
• the probability of occurrence of each input pattern
• the law of association between input patterns and correct categories.

On intuitive grounds, one would expect that, in situations in which the environment behaves in a simple and predictable way, the behavior of the model should differ from a DYSAM-like one. These situations occur when the number of possible different input patterns is very small or, more frequently, when their probability of occurrence is described by a distribution which is sharply "peaked" over a small number of patterns. Besides, the law of association between input patterns and correct categories should not change with time. In such situations, the input patterns are always more or less the same and the model does nothing but to *learn* the statistical structure of a fairly simple environment. No DYSAM strategy is required, except in the initial phases, because a single network (that is a single model) is sufficient to capture the environmental regularities. In the opposite situation, however, in which the probability distribution for the occurrence of input patterns tends to be nearly flat, a DYSAM-like behavior should develop, and such an effect should increase in the presence of a variation with time of the law connecting the input patterns with the correct categories.

The correctness of these arguments were tested using a number of computer simulations of model behavior. Within all simulations the number *nc* of categorizing units was the same for all networks and kept constant over the whole model evolution. Besides, all input and weight vectors were *normalized* in such a way as to keep their modulus always equal to 1. The initial values of weight vectors were chosen randomly between + 0.1 and − 0.1. A further constraint was that the minimum fitness threshold *fth* was always given by *fmax* /2.

A first interesting example of the outcome of a simulation was obtained in correspondence with an evolutionary history of the environment of 2000 steps, based upon 100 different possible inputs, each one of them represented by a randomly chosen 5-component vector with unit modulus. The whole duration of the evolutionary history was subdivided into 20 different *epochs*, each one lasting for 100 steps. The probability distribution for the occurrence of each pattern changed from one epoch to another in such a way as to tend toward uniformity, as a consequence of the relationship:

$$P(i, t + 1) = P(i, t) + eps \,[PM(t) - P(i, t)]$$

where $P(i, t)$ denotes the probability of occurrence of the i th input pattern at epoch t, $PM(t)$ is the average probability of occurrence at the same epoch, and *eps* is a parameter. We set *eps* = 0.1. The probability distribution for the occurrence of input patterns in the first epoch was chosen as Gaussian, with mean = 50 and variance =1. Moreover, the number of possible categories was 5 and the correct classification of input patterns was achieved through a particular categorizing neural network, whose weight vectors were chosen randomly. This network lasted unchanged for all epochs, so the law of association between input vectors and correct categories did not change with time.

Regarding the model evolution corresponding to the evolutionary history of the environment resulting from the above choices, we started with an initial repertoire of 10 networks. The values of the other parameters were: $fmax = 1$; $rth = 0.45$; $df = 0.1$; $p = 0.4$; $fmin = 0.3$; $dmax = 0.2$; $tapp =$ 100 steps ; $\eta_0 = 0.1$; $\beta_0 = 0.01$. The first interesting result of the simulation is that the number of active networks (that is, of effectively used ones) does not change with time in an ordered way, as shown in the graph reported in Figure 2.4 below.

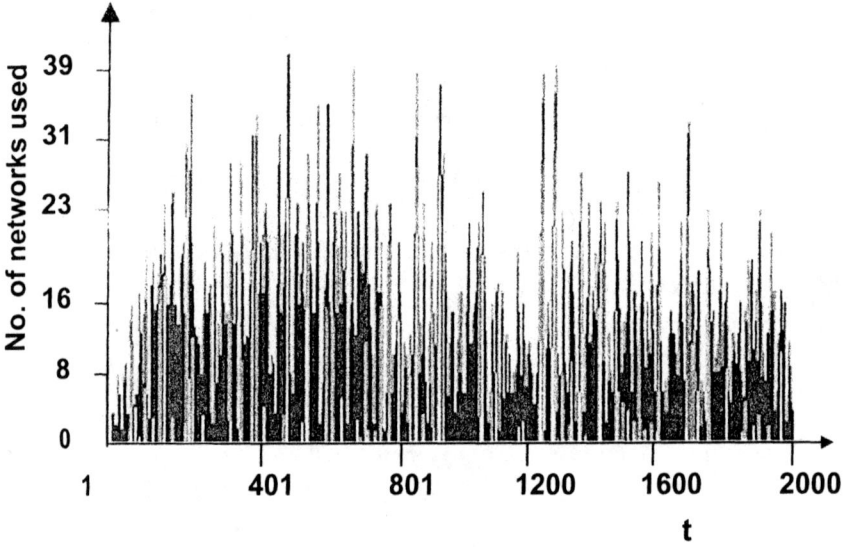

Figure 2-4. Number of networks used vs. time

As we can see, sometimes the model behavior is DYSAM-like and sometimes it is not. This could, of course, depend upon the complexity of some environmental features. To detect the possible causes of this behavior we resorted to a suitable measure of a particular kind of complexity, that associated with the degree of uniformity of the probability distribution for input pattern occurrence. Namely, if this distribution is highly "peaked", then only a very few patterns are associated with a high probability of occurrence, and the environmental behavior is, in some sense, more predictable. If, on the contrary, the distribution is "flattened", then all input patterns are, more or less, associated with the same probability of occurrence and the world becomes unpredictable and, therefore, more complex. As a measure of the "flatness" of the distribution we chose the so-called *Excess*, defined through the usual formula contained in Statistics textbooks:

Excess = $[\mu_4/(\sigma^2)^2] - 3$

where μ_4 is the fourth-order centered moment of the input pattern distribution, given by:

$\mu_4 = \{\Sigma_j [j\, freq(j) - me]^4\}/ftot$

and *freq(j)* denotes the frequency of occurrence of the *j*-th input pattern within a given epoch, whereas:

$me = [\Sigma_j j\, freq(j)]/ftot, \quad ftot = \Sigma_j freq(j)$

Besides, σ^2 , as usually, is defined by:

$\sigma^2 = \{\Sigma_j [j\, freq(j) - me]^2\}/ftot$

It is well-known that when *Excess* = 0 the distribution is perfectly Gaussian, whereas when *Excess* is greater than zero, the distribution is more "peaked" than a Gaussian one. In general, a diminution of the value of *Excess* is to be interpreted as an increase in the "flatness" of the distribution.

Within our evolutionary history we can compute the *Excess* for each epoch, and it is possible to show that its value decreases with time (in a fluctuating way), as shown in Figure 2.5 below.

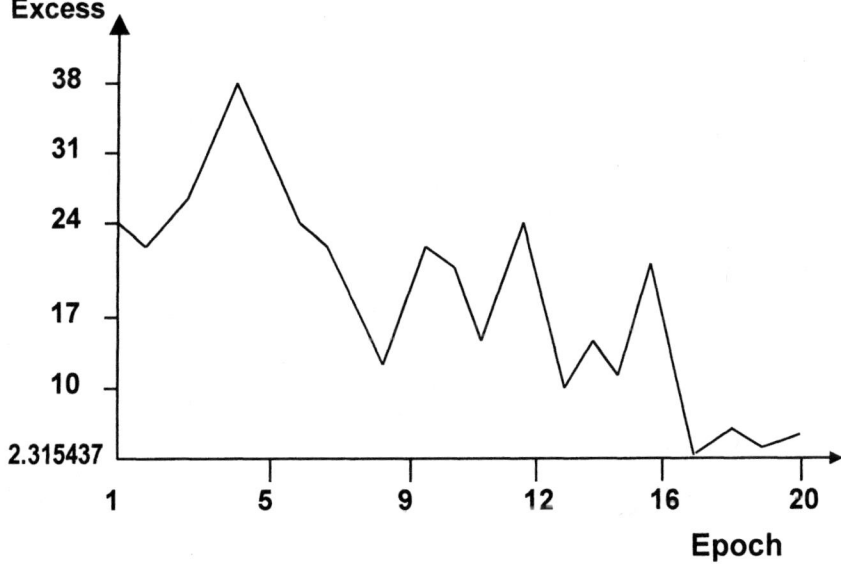

Figure 2-5. Excess *vs.* Epoch

If now we plot the number of used networks (averaged over each epoch) vs. the *Excess*, we find that there is no clear indication of a decrease in the number of used networks with a decrease in the *Excess* (see Figure 2.6).

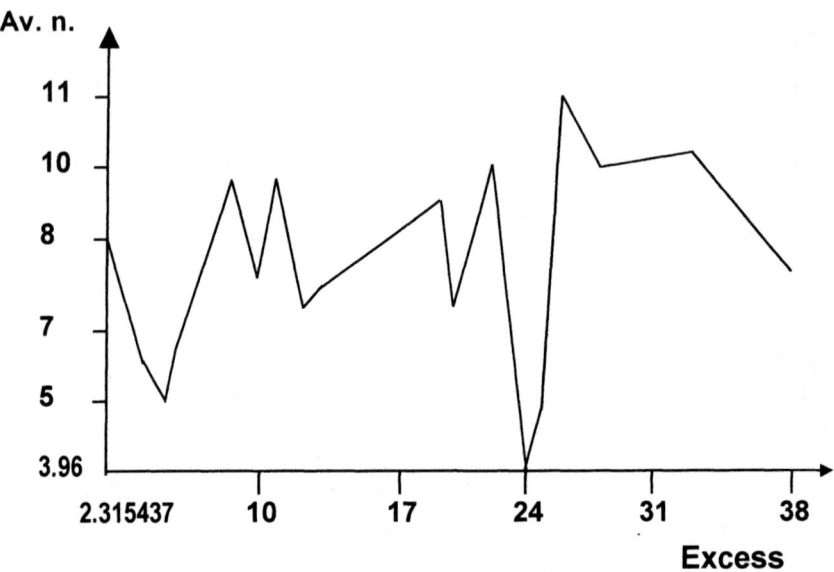

Figure 2-6. Average number of networks used *vs.* Excess

In other words, the DYSAM-like behavior does not seem to be due to the flattening of the distribution of input patterns. This behavior could rather be due to the high number of different possible input patterns.

A surprise came with a second simulation, in which, by keeping constant all features and parameter values used in the previous simulation, we allowed for a random change of the law of association between input patterns and correct categories for different epochs. First of all, we obtained the disappearance over time of DYSAM-like behavior, as shown by the graph giving the number of used networks *vs.* time, reported in Figure 2.7.

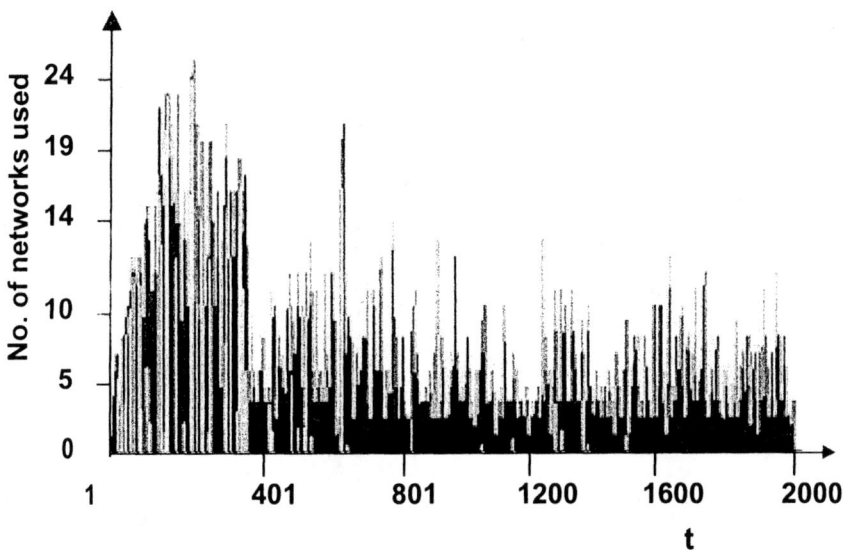

Figure 2-7. Number of networks used vs. time

As, even in this case, the *Excess* is decreasing with time, it is instructive to plot the number of used networks (averaged over each epoch) *vs.* the *Excess* value. The graph is shown in Figure 2.8 below.

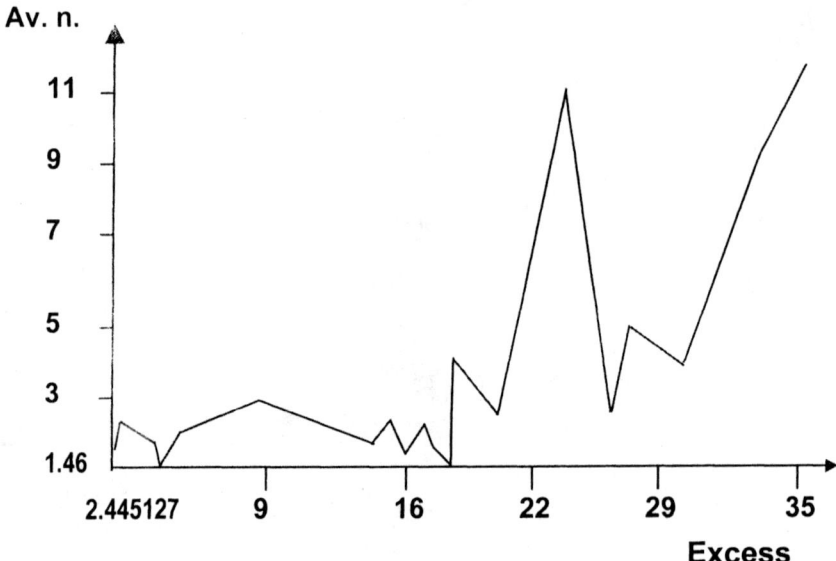

Figure 2-8. Average number of networks used *vs.* Excess

Here there is a clear and strong decrease in the number of networks used with decreasing *Excess* values. In other words, when the rules underlying the world change with time (as in our case), there is a strong tendency toward a sort of *rigidity*, avoiding a DYSAM-like behavior, by adopting, when the complexity of the world is low enough (low *Excess*), the smallest possible number of models. Paradoxically, as the unpredictability of the world increases (as induced by a change with time of the rules underlying the environmental behavior), we observe, rather than a more frequent resort to a DYSAM-like strategy, a sort of *obsessive focusing* on the smallest possible number of models, as if we were blocked by the *fear* of an environment which we assume to be unmanageable.

Even the small number of results emerging from these computer simulations indicate just how friutful a study of models of DYSAM-like behaviors could be for those interested in the study of human behavior, of decision and game theory, of the foundations and the development of science. Without pursuing further this line of argument the need for a greater engagement of the systems community in building and studying models of this kind should be stressed. The latter constitute a first step in modeling observer-environment inter-relationships.

References

Ackley, D. H., and Littman, M. S., 1990, Learning from natural selection in an artificial environment. In: *Proceedings of the International Joint Conference on Neural Networks (IJCNN)*, Washington, DC, vol. I , Erlbaum, Hillsdale, NJ, pp. 189-193.

Allen, R. B., 1990, User model: theory, methods, and practice. *International Journal of Man-Machine Studies* **32**:511-543.

Andreewsky, E., and Bourcier, D., 2000, Abduction in Language interpretation and Law making, *Kybernetes* **29**:836-845.

Aumann, R. J., 1987, Correlated Equilibrium as an Expression of Bayesian Rationality, *Econometrica* **55**:1-18.

Axelrod, R., 1984, *The evolution of cooperation,* Basic Books, New York.

Axerold, R.,1997, *The Complexity of Cooperation: Agent-Based Models of Competition and Cooperation.* Princeton University Press , Princeton, NJ.

Bauer, E., and Kohavi, R., 1999, An Empirical Comparison of Voting Classification Algorithms: Bagging, Boosting and Variants, *Machine Learning* **36**:105-139.

Bayes, T., 1763, An Essay Toward Solving a Problem in the Doctrine of Chances. In: *Philosophical Transactions of the Royal Society of London* **53**:370-418; reprinted in *Biometrika* **45**:293-315 (1958), and in *Two Papers by Bayes* (W. E. Deming, ed., 1963) Hafner, New York.

Berger, P. L., and Luckmann, T., 1966, *The Social Construction of Reality.* Penguin Books, New York.

Bishop, C. M., 1995, *Neural Networks for Pattern Recognition.* Oxford University Press, Oxford, UK.

Breiman, L., 1996, Bagging predictors , *Machine Learning* **24**:123-140.

Bretthorst, G., L., 1994, An Introduction to Model Selection Using Probability Theory as Logic. In: *Maximum Entropy and Bayesian Methods* (G. Heidbreder, ed.), Kluwer, Dordrecht, pp. 1-42.

Brody, D. C., and Hughston, L. P., 1997, Generalised Heisenberg relations for quantum statistical estimation, *Physics Letters A* **236**:257-262.

Bury, K. V., 1976, *Statistical Models in Applied Science.* Wiley, New York.

Butts, R., and Brown, J., (eds.), 1989, *Constructivism and Science.* Kluwer, Dordrecht.

Butz, M., V., 2002, *Anticipatory Learning Classifier Systems. Genetic Algorithms and Evolutionary Computation.* Kluwer, Boston, MA .

Carroll, J. B., (ed.), 1956, *Language, Thought and Reality: Selected Writings of B. L. Whorf.* Wiley, New York

Carroll, J. M., (ed.), 1989, *Interfacing thought: Cognitive Aspects of Human-Computer Interactions.* MIT Press, Cambridge, MA.

Cheng, J., and Ushijima, K., 1991, Partial Order Transparency as a Tool to Reduce Interference in Monitoring Concurrent Systems. In: *Distributed Environments* (Y. Ohno, ed.), Springer, Berlin-Heidelberg-New York, pp. 156-171.

Cho, I.-K., 1995, Perceptrons Play the Repeated Prisoner's Dilemma, *Journal of Economic Theory* **67**:266-284.

Churchland, P. M., 1979, *Scientific realism and the plasticity of mind*, Cambridge University Press, Cambridge, UK.

Churchland, P. M., 1995, *The engine of reason, the seat of the soul. A philosophical journey into the brain*, MIT Press, Cambridge, MA.

Cruchtfield, J. P., 1994, The Calculi of Emergence: Computation, Dynamics and Induction, *Physica D* **75**:11-54.

Deneubourg, J. L., Goss, S., Franks, N., and Pasteels, J. M., 1989, The blind leading the blind: Modeling chemically mediated army ant raid patterns, *Journal of Insect Behavior* **23**:719-725.

Dietterich, T. G., 1997, Machine Learning Research: Four Current Directions, *AI Magazine* **18**:97-136.

Dietterich, T., 2000, Ensemble methods in machine learning. In: *Proceedings of the First International Workshop on Multiple Classifier Systems* (J. Kittler and F. Roli, eds.), Springer, New York, pp. 1-15.

Diettrich, O., 2001, A Physical Approach to the Construction of Cognition and to Cognitive Evolution, *Foundations of Science*, **6**:273-341.

Feng. C., Sutherland, A., King, R., Muggleton, S., and Henery, R., 1994, Comparison of machine learning classifiers to statistics and neural networks. In: *Selecting Models from Data: Artificial Intelligence and Statistics IV* (P. Cheesemnan and R. W. Oldford, eds.), Springer, Berlin, pp. 41-52.

Flood, R. L., and Jackson, M. C., 1991, *Creative Problem Solving: Total Systems Intervention.* Wiley, Chichester, UK.

Franks, N. R., Gomez, N., Goss, S., and Deneubourg, J. L., 1991, The blind leading the blind: Testing a model of self-organization (Hymenoptera: Formicidae), *Journal of Insect Behavior* **4**:583-607.

Gintis, H., 2000, *Game Theory Evolving: A Problem-Centered Introduction to Modeling Strategic Interaction.* Princeton University Press, Princeton, NJ.

Hall, M. J. W., and Reginatto, M., 2002, Schroedinger equation from an exact uncertainty principle, *Journal of Physics A* **35**:3289-3303.

Heisenberg, W., 1971, *Physics and Beyond*. Harper & Row, New York.

Herbrich, R., 2001, *Learning Kernel Classifiers: Theory and Algorithms.* Adaptive Computation and Machine Learning, Series, MIT Press, Cambridge, MA.

Hines, W. G., 1987, Evolutionary Stable Strategies: A Review of Basic Theory, *Theoretical Population Biology* **31**:195-272.

Hinton, G. E., and Van Camp, D., 1993, Keeping neural networks simple by minimizing the description length of the weights. In: *Proceedings of the Sixth Annual Conference on Computational Learning Theory* (L. Pitt, ed.), ACM Press, New York, pp.5-13.

Hofbauer, J., and Sigmund, K., 1998, *Evolutionary Games and Population Dynamics.* Cambridge University Press, Cambridge, UK.

Holland, J. H., Holyoak, K. Y., Nisbett, R. E., and Thagard, P. R., 1986, *Induction*. MIT Press, Cambridge, MA.

Huberman, B. A. and Hogg, T., 1988, The behavior of computational ecologies. In: *The Ecology of Computation* (B. A. Huberman, ed.), Elsevier North Holland, Amsterdam, pp. 77-115.

Huberman, B. A., and Hogg, T., 1993, The Emergence of Computational Ecologies. In: *Lectures in Complex Systems*, (L. Nadel and D. Stein, eds.), Addison-Wesley, Reading. MA, pp.163-205.

Kay, S. M., 1993, *Fundamentals of Statistical Signal Processing: Estimation Theory*. Prentice-Hall, Englewood Cliffs, NJ.

Kobsa, A., 1993, User Modeling: Recent work, prospects and hazards. In: *Adaptive User Interfaces: Principles and Practice* (M. Schneider-Hufschmidt, T. Kuhme, and U. Malinowski, eds.), Elsevier Science Publishers B. V., Amsterdam, pp. 111-128.

Kohonen, T., 1984, *Self-organization and Associative Memory*, Springer, Berlin.

Lanzi, P. L., Stolzmann, W., Wilson, S. W., (eds.), 2002, *Advances in Learning Classifier Systems*. Springer, Berlin.

Laplace, P. S., 1814/1951, *A Philosophical Essay on Probabilities, unabridged and unaltered reprint of Truscott and Emory translation*. Dover, New York (original work published in 1814).

Maynard-Smith, J., 1982, *Evolution and the Theory of Games*. Cambridge University Press, Cambridge, UK.

McTear, M., 1993, User modeling for adaptive computer systems: A survey of recent developments, *Artificial Intelligence Review* 7:157-184.

Minati, G., 2001, Experimenting with the DYnamic uSAge of Models (DYSAM) approach: the cases of corporate communication and education. In: *Proceedings or the 45th Conference of the International Society for the Systems Sciences* (J. Wilby and J. K. Allend, eds.), Asilomar, CA, 01-94, pp. 1-15.

Minati, G., and Brahms, S., 2002, The DYnamic uSAge of Models (DYSAM). In: *Emergence in Complex Cognitive, Social and Biological Systems* (G. Minati and E. Pessa, eds.), Kluwer, New York, pp. 41-52.

Minati, G., Penna, M. P., and Pessa, E., 1998, Thermodynamic and Logical Openness in General Systems, *Systems Research and Behavioral Science* 15:131-145.

Mingers, J., and Gill, A., (eds.), 1997, *Multimethodology: Towards Theory and Practice and Mixing and Matching Methodologies*. Wiley, Chichester, UK.

Nash, J., 1950a, The bargaining problem, *Econometrica* 18:155-162

Nash, J., 1950b, Equilibrium points in n-person games. In: *Proceedings of the National Academy of Sciences of the USA* 36:48-49.

Nash, J., 1951, Non-Cooperative Games, *Annals of Mathematics* 54:286-295.

Peirce, C. S., 1998, Harvard Lectures on Pragmatism. In: *The Essential Peirce: Selected Philosophical Writings, 1893-1913*, (N. Houser, J. R. Eller, A. C. Lewis, A. De Tienne, C. L. Clark and D. B. Davis, eds.), Indiana University Press, Bloomington, IN, Chapters 10-16, pp. 133-241.

Pessa, E., 1994, Symbolic and sub-symbolic models, and their use in systems research, *Systems Research and Behavioral Sciences* 11:23-41.

Pessa, E., 1998, Emergence, Self-Organization, and Quantum Theory. In: *Proceedings of the First Italian Conference on Systemics* (G. Minati, ed.), Apogeo scientifica, Milano.

Pessa, E., Penna, M. P., Montesanto, A., 1998, A systemic description of the Interactions between Two Players in an Iterated Prisoner dilemma Game. In: *Proceedings of the First Italian Conference on Systemics* (G. Minati, ed.), Apogeo scientifica, Milano, Italy, pp. 59-79.

Raymer. M. G., 1994, Uncertainty Principle for Joint Measurement of Noncommuting Variables, *American Journal of Physics* 62:986-993.

Rojas, R., 1996, *Neural networks. A systematic introduction*. Springer, Berlin-Heidelberg-New York.

Roy, F. B., 1998, *Physics from Fisher Information, a Unification*. Cambridge University Press, Cambridge, UK.

Runeson, P., and Wohlin, C., 1982, Usage Modeling; The basis for statistical quality control. In: *Proceedings of 10th Annual Software Reliability Symposium*, IEEE Reliability Society, Denver, CO, pp. 77-84.

Samuelson, L., 1997, *Evolutionary Games and Equilibrium Selection*. MIT Press, Cambridge, MA.

Schapire, R. E., 1990, The strength of weak learnability, *Machine Learning* 5:197-227.

Schapire, R. E., Freund, Y., Bartlett, P., and Lee, W., 1997, Boosting the margin: A new explanation for the effectiveness of voting methods. In: *Proceedings of the Fourteenth International Conference on Machine Learning (ICML '97)*, (D. H. Fischer, ed.), Morgan Kaufmann, San Francisco, CA, pp. 322-330.

Schuster, P., 1998, Evolution at molecular resolution. In: *Nonlinear Cooperative Phenomena in Biological Systems*, (L. Matsson, ed.), World Scientific, Singapore, pp. 86-112.

Schwenk, H., and Bengio, Y., 2000, Boosting naural networks, *Neural Computation* **12**:1869-1887.

Shubik, M., 1959, *Strategy and Market Structure*. Wiley, New York.

Taylor, P. D., and Jounker, L. B., 1978, Evolutionarily Stable Strategies and Game Dynamics, *Mathematical Biosciences* **40**:145-156.

Umezawa, H., 1993, *Advanced Field Theory. Micro, Macro, and Thermal Physics*. American Institute of Physics, New York.

Varela, F., Thompson, E., and Rosch, E., 1991, *The Embodied Mind: Cognitive Science and Human Experience*. MIT Press, Cambridge, MA.

Von Bertalanffy, L., 1968, *General System Theory. Development, Applications*. George Braziller, New York

Von Foerster, H., 1974, Notes pour une épistémologie des objets vivants, In: *L'unité de l'homme: Invariants biologiques et universaux culturels*, (E. Morin and M. Piattelli-Palmerini, eds.), Seuil, Paris, pp. 139-155.

Von Foerster, H, 1979, Cybernetics of Cybernetics. In: *Communication and Control in Society* (K. Krippendorff , ed.), Gordon and Breach, New York, pp. 5-8.

Von Foerster, H., 1981, *Observing Systems*. Intersystems Publications, Seaside, CA.

Von Foerster, H., 2003, *Understanding Understanding: Essays on Cybernetics and Cognition*. Springer, New York

Von Glasersfeld, E , 1991, Knowing without metaphysics. Aspects of the radical constructivist position. In: *Research and reflexivity* (F. Steier, ed.), Sage, London-Newbury Park, CA, pp. 12-29.

Von Glasersfeld, E , 1995, *Radical constructivism: a way of knowing and learning*. Falmer Press, London.

Von Neumann, H., 1996, Mechanisms of neural architecture for visual contrast and brightness perception, *Neural Networks* **9**: 921-936.

Watzlawick, P., (ed.), 1983, *The Invented Reality*. Norton, New York.

Watzlawick, P., Paul, J. H., Janet, H., and Jackson, D., 1967, *Pragmatics of Human Communication: A Study of International Patterns, Pathologies, and Paradoxes*. Norton, New York.

Watzlawick, P., Weakland, J., H., and Fisch, R., 1974, *Change-Principles of Problem Formation and Problem Resolution*. Norton, New York.

Weibull, J. W., 1995, Evolutionary Game Theory. MIT Press, Cambridge, MA.

Wigner, P. E., 1960, The unreasonable effectiveness of mathematics in the natural sciences, *Communications in Pure and Applied Mathematics*, **13**:1-14.

Web Resources

Stewart, S., and Davies, J., 1997, User Profiling Techniques: A Critical Review, In: *Information Retrieval Research* (Furner, J., and Harper, D., eds.), Springer
http://ewic.bcs.org/conferences/1997/irsg/papers/paper10.pdf

Chapter 3

EMERGENCE

In this chapter the reader is introduced to the concept of emergence through a short history of the concept, its relevance for Systemics and references to various definitions and conceptual representations. The references to the emergence of collective behaviors introduce the central subject of the chapter: the concept of *Collective Being.*

3.1 A short history of the concept

The concept of emergence and of "emergent evolutionism" were first introduced by C. L. Morgan in 1923 (Morgan, 1923) and during the same period the philosopher C. D. Broad introduced *the concept of emergent properties present at a certain levels of complexity but not at lower ones.* For a long time the topic of emergence was considered as belonging particularly to the context of biology. The attribute "emergent" was considered as a synonym of "new", "unpredictable", to underline that within the framework of biological evolution it is often possible to detect the "becoming" of certain characteristics in a discontinuous way, unpredictable on the basis of the previously existing ones. This rough and ready concept of emergence was already implicit in the proposal put forward by L. von Bertalanffy, also a biologist, of a General Systems Theory.

Corning (Corning, 2002) dates the introduction of the term of emergence even further back. He introduces an approach for defining emergence as well as for the problem of building a theory of emergence. The theoretical contribution of this paper is discussed below in this chapter. For the moment, in this initial introduction to the subject, we recall his quotation of the philosopher G. H. Lewes (Lewes, 1877), who published a collection of 5 books between 1874–79 with the overall title: *Problems of Life and Mind*:

> Every resultant is either a sum or a difference of the cooperant forces; their sum, when their directions are the same – their difference, when their directions are contrary. Further, every resultant is clearly traceable in its components, because these are homogeneous and commensurable ... It is otherwise with emergents, when, instead of adding measurable motion to measurable motion, or things of one kind to other individuals of their kind, there is cooperation of things of unlike kinds... The emergent is unlike its components in so far as these are incommensurable, and it cannot be reduced to their sum or their difference (p.414).

It is important to mention the conceptual difference between the processes of *composition* and of *emergence*. In the case of composition, acting upon the components gives rise to a new component with new properties, as in chemistry. This is a process of *assembling* components in a reversible or irreversible manner. In the case of emergence, we refer to three main features of the concept (Beckermann *et al.*, 1992; Goldstein, 1999; Minati, 2001a; 2001b; Pessa, 2002):

1) emergence is *observer-dependent*; within the scientific context the problem is not the existence of emergence *per se*, but rather its detection within the models used by a given observer, equipped with suitable detecting devices, with given goals and cognitive schemata;

2) emergence is associated with the existence of different *observational* or *descriptional* levels which are usually arranged in a hierarchical way

with the higher levels emerging as a consequence of the interactions taking place within the lower ones; a general feature seems to be the existence of both *bottom-up* and *top-down* influences inter-relating different levels, in the sense that, on one hand, the higher levels arise, and are influenced, from what is occurring within the lower ones and, on the other hand, the lower levels themselves are influenced, in turn, by what is occurring within the higher ones;

3) in all cases considered so far, emergence results in the formation of new *coherent entities*, behaving as a whole at the higher levels; such coherence, of course, is not an objective property but depends on the *coherence detectors* available to the observer (Bonabeau and Dessalles, 1997).

These three aspects will now be considered as a function of their dependence upon the observer. It must be stressed that the **concept of emergence cannot, *in principle*, be defined in an *objective* way. Emergence can only be defined relative to a given observer**. Particular instances of this feature include:

- **the establishment of unexpected effects by using a specific model** (e. g., deterministic chaos, the *Three Body Problem*). It was not expected, using the Newtonian paradigm, that the problem of computing the mutual gravitational interaction of three masses was unsolvable, even when they are constrained to lie in a common plane. This may be called *computational emergence*.
- **the need of the observer to *change* conceptually the model or the representation used during observation of an evolving phenomenon; typical cases involve the occurrence of *collective behaviors***; an example is given by the occurrence of phenomena such as *superconductivity, ferromagnetism* or *lasers* which, being manifestations of collective effects, can no longer be described by resorting to the traditional models of classical Physics. Another example is provided by the cooperative phenomena underlying the operation of *industrial districts* (see Chapter 7) which cannot be described by using the classical models of *Economics*. These cases may be described as *'phenomenological emergence'*.

In both cases the role of the *observer*, equipped with a suitable model for detecting discontinuity and discrepancy according to his expectations, is crucial. In principle, reference should not be made to a generic observer, but to a specific one with a particular model. Attention should be paid, however, to the dangers of extreme subjectivism. If it were assumed that every particular observer is intrinsically different from, and irreducible to, any other observer, we would end up with the impossibility of any form of a theory of emergence. Even the term 'emergence' itself should be banned from the scientific vocabulary: 'my' emergence would be irreducible to 'your' emergence, and therefore not communicable. To avoid this trap we will assume that the reference is to *classes* of observers, having common properties, which can somehow be defined and detected. In this way we can build a theory of emergence embodying the observer itself, provided the definition of the class of allowable observers be stated at the very beginning of the theory we try to develop. This is especially important within a framework of *multi-modelling*, as when we adopt the DYSAM strategy.

Once a class of observers has been defined, it is possible to discuss the important topic of *cues* for emergence, which signal that a process of emergence is in *progress* (see Chapter 5, on emergence and ergodicity) even when no unexpected effects or needs for changing the model occur. The simplest example is offered by phase transitions in Physics. Here the class of observers is constituted by average physicists, equipped with standard measuring and computing devices and with a good knowledge of classical physics and of standard macroscopic thermodynamics. As is well known, near the critical point of a phase transition, we have the appearance of a number of cues which show that some profound structural change is occurring: the curve of specific heat *vs.* temperature shows sudden variations for very small temperature changes, the same occuring for other curves, such as that of susceptibility *vs.* temperature. Standard thermodynamics, of course, which can describe the behavior of a sample of water at, say, 40°C continues to hold at 99°C. No discrepancy is present and the physicist using classical theory can sleep easy. Even here, however, there are some cues which could induce a state of alarm. Something is happening! An important contribution offered by a theory of emergence would be simply its ability to interpret and elaborate upon such cues.

Let us now deal with the second feature of emergence: the existence of different levels. Some simple examples, taken from everyday life, show that we cannot speak of emergence without, at one and the same time, speaking of levels of observation and of description. A typical case is that of an Italian garden, surrounding a beautiful sixteenth-century villa. An observer lying within the garden, looking on a small scale, will perceive a number of irregularities in the disposition of plants, flowers, trees and so on. But, if the observer is in a helicopter flying over the garden a very regular geometrical structure of the garden on a much larger scale will be perceived. In a sense, such a structure is emergent, at least with respect to the means at the disposal of an ordinary human being on foot. It can be detected only if the observation is from a new higher level and the emergent structure itself is present only at this new level.

Such a simple example should not induce us to neglect that a consistent theory of what the term 'level' means is still lacking. In many cases, such as the previous one, it could be easily defined by the spatial scale of observation and description. In other cases the level could merely be associated with the choice of a particular time scale. However, even in the case of simple physical systems, both spatial and time scales need to be specified. How are the different levels related? In many other cases, and very frequently in the domain of social and cognitive sciences, resorting to notions of space and time could be useless, for a suitable description, other different kinds of variables (which could still be measurable) are required. How can the scales of these variables be defined? Can levels associated with them be introduced? Or can the 'level' not be defined from the outside but only from the very structure of the theory of emergence adopted? It is found that this is exactly the case in most theories of emergent phenomena within the domain of physics (as will be shown later). But it is currently unknown whether this is or is not a general rule.

The existence of different levels entails the existence of different observational devices suited to each of them (or, briefly, to different observers). Such a circumstance has been used to introduce some formal definitions of emergent properties, such as that proposed by Baas (Baas, 1994; Baas and Emmeche, 1997; see also Mayer and Rasmussen, 1998). The definition is as follows:

"Let $\{S_i\}_{i \in I}$ be a family of general systems or "agents". Let Obs^1 be "observation" mechanisms and Int^1 be interactions between agents. The observation mechanisms measure the properties of the agents to be used in the interactions. The interactions then generate a new kind of structure $S^2 = R(S_i^1, Obs^1, Int^1)$ that is the result of the interactions. This could be a stable pattern or a dynamically interacting system. We call S^2 an *emergent structure* that may be subject to new observational mechanisms Obs^2. This leads to the following definition:

P is an emergent property

$$\Updownarrow$$

$$P \in Obs^2(S^2) \text{ and } P \notin Obs^2(S_i^1)$$

For instance, while observing the behavior of a group of people, of cars, or the flight path of a group of birds, one might conclude that they respectively form a crowd, traffic jam, and a flock. But if property P is not observable by looking at individual behavior, the property is said to be an emergent property of the group. Here the notion of "emergence" does not mean "unpredictability". One could use a definition such as:

"Property P of S is emergent if, and only if, it is observable by Obson S with a positive, but less than 1, probability and is not observable at a lower level, i.e. at level S."

The reference to probability is related to the fact that by observing at a higher level of description, say S, it is *possible* to detect new emergent properties, although not *certain*. This is because activity at a lower level is a necessary, but not sufficient, condition for the establishment of emergence.

Turning now to the third feature of emergence, the appearance of coherent entities, it should be stressed that their occurrence entails their detectability and that the latter property, in turn, entails their *stability*. Namely a coherent entity behaves like a whole only when it is stable, at least to a certain degree, with respect to external perturbations,. This poses a new problem for every theory of emergence, as the latter should specify the mechanism leading to the stability of the emergent coherent entities. On this point, the only theories developed so far, containing such a mechanism, lie within the domain of physics (a prototypical example being Quantum Field Theory, which will be discussed below).

The topics briefly covered within this Section have been the subject of an intense debate amongst researchers in Systemics. Namely, since the origins of General Systems Theory, efforts have been made to discover how holistic, systemic properties might *emerge, be established,* from interactions among individual system components.

In the 1970s, mainly due to the work of **Ilya Prigogine** and his school (Nicolis and Prigogine, 1977) and **Hermann Haken** (Haken, 1983), emergence was identified with so-called order-disorder transitions, bringing ordered frameworks to occur within systems fulfilling certain suitable boundary conditions. Such processes were denoted as 'self-organization' processes (Holland, 1998) and, from then onwards, the terms of emergence and self-organization have been considered as synonyms.

Physicists focused on the topics of emergence during the 1980s mainly due to the intense activity in the field of deterministic chaos and of Cellular Automata. One example is the impossibility of finding an algorithm able to deterministically compute the final 'fate' of the evolution of a Cellular Automaton, endowed with fully deterministic evolution rules, without computing all discrete intermediate states. Deterministic chaotic behaviors look "new" when considering their simple computational generating rules (designed to act in a classical deterministic way), which, nonetheless, make them "unpredictable" by definition. At the beginning chaotic behaviors seemed to be good examples of emerging phenomena, in the sense mentioned above. Only when researchers in the field of chaos theory began to experiment with models of Artificial Life (Bonabeau and Theraulaz, 1994) did the need for a more precise definition of emergence arise. Within this framework the expression "emergent computation" was introduced and refers to the occurrence of new and always increasing computational power generated by the cooperation among a certain number of elements interacting through simple rules, as in Neural Networks (Forrest, 1990; Lundh *et al.*, 1997).

In the 1980s a new framework for dealing with emergence was introduced by physicists such as **P. W. Anderson** (Anderson, 1981; Anderson and Stein, 1985). The focus was on processes of *spontaneous symmetry breaking*. This concept is widely used in physics for the description of phenomena such as superconductivity and superfluidity. Moreover *spontaneous symmetry breaking* carries certain special features when occurring within the quantum-mechanical framework.

If the arguments introduced by Anderson and others were correct, then they would entail the paradoxical conclusion that a description of emergent phenomena is possible only within the theoretical framework of Quantum Theory (on this point, see also Sewell, 1986). Consequently all other descriptions previously mentioned, including those introduced by Von Bertalanffy, Prigogine and Haken, would have to be considered as inadequate. In this case, it becomes clear that a large part of the theoretical content of the systemic approach would have to be reconsidered.

Moreover, this conclusion seems to contradict the fact that other researchers, such as those designing models for Artificial Life, have recently discovered real examples of emergence, using algorithms which do not seem to be based upon Quantum Theory (Pessa, 1998).

With reference to computer programs, emergent properties are defined as properties which are a consequence of the interactions between the behaviors programmed into the simpler parts of the system. If these properties are different from the set of properties of single programs they are said to be *synergic*. The terminology is also useful for identifying fuzzy properties of complex systems for which formal descriptions are difficult. This is what usually happens in so-called computational ecologies, produced by building artificial environments for computer programs to live in. A natural ecology is produced by the interaction of many organisms. A computational ecology is the result of the interaction of many independent programs. A general overview on the concept of emergence in different disciplinary fields may be found in (Johnson, 2001).

A further important contribution to the subject is that proposed by Crutchfield (Crutchfield, 1994) who distinguished three kinds of emergence:

1. *Intuitive definition,* corresponding to an initial identification of "emergence" with novelty, as introduced at the beginning of this chapter;
2. *Pattern formation,* referring to a process in which a pattern is said to be emergent when it occurs as a non-trivial consequence of the structure of the adopted model but nevertheless it is *predictable* in advance on the basis of sophisticated mathematical analyses of the model itself. This is, for instance, the case of the so-called *Dissipative Structures* (Nicolis and Prigogine, 1977; Prigogine, 1967; Prigogine and Glansdorff, 1971);
3. *Intrinsic emergence,* referring to a process in which the occurrence of a certain behavior is not only unpredictable, but its occurrence gives rise to profound changes in system structures such as to require a formulation of a new model of the system itself.

In accordance with these definitions only *intrinsic emergence* should be considered as a correct specification of what is intended when speaking about systemic properties, not reducible to properties of the specific single components of a given system but, on the contrary, emergent only from interactions among them (Pessa, 1998; Genakopolos and Gray, 1991). The importance of this concept of intrinsic emergence has led to, over the past few years, a search both for the conditions under which it arises and for the best ways to describe the time evolution of a system possessing it. Without entering into detail here (it will be dealt with in the next chapter), we limit ourselves to quoting the main ingredients which came to light during a comparative analysis of different processes of emergence and which were deemed to be necessary (but not always sufficient) for the occurrence of intrinsic emergence. These ingredients can be briefly listed as:

a) existence of the possibility of specifying and describing the *interactions* between the components of the system under study (in physics they can be simply described and measured through concepts such as *force, energy,* and so on);
b) existence of intrinsic *fluctuations* of the behaviors of these components; these fluctuations may have different origins: they may be due to *stochastic noise,* to *chaotic behaviors* on a smaller scale, or to *quantum-like* phenomena;
c) the system must be *open* with respect to the external environment; such an openness can be of many different kinds: *thermodynamical, logical, parametric* and so on (for a discussion on this topic see Minati *et al.,* 1998); such openness is needed for the survival of the intrinsically

emergent coherent entities, owing to contributions from the outside world.

A more detailed description of the roles played by these ingredients will be made in Chapter 4.

3.2 Collective Beings

3.2.1 Aggregations, Congregations

Aggregation (Parrish and Hamner, 1998) is a pervasive phenomenon in nature. Generally speaking an aggregate may be intended as a collection of elements. Inanimate objects aggregate to constitute landscapes and beaches; molecules aggregate to form the basic building blocks of matter. Inanimate objects may be not only aggregated but sorted by physical gradients: e. g., the sand grains of a beach sorted by wave action. A beach is characterized by a spatial gradient of grain sizes and not by a random spatial distribution of them. Aggregates and processes of aggregation take place in nature at very different levels. While a set is an abstract collection of elements for which we have a rule of belonging, an aggregate is a set containing, implicitly, the history of its aggregation and of the process producing that aggregate. Our interest is in the process of aggregation.

A primary distinction between *active* and *passive* aggregation must be made. The sand of a dune is constituted by grains *accumulated*, for instance, by the wind. The elements of this kind of aggregate do not play an active role in constituting the aggregate itself. In other cases, however, the elements aggregate *actively*, such as, for instance, in the presence of attractive or repulsive forces between the elements themselves. The balance between repulsion and attraction allows atoms and molecules to give rise to different phases of matter (i. e., liquid, solid, gas).

Aggregates and processes of aggregation refer to both *inanimate* and *animate* elements. The term 'animate' will refer to elements identifiable with living units. The aggregates made by animate elements, such as assemblages of plants or animals may also result from a physical sorting based on *passive* aggregation: when they are passively shaped or driven by physical processes. In other cases the members of the (active) aggregate may interact within the group, once formed (Wilson, 1962; 1971, 1975).

Even when localised concentrations of any resource (food, space, light) or aggregates themselves act as attractive sources it is possible to speak of *active* aggregation. An active aggregation may disappear if the source of attraction wanes. In case of active aggregation the elements may join and leave the aggregate and not always be part of that aggregate. *Repulsion* also plays a crucial role (Payne, 1980) keeping a balance between different driving forces.

When the source of attraction is the group itself, the term *congregation* can be used (Turchin 1989; 1998). In a *passive congregation* the members are attracted by the group itself without displaying any social behavior: a typical example is the formation of a fish school, considered as a "selfish herd", where each member simply tries to obtain the maximum advantage from the very existence of the group itself (Pitcher and Parrish, 1993).

"Although a large variety of animals congregate, interactions within a group differ markedly across species. Spatially well-defined congregations, such as fish schools, may be composed of individuals with little or no genetic relationships to each other (Hilborn, 1991). ... Social congregations display a variety of inter-individual behaviors, necessitating active information transfer." (Parrish and Hamner, 1998).

3.2.2 Collective behavior: an introduction

The expression *collective behavior* (Theraulaz and Gervet, 1992; Millonas 1992; 1994) denotes a non-linear phenomenon involving a macroscopically large number of particles or agents. This phenomenon consists in the occurrence of coherent wholes (generally associated to scales beyond those typical of the single elements), resulting from a sort of coordination between the individual behaviors of the single particles or agents. Such coordination is, of course, associated with the appearance of some sort of long-range, or medium-range, correlations between the individual behaviors themselves. In physics, examples of collective behaviors (Iberall and Soodak, 1978) have been observed in condensed matter, especially as a consequence of phase transitions (e. g., magnetism and superconductivity), and in biophysics, such as in DNA replication (see, e.g., Bieberich, 2000).

We may distinguish between two different kinds of *collective behaviors*:
- Those taking place, for instance, as a consequence of cooperative/competitive processes among populations of different types (such as equilibrium between prey and predators), the synergic effects of these kinds of collective behavior give rise to the formation of ecosystems (for models of this kind see, for example, Yokozawa and Hara, 1999; Sakaguchi, 2003);

- Those emerging from synergic effects due to interactions between *agents of the same type*, presumably behaving with *the same cognitive model*, which will be introduced below.

Research in the field of collective behaviors **deals with processes** (e.g., the formation of ecosystems, see, for instance, Golley, 1993**) which occur not due to an explicit design regarding roles and functions of agents (e.g. predator or prey), but as a consequence of their interaction, even when they are structured in specific functions and roles (as required, for example, by ant-nests).**

A great number of papers have appeared in the literature regarding the development of mathematical models and of simulations in the field of Artificial Life (Cariani, 1992) and in Synthetic Biology, the "techno-science" of Artificial Life (Ray, 1991; 1994).

Examples of beings having an artificial life, marked by computational cycles, and whose evolution is studied by Synthetic Biology, are given, for instance, by computer viruses. Other examples are given by flocks of *Boids* used, for instance, in simulation programs for computer-animated cartoon production (Reynolds, 1987). Focus in this case is on *phenomenological emergence*, even though it may occur through artificial agents behaving in such a way as to run a mathematical model which may produce unexpected effects, known as *computational emergence*.

In social systems various processes of emergence of collective behaviors may occur (Theraulaz and Gervet, 1992; Millonas 1992; 1994), giving rise to the appearance of new coherent systems such as flocks, herds, fish schools and swarms from interactions among individual living beings (Anderson, 1980; Aron *et al.*, 1990; Aoki, 1984; Belic *et al.*, 1986; Breder, 1954; Parrish and Hamner, 1998; Parrish and Edelstein-Keshet, 1999). The *new emerging being* does not possess the same properties as those of its interacting components. In the previous examples, it was possible to assume that this *emerging being* was not a living being, even though it was able to learn, to die and to be born (i.e., to disintegrate and then to reconstitute itself). This kind of collective behavior is characterized by the fact that the single components continue to follow their own *autonomous* behavioral rules, rather than to follow new rules dictated by the appearance of the collective system itself. In this sense they lack organization and the fulfillment of different and specific roles, as is assumed to occur, for instance, in Living Systems Theory, generalized in Living Systems Analysis (LSA) (Miller, 1978).

The same kinds of processes may be observed when considering phenomena such as crowds or traffic jams. Within this kind of emergence the components have no specific and different roles. From their interactions a new and unexpected behavior emerges, unexpected of course, with respect to the behavior of the individual components.

A specific example of this kind of collective behavior is that based upon the concept of *swarm*.

The theoretical concept of *swarm* has been introduced by many authors (see, for example, Bonabeau *et al.*, 2000; Mikhailov and Calenbuhr, 2002) and is intended as a collection of elementary agents, interacting locally but having a globally adaptive behavior. For instance a set of mobile agents able to directly or indirectly communicate with each other, acting on their local environment and supporting a process of *distributed problem solving*, constitutes a *swarm*.

The concept of swarm gave rise to the further concept of *swarm intelligence* (Bonabeau *et al.*, 1999), which may be considered as a property of non-intelligent agents having collectively an intelligent behavior (Beckers *et al.*, 1990). The equivalence between *collective intelligence* and *swarm intelligence* will be discussed below.

It must be remembered that collective behaviors, arising from agents locally characterized by very simple behaviors, are linked, as we saw for all *emergent* phenomena, in an indissoluble way to the existence of an observer able to detect them.

In recent times there has been a resurgence of interest in the study of emergent phenomena in the field of social systems. Research has tried to apply the methods of theoretical physics in order to describe the emergence of collective behaviors as flocks (Darling, 1938), swarms (Millonas 1992; 1994; Bonabeau *et al.*, 1996; Lindauer, 1961), fish schools (Breder 1954; Echelle and Kornfield, 1984), herds, ant colonies (Deneauburg *et al.*,1983; 1987; 1989; 1990; 1991; Deneauburg and Goss, 1989; Franks *et al.*, 1991; Goss *et al.*, 1990; Holldobler and Wilson, 1990; Millonas, 1992), organizations and industrial districts (see Chapter 7 for the pertinent references). Moreover, this approach may be considered as somewhat limited (Pessa *et al.*, 2004) as it is based on the use of models originally designed for physics and assumed to be valid for describing social systems.

This induced us to introduce a further distinction between two kinds of collective behaviors:
- That in which the *cognitive* interactions among elements are not essential to account for the observed collective behavior;
- That in which such interactions are fundamental for the emergence of collective behaviors.

It will be shown that the classical modelling approach, based on the use of single models, is not sufficient to describe this second kind of collective behavior: the following Sections will introduce the concept of *Collective Being* and link it to the concept of *DYnamic uSAge of Models* (DYSAM), already introduced in Chapter 2.

So far collective behaviors have been studied almost exclusively by Physics. In physics, the focus has been mostly on collective phenomena such as ferromagnetism, superconductivity, superfluidity, collective oscillations in crystal lattices, the laser effect and so on. These phenomena are the main topics studied by Synergetics (Haken 1981; 1987).

Intuitively, a *collective behavior* is established when mutually interacting agents behave as if they agreed to act in such a way to produce some desired effect. This is what happens, for instance, in social systems when massive purchases or sales occur because of bull or bear influences in the Stock Exchange. Effects of the kind *'the more someone sells and the more the others sell'* occur. By using conceptual categories of Synergetics (Bushev, 1994) we may say that *order parameters* emerge.

As mentioned above, this is very well described in physics and by mathematical models. However, when dealing with social systems, the theoretical models based on considering all agents as particles suffer from many drawbacks. This situation led to a search for phenomenological models able to simulate, in a more realistic way, aspects of the considered collective phenomena. As a first approximation, many of the models introduced for flocks, swarms, insect colonies, are based upon the common assumption that the collective behavior considered is **independent of the complex cognitive abilities** of the single living beings composing the system showing the collective behavior under study[3]. In this case the cognitive model, reduced to a description of a simple stimulus-response device, is represented through a simple computational model, as discussed in Chapter 8. As in some models in the field of robotics, it is assumed that system components are only able to react to local stimuli, in real time. Another context in which the same approach was taken for many years is that of models of pattern formation, of pattern recognition (Haken, 1979) and of the emergence of order (Kauffmann, 1993), where the concepts themselves are in reality related to the cognitive aspects of the observer. This

---

[3] It is possible to hypothesize very simple interaction rules for flock components, applied algorithmically, for instance:
1. max distance < M;
2. min distance > m;
3. distances always change with time;
4. different directions among agents, but with angle $< \alpha$.

recalls the old behaviorist approach. This assumption tends to limit a correct description and understanding of collective behaviors detected in social systems. The crucial consideration introduced in the new approach is that models must take into account the **cognitive system** and the **cognitive model** of component agents (Pessa, 2000; Pessa *et al.*, 2004).

3.2.3 Cognitive System and cognitive models

For many years the study of human and animal behavior has been influenced by the **Behaviorist** approach (see, for example, Skinner 1938; 1953). According to such a conception, the observable behavior is explained merely in terms of the association mechanisms between stimuli and responses, without an active role of the organism.

The novelty presented by the **Cognitive approach** (see, for example, Anderson, 1985; Benjafield, 1992), formally introduced for the first time in the celebrated book *Cognitive Psychology* (Neisser, 1967), consists of considering the processes internal to the organism as fundamental mediators between stimuli and responses. Such processes may be *studied* by formulating models of their operation and comparing their predictions with the results of suitable experiments.

According to the Cognitive approach, these internal processes consist of information processing activities and as such they may be simulated by the activities of computer systems running programs simulating the observed behaviors. Even if this view is fascinating and gave birth to disciplines such as Artificial Intelligence, its practical implementation was plagued by a number of shortcomings arising from the fact that it only works within very limited domains, such as those of visual perception or memory, as studied within laboratory experiments. Many other processes, mainly those connected to motivation and emotional-affectionate aspects of life, are completely neglected. On the contrary, it is essential that the description of a *cognitive system* should take into account all these aspects, and not merely the strictly cognitive ones. *A Cognitive System is intended as a complete system of interactions among activities, which cannot be separated from one another, such as those related to attention, perception, language, the affectionate-emotional sphere, memory and the inferential system*, as described by Pessa (Pessa, 2000).

A distinction is made here between agents provided with a non-artificial cognitive system, identified with living beings, and agents provided with an artificial cognitive system, deriving from the computational modelling of cognitive processes. The cognitive system of the former can be described, although only to a first approximation, as having a general structure as depicted in Table 3.1.

Table 3-1. Schema of a cognitive system

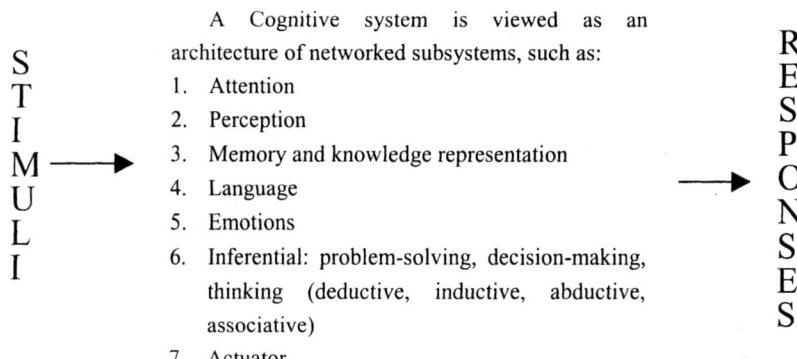

S
T
I
M ⟶
U
L
I

A Cognitive system is viewed as an architecture of networked subsystems, such as:
1. Attention
2. Perception
3. Memory and knowledge representation
4. Language
5. Emotions
6. Inferential: problem-solving, decision-making, thinking (deductive, inductive, abductive, associative)
7. Actuator

⟶

R
E
S
P
O
N
S
E
S

Examples of these are found in the emergence of computational ecologies (Huberman and Hogg, 1993). This approach is peculiar to studies in Artificial Life (Brooks, 1992; Hemerlijk, 2001) and especially in robotics (Cao *et al.*, 1977; Cliff *et al.*, 1993).

The *cognitive approach* has exerted a strong influence on most domains of scientific psychology, by introducing **Cognitive models** referring to specific processes belonging to general cognitive activity (divided into several subsystems in Table 3.1). The cognitive subsystems should, in principle, be considered not as being separated but as mutually interacting. However, for research purposes, such an assumption is often neglected and, in order to reach short-term goals, a number of models of a given particular cognitive process completely neglect the influence of other processes or treat them as external noise. Chapter 8 will focus upon the crucial question: "Is Cognition equivalent to Computation?" By assuming this equivalence, **a cognitive system may be assumed as a system of models interacting within a cognitive architecture** (Anderson, 1983; Anderson and Lebière, 1999).

3.2.4 Collective behavior

Different approaches for simulating the emergence of collective behaviors have been introduced in the literature. Some of those based on single agent behaviors giving rise to flocks or swarms are listed below.

a) Boids (Reynolds, 1987)
This term denotes artificial agents designed to reproduce a swarm-like behavior in a virtual world; this effect is obtained as a consequence of the following behavioral rules:
- separation rules: individuals must control their motion in order to avoid the crowding of locally adjacent components;
- alignment rules: individuals must control their motion so as to point towards the average motion direction of locally adjacent components;
- cohesion rules: individuals must control their motion so as to point towards the average position of locally adjacent components;

b) Lattice models
This term refers to a number of different models having in common the representation of individual agents as moving entities localized within a discretized spatial lattice. The motion of the single agents and, in general, the evolution of the whole system of interacting agents is a consequence of suitable local evolutionary rules, giving the state of a lattice point at a given time instant as a function of the state of the neighboring points at the previous time instant. Such a framework enables both computer simulations and an analytical treatment based on Statistical Mechanics. Among these models we will quote *Fluid Neural Networks* (Delgado and Solé, 1998; Manrubia *et al.*, 2001; Montoya and Solé, 2002; Solé *et al.*, 1993; Solé and Miramontes, 1995; Solé and Montoya, 2001;), and the *Fixed Threshold Model* (Bonabeau *et al.*, 1996; Bonabeau, 1997; Bonabeau *et al.*, 2000; Dorigo *et al.*, 2000), as well as *Swarm Networks* (Millonas 1992; Rauch *et al.*, 1995). This approach does not only model swarms or flocks, but even more complicated collective pattern building processes, such as, for example, nest building (Bonabeau, 1997; Theraulaz and Bonabeau, 1995b).

c) Hydrodynamic-like models
This term refers to models based on a macroscopic continuum-like description of a system of agents, viewed as components of a suitable "fluid". This latter is described through evolution equations resembling those of hydrodynamics. It is, of course, possible to derive these equations by starting from a particle-like description of the motions of the individual

agents and then performing a suitable continuum-limiting procedure. The emergence of collective behavior is identified by the appearance of large-scale spatially dependent stationary stable or metastable solutions of the macroscopic evolution equations. The advantage of this approach is to make explicit the influences controlling the onset and the maintenance of these "collective" solutions.

These influences include (Minati *et al.*, 1997; Pessa *et al.*, 2004):

- *local inhibition*; this comes from the fact that the operation of each single cognitive system requires the processing of information arriving from external stimuli; the information to be processed may be assumed to come from adjacent individuals; however, such information processing requires a significant amount of energy; it may therefore be viewed as equivalent to the effect produced on individuals by a force, opposing individual movement; such a force, then, may be considered as dependent only upon nearest neighbours;
- *medium range activation*; this comes from the very existence of a tendency towards cooperation or behavioral imitation between different agents; it is medium range, owing to the fact that the perceptual abilities of every agent can cover only a limited spatial range, although greater than the shorter range between immediate neighbors;
- *long range inhibition*; this is due to the existence of forces which counteract the control abilities of each agent (e.g., friction); typically their effect manifests itself only over very large spatial and/or temporal scales.

Such a schema is a generalization of that already proposed by Gierer and Meinhardt (Gierer and Meinhardt, 1972). Within it the collective behaviors emerge and are maintained as a consequence of a suitable balance between these various influences. A similar framework was adopted in other interesting hydrodynamic-like models (e.g., Toner and Tu, 1995). The *local, medium or long-range* interactions (Barrè *et al.*, 2001) between agents are a very important issue in modelling emergence. We recall, on this point, that even in brain models the long-range interactions explain the robustness of memory activity (Vitiello, 2001).

So far focus has been placed upon processes of emergence of collective behaviors in which the observer does not detect specific roles or functions for the interacting agents (such as, for instance, belonging to a specific flock) or those in which specific roles or functions are not the reason for the collective behavior of agents (contrary to that occurring, for instance, in real ant-hills).

On the basis of what has been said, emergent collective behavior can not be adequately described using only rules of interaction among

agents following simple stimulus-response, action-reaction mechanisms, adopting a behaviorist conceptual framework (Therelauz and Gervet, 1992; Therelauz and Deneubourg, 1994). **In order to model collective behavior in a more useful and realistic way, the *cognitive model* used by autonomous, interacting agents must be taken into account.**

However, in nature, collective behavior phenomena normally emerge from autonomous agents *all of the same type*. When considering living components, such as birds, fishes and insects, it appears that collective phenomena such as flocks, schools of fishes and swarms are produced by individuals all belonging to the same *species*. The particular kind of emergent collective behavior observed does indeed characterize different species; for instance flocks composed of geese are different from flocks composed of swallows, and so on.

At this point more precision is required. First of all the concept of species is not formally well established. A simple definition of species may be the following: a species is intended as a systematic category comprising one or more populations of individuals able to mate and reproduce. Moreover, the inter-fertility criterion, fundamental for this definition of species, does not apply to organisms having parthenogenetic (egg develops without fecundation) or agamic (asexual, i.e. per fission, gemmation, fragmentation) reproduction. The definition of *species* is a very controversial and open subject in biology. Another definition of species is the genotypic cluster definition: a species may be intended as a group of organisms or beings which share a common genetic make-up.

The need to be more specific also derives from the fact that there are examples of flocks of parrots formed by more than one species, called multi-species flocks.

Multi-species flocks have been observed in South America, including flocks of parrots comprising Amazons, Conures, Macaws and other species. Multi-species flocks have also been seen in Australia, where it is quite common to see Galahs (Rose-breasted), Greater Sulfur-crested Cockatoos and Corellas eating and flying together. They are considered as multi-species flocks even though composed of parrots because parrots comprise many different species within the *psitticine* genus (Hino, 2002; Jullien and Thiollay, 1998; Munn and Terborgh, 1979). Piping plovers also appear to roost in multi-species flocks (Nicholls and Baldassarre, 1990). Thus, when attempting to indicate that autonomous agents establishing collective behaviors are *all of the same type* instead of stating that they belong to a particular species, we may say that they belong to a particular genus.

When describing the common shared properties of such autonomous agents, one of the implications of their belonging to the same species or genus is their ability to behave in such a way as to induce the emergence of a specific collective behavior.

Given the fact that autonomous agents inducing the emergence of collective behavior are *all of the same type* (i.e. same species or same genus), one can say that their behavior is driven by the *same* cognitive systems using the *same* cognitive model.

In any case, the sharing of the same cognitive system using the *same* cognitive model may be assumed as a *necessary* but not a *sufficient* condition to establish collective behaviors amongst agents.

Why a *necessary* condition? Because a collective behavior is different from symbiosis, that is the existence of two or more dissimilar organisms in intimate association with mutual advantages. A collective behavior of autonomous agents is established by reciprocal interactions sharing the same processing of information leading to synchronous behavior.

On the other hand, a *sufficient* condition for the occurrence of a collective behavior is that the agents process information using the *same cognitive model* in such a way as to induce collective behavior, that is, by assuming the **conceptual interchangeability of agents, playing the same roles at different times.**

The concepts of *cognitive system* and *cognitive model* were introduced many years ago. It is now possible to introduce the concept of *cognitive space*. The geographical literature offers many different cognitive categorizations of *space* (Couclelis and Gale, 1986; Freundschuh and Egenhofer, 1997) distinguishing between six different kinds of spaces: pure Euclidean, physical, sensorimotor, perceptual, cognitive, and symbolic space. Within the cognitive science literature this concept is denominated as *conceptual space* (Gärdenfors, 2000). The latter can be characterized both theoretically and experimentally (through experiments on human subjects).

An important difference is that between *perceptual space* (related to the fact that objects are perceived through the senses at one place and one time) and *cognitive space* (dealing with sensory images of objects themselves and the set of allowable movements in the space accessible to the agent in relation, of course, to its own cognitive model). The cognitive space is therefore characterized by the cognitive model of the considered agent even if it is not coincident with the cognitive model itself.

Consequently, by observing collective behaviors emerging from agents sharing the *same cognitive model*, one can consider that single agents, when using the same (suitable) cognitive model, *can only* make emergent a collective behavior when acting together.

Therefore a system composed of agents possessing the same cognitive model, by mutual interactions, should manifest only a collective behavior, of course within their (shared) cognitive space. Thus, birds can only fly in flocks when they fly together, insects in swarms, etc.

Some further remarks must be added, in order to place what has been said within a wider context.

Reference has already been made to the importance of the observer and his/her models used to recognize and detect processes of emergence.

One must then ask exactly when does an observer recognize, for instance, the emergence of a flock. Observations, measurements and considerations are possible as long as the interaction does not perturb the process of emergence itself. Besides, do we have sufficient systemic categories to always recognize a flock without considerations regarding the quantity of interacting agents, their configuration, speed and distance? As mentioned above a flock, composed of birds flying with very large distances between them would be difficult to detect by a human observer lying on the earth without special equipment, whereas it is more likely to be detected by an observer flying in an aircraft, at a distance such as not to perturb the flock itself. *When is a flock no longer a flock?* Reference must be made to the models of the observer.

It is possible now to consider whether cognitive interactions can occur among agents (that is, those involving cognitive meanings, interpreted respectively by their various cognitive models), such as living beings, possessing *different* cognitive models. The answer is positive if we take into account how, for instance, communication is possible among different species. A species may even try to *represent* the other's cognitive system. Examples include relationships such as prey-predator; hunting, defense and camouflage, animal training, and so on.

In the same way small animals, by acting collectively, give rise to behaviors which can induce perception of shapes which can frighten predators. In this case individuals sharing the cognitive model of a particular species are able to process at least some aspects of other cognitive systems.

The cognitive space in which an autonomous agent performs may have non-empty interactions with that of agents possessing a different cognitive model. This happens due to the conjectures, made by one of the observers, about the *supposed* overlapping between their respective cognitive spaces.

These considerations and hypotheses have been introduced here to consider cognitive spaces generated by human beings, in order to understand how the space is generated in which social systems, emergent from their interactions, work.

As seen below, it is particularly interesting to apply such considerations and hypotheses to social systems such as companies and corporations.

In order to generalize the concepts underlying the theory of collective behaviors from the domain of swarms and flocks to that of human groups and societies, however, the introduction of another concept is necessary: that of *macroscopic* (or *large-scale*) *performance* resulting from those collective behaviors. This was often used in the scientific literature to characterize the very nature of these collective behaviors, such as for swarms intended as collections of locally interacting elementary organisms having a comprehensive adapting behavior. On this point, authors such as Bonabeau, Chialvo, Deneubourg, Dorigo, Gervet, Millonas, Theraulaz, Toner, et al., introduced definitions such as:

"We define a swarm as a set of (mobile) agents which are liable to communicate directly or indirectly (by acting on their local environment) with each other, and which collectively carry out a distributed problem solving..." (Theraulaz *et al.*, 1990). See also (Theraulaz and Gervet, 1992; Theraulaz and Deneubourg, 1994; Theraulaz and Bonabeau, 1995a).

"*Collective intelligence* or *swarm intelligence* may be intended as the property of non intelligent agents to manifest an intelligent **collective** behavior" (Franks *et al.*, 1991). See also (Bonabeau, *et al.*, 1999; Millonas, 1992; 1994; Theraulaz *et al.*, 1990; Theraulaz and Deneuburg, 1994).

This is the opposite of crowd behavior, characterized by reactions caused by terror and panic. In the case of the emergence of *collective intelligence*, behaving collectively provides solutions to problems which the individual components are unable to solve.

Consider now the occurrence of a *collective representation with individuals unable to formulate an abstract representation.*

An example is the behavior of ants looking for food. When an ant detects a source of food, it marks the followed path with a chemical track (pheromone). This mark induces other ants to go towards the source of food. When the ants return they leave a further track of pheromone reinforcing the original one. This kind of behavior amplifies the importance of the discovery by organizing a kind of communication, but it also allows the ant to evaluate which path is more interesting than another. The closest source of food is the one having the strongest chemical track, the others being associated with a weaker signal because of the decay of the chemical signal along the path due to its lesser frequentation.

Space representation is essential to describe and represent swarm intelligence, since *intelligence is more in the representation of cognitive space* rather than in the individual agents: it is an emergent property of the space structured by their behavior. It relates to the concept of spaces of rules, such as ethics, metaphorically intended as *social software.*

These references to collective behaviors can be summarized by stressing the need to share the same cognitive model, in order to obtain the emergence of collective intelligence, and by quoting, on this point, a very interesting subject, for future consideration: the emergence of a *collective mind* (Minsky, 1985).

Table 3.2 below contains a concise schema of the different conceptual layers encountered along the path leading from Sets to *Collective Beings.*

Table 3-2. Conceptual systemic levels

Sets	as introduced in Section 1.2
Systems	whose properties are *all emergent* (see Section 1.2.4):
	a) *organized systems* having identifiable roles of components, such as devices, companies, living beings;
	b) *non-organized systems* having emergent *properties,* such as crystals showing superconductivity or ferromagnetism; at this level elements are *interchangeable*;
	c) *non organized systems* having *emergent roles* of components, such as leadership in social systems, in devices depending on their usage, in the brain when an area may functionally substitute another damaged one;
	d) *emergent systems* showing collective behavior, such as the building of ant-hills or bee-hives, where properties such as those in a), b) or c) may be *simultaneous*ly present;
Collective Beings	introduced in the following Section.

3.2.5 Collective Beings

The concept of Collective Being refers in some way to the context of *constructivism* introduced by authors like H. von Foerster, H. Maturana, F. Varela, P. Watzlawick, E. von Glasersfeld (Von Foerster, 1979; Maturana and Varela, 1980; 1992; Varela *et al.*, 1991; Watzlawick, 1983; Von Glasersfeld, 1996). Introducing this concept is an attempt at applying an approach based on *how it is more convenient to think that something is rather than trying to find out how something really is.* By the way, we are convinced that both approaches should be considered to different extents, empirically and not ideologically: sometimes, in different contexts, one may be more inspiring than the other. Sometimes it may be more *effective* to think that something *really* exists: in fact the second approach may be considered as simply a particular case of the first one.

Before discussing the concept of Collective Being the concept of **multiple-system** has to be introduced. This concept is used, for instance, in *systems engineering* when dealing with networked interacting computer systems having cooperative tasks to perform, as in corporate information systems and on the Internet. The concept of multiple-systems was introduced several years ago in disciplinary studies such as psychology where there has been the introduction of multiple-memory-systems models, partly to account for findings of stochastic independence of direct priming effects and recognition memory effects (Tulving, 1985).

The concept of **Collective Being** makes reference to the fact that the same components of a system may *simultaneously* or *dynamically* give rise to *different* systems. It occurs when the elements involved in making emergent a system by interacting according to specific rules, are also involved in interacting, at the same time (*simultaneously*) or at different times (*dynamically*), by following other rules with the same and/or different elements. The concept of Collective Being is particularly suitable for all the cases where elements are agents equipped with a cognitive system and with the ability to use different cognitive models depending upon contextual situations. In some ways it recalls the concept of *logical openness* mentioned in Chapter 2 and introduced in Appendix 1, with reference to the ability of the system to move among *different levels of openness.*

Consider two different kinds of processes of emergence from agents *equipped with a cognitive system,* as illustrated in the following cases 1 and 2:

As introduced above, collective behaviors are assumed to be emerging from interacting agents equipped with the *same* cognitive model in such a way as to establish synergic (cooperative/competitive) effects in a context having *fixed* evolutionary environmental rules (i.e. ant-hill building).

There are, however, collective behaviors emerging from interacting agents equipped with the *same cognitive system* and *simultaneously* belonging to multiple-systems (i.e. deciding by considering different rules at the same time) or *dynamically* belonging to different systems in time (i.e. deciding by considering different rules at different times).

In this way they may be modeled as possessing *different* cognitive models used *simultaneously* or *dynamically* to perform cognitive processing, i.e. to take decisions. They establish synergic (cooperative/competitive) effects in a context having evolutionary environmental rules partly *fixed* and partly *self-designed*. Examples are *Human Social Systems* for cases in which: (a) agents may *simultaneously* belong to (that is to decide and behave as components of) such as a traffic system, consumer market, family, mobile telephone network, music listeners; *swarms*, (b) agents may *dynamically* give rise to different systems, such as bird flock configurations for food acquirement, defense and migration or bird communities on the ground when eating, drinking and sleeping.

With reference to point (a) we can observe how a human being reaches a decision by simultaneously taking into account multiple perspectives, such as being a parent, belonging to a company, a university, a political party, a sports team and so on. We are *not* considering roles in different social communities, but the act of belonging to them, that is the use of multi-perspective, multiple approaches when facing problems and taking decisions. In case (b) the agents, by using multiple cognitive systems, are able to dynamically make emergent different systems depending upon the contextual conditions. When the agents take on different roles at one and the same time or dynamically (see Chapter 5 for ergodic behavior) there is the emergence of a Collective Being.

However, multiple-systems may **degenerate** into single systems when the components are no longer subject to *concurrent* interactions. Collective Beings degenerate in single systems when the components behave by using the **same** cognitive model.

In Collective Beings we speak of *different cognitive models* and not just of different ways of processing data by the same cognitive model because the cognitive aspects are involved in different ways, involving attention, perception, language, the affectionate-emotional sphere, memory and an inferential system.

In case 1 considered above the natural environment is assumed to have fixed evolutionary rules, while in the second there is a mixture of natural rules and self-established, local, social rules.

The concept of Collective Being allows an approach for modeling systems emerging from interacting systems constituted by autonomous agents possessing a Cognitive System, behaving with the *same* cognitive model in time, simultaneously or dynamically belonging to different systems. The emergent higher systems (Collective Beings) have a behavior which is not detectable by an observer looking only at the agents or at the different interacting component systems. At a higher level of abstraction a

Collective Being may emerge from a system of Collective Beings, etc. It is possible to manage, influence, control and perhaps even design Collective Beings by using a methodology such as the DYnamic uSAge of Models (DYSAM) introduced in Chapter 2.

Different simultaneous or dynamical systems within a Collective Being do not only influence each other by interacting, as systems usually do. Actually in this case a system does not only become emergent through its components, but it also affects those components, as will be seen in Section 3.4.7. *Within Collective Beings the various systems influence each other due to their effects on common elements.*

3.2.6 Components contemporarily *belonging* to many systems

As introduced in the previous paragraph, it is possible to identify systems having components belonging to **more than a single system**. *Belonging* in this case means that the same components can play different roles when interacting with other components; thus, "belonging" is not to be identified only with traditional *set inclusion*. In this case, we have situations in which the *same* components can give rise to the *emergence* of different systems, and the different roles played by the same component may take place **at different times or even simultaneously**.

It must be noted that the case under discussion is different from coping with *systems and their subsystems* (Chapter 1) performing specific functions within the context of the organization of the whole system. In this case a component may have *shared* involvement in making emergent different systems. This occurs, for instance, when a computer, while multitasking or running different work sessions, can also be part of various networks performing the same or different functions. In this case the component has a shared, simultaneous, participation in making different systems emergent. In many systems the components play exclusive, fully dedicated roles, making only *one* specific system emergent.

The concept of **multiple-system** refers to elements *simultaneously or dynamically* making **different** systems emergent with reference to the four cases considered above.

In the context of components belonging to different systems, is a component still identifiable as the *same* one when performing different roles in different processes (of emergence, in this case)? How is it possible to say that a component of a system is equal to another one? The answer is related to the fact that by acting upon a component in a system it is possible to produce effects on its way of interacting in another system. The reference is to the concept of *interchangeability* between components.

Because of the crucial role of the observer it is not possible to distinguish between multiple-systems and the need to dynamically use different models to model a system. DYSAM and Collective Beings are two aspects of the same phenomena.

The expression 'belonging to a system' may have two different meanings:

The components belong to different organizations, different structures, giving rise, even simultaneously or dynamically, to different systems. From this point of view such a situation is very unlikely. How can an electronic component contemporarily be component of different electronic systems? Or a wheel be a component of two cars? Or a molecule, a cell, an organ, be part of different organisms at the same time? But different components may be simultaneously part of different systems if they play *different independent roles*. The behavior of an electronic component may be a source of information for a security monitoring system. The reaction of a cell, while being a component of a living system, may also be a source of information for a physician using diagnostic techniques.

A different situation occurs when components belonging to different systems are endowed with a cognitive system. The components have *memory* of the different roles, making them aware of the roles themselves; this implies that they have a cognitive model appropriate for each role. In this case they are obviously autonomous agents provided with a cognitive system, and, as such, are able to select the cognitive model to be used, each model corresponding to the emerging system to which they belong. Likewise a person has multiple-simultaneous-roles in a society: worker, family component, buyer, supporter, etc.

Consider, for instance, a company. It consists of different kinds of resources, the human ones being essential. Each of the latter is endowed with a **cognitive system and with its own cognitive models.** And each of these resources belongs to or, better, contributes *simultaneously* or *dynamically* to make different social systems emergent, such as:

- the company itself;
- families;
- markets of given products (e.g., consumer products such as clothes, cars, electronics, etc.) and services (e.g., banking, transporting, telecommunications, etc.);
- public (attending, for instance, TV programs, movies, theatrical representations, music concerts);
- religions, by considering communities and rites;
- political parties, unions and their conventions;
- sports teams;

- voluntary associations;
- etc…

Traditional approaches to the study of these social systems used to model an *agent*, a customer in a supermarket, for instance, simply as a buyer, a wealth-allocating operator to be studied according to models considering optimization, timing, effects of stimuli such as advertising, colors, positions of goods in the store, etc.

In most cases, modern Cognitive Science showed how the modeling of behaviors by considering only particular aspects of them is not only reductive, but makes it impossible to take advantage of the interactions, of the interdependence between various simultaneous aspects which can be used not only to predict particular outcomes but to also influence them.

To model collective behaviors in living systems giving rise to the emergence of multiple-systems, i.e., Collective Beings, it is not sufficient to look for the optimal model, of an adequate complexity, in order to be effective in any given context. A different strategy is needed, based on using more than one model, depending upon the context being considered. This is the approach introduced with DYSAM.

Why can different models not be reduced to a single, adequately articulated model? Referring to Figure 2.2, in order to find some answers to this question:

- One general reason is related to the ability to deal with *different levels of representation*. Different levels of representation call for different processing, using different models. It refers to the cognitive processing of different kinds of inputs in living beings at different levels of complexity: from optical, acoustic and sensorial to semantic ones.
- Another reason is to avoid rigidity of approach, of the architecture: this is why, since the origins of computer science, the most often adopted strategy has been that of writing many different subprograms (corresponding to specific procedural needs) rather than only one capable of solving all the processing needs of a given procedure.
- A further reason is the need to be able to act without being forced to select, each time, *the* model to be used, but to apply strategies based upon the *simultaneous* use of different models. That is, to allow the best and most effective **decision and learning** processes when dealing with behavior simulation and explication. Besides, the need to learn allows the opportunity of having various processing and representing techniques (symbolic, neural) available, with different levels of effectiveness according to the problem in hand. It corresponds to the difference between seeing all problems in the same way when one only tool is available and inventing different usages of the same tool

according to the problem. *Of course, the consumer society privileges the first approach.*

- The level devoted to learning is specific for each system-agent problem. Whereas some information processing functions may be assumed to be conceptually standardized, at other levels, such as those devoted to learning and semantic processing, are specific to that processing. Learning refers to how the output from different models is processed in parallel by the cognitive system of the agent.
- As DYSAM is a meta-model, **selection** is not the outcome of a single model, but the output of a dynamic system of models. Such a system of models is specific for any agent, depending dynamically upon problems, context, resources, etc.

But is the use of the dynamic modelling approach not simply a model in itself? A *Multi-modelling* approach is a *modelling strategy* based upon using various models, or *architectures of usage.* To model a Collective Being focus is placed upon the need to adopt a dynamical use of models, *able to select among the results of previous/parallel processing based upon various models and depending upon the results themselves, on context, on strategies, and on learning.*

3.2.7 Modelling issues of *Collective Beings*

The concept of *Collective Being* (Minati, 2001a; 2004a) has been introduced to model, with DYSAM, multiple-systems emerging from the behavior of agents having the *same* cognitive system and using dynamically different cognitive models. A Collective Being is intended to be a system oscillating among different systemic aspects, having components simultaneously or dynamically belonging to different systems and which can be suitably modelled with the DYnamic uSAge of Models (DYSAM) (Minati 2001a; 2001c; Minati and Brahms, 2002).

A good example of systems oscillating among different systemic aspects can be obtained when considering *openness and closeness.* As introduced by Minati (Minati *et al.*, 1998) a system may be assumed to be not just *open or closed*, but to behave sometime as open and some time as closed and to different degrees (see Appendix 1).

Agents, elements of a Collective Being, are not characterized by different functions, as in the case where they were considered as subsystems. An example is a *beehive* (organized into fixed roles and subsystems), which is not a Collective Being, when compared to a *swarm* (Millonas 1992; 1994) (assuming different roles depending on the context), which is a Collective

Being (bees make emergent different types of systems including swarms for defense, or when looking for food, and beehives). The behavior of agents making collective behaviors emergent and a Collective Being is not related to their following standardized rules (Deneubourg *et al.*, 1987; Pasteels *et al.*, 1987) deriving from fixed features of their physical or biological components, but to their being equipped with the *same cognitive system* and dynamically using the *same cognitive models*. On the contrary, they are able to modulate, to change, to design their way of interacting (Anderson, 1980; Benjafield, 1992), possessing, at high levels of complexity, *awareness* (Vitiello, 2001). For instance, a social system may be very different over time because of the change in the way the components interact.

A multiple-system may be detected (1) by different observers considering the system from different points of view coincident with the interests of individual interacting agents, or (2) by recognizing the emergence of different systems.

The example of the *crowd* is illuminating. It may be constituted of people walking, standing, silent, crying, clapping, etc.

Different observers model from different points of view (1): the system may be modeled with reference to security, energy consumption, pollution generation, as a market, etc.

Multiple-systems, a Collective Being emerges from the *Crowd* (2): from where different systems may emerge, including traffic, political demonstrations or religious processions.

In many cases the observer is interested in different aspects of the phenomena, considered as physically invariant and can thus be suitably dealt with by using a single model. A system may be interesting for its physical aspects, such as temperature, weight, energy consumption, etc.

However, different systems emerging from the same agents must be described using suitable modeling tools. Thus the lava from a volcanic eruption may be interesting, from the point of view of its chemical composition, speed, temperature, and from another for its effects on atmosphere, territory, buildings. Analogously a tide can be interesting for its amplitude, timing, duration but also for fishing, or for its damaging impact on buildings.

The whole *system-observer* system to be modeled may changing so much that the observer may need to have different models:
- depending on the approach required;
- until a more complete and effective one is identified, that is, when the most suitable model is not easily identifiable, but the dynamic use of models is expected to **converge** to the most effective and suitable one. In

this case DYSAM is used to *approximate* a model which is not yet available;

- which can be *dynamically* used as a consequence to continuous changing, when the problem is not to identify the most suitable one because of the continuous changing of the phenomenon to be modelled. The problem in this case is to adapt as quickly as possible to a changing context. It may be an effective approach when emergence processes with unexpected evolutions have to be modelled;
- which can be processed *in parallel* by using strategies to decide the one to be used depending upon the time available to react;
- *interacting* which can be processed *in parallel* and *simultaneously* for making decisions.

Decisions made by a Collective Being, such as to follow a specific model over time, never result from a single computational process. Decisions come from emergent computation. The Collective Being's mind is not computational, and this suggests some analogies with the concept of computational mind in cognitive science (Newell and Simon, 1976). The decision-making process of a Collective Being is to be intended as not simply computational (Beckers *et al.*, 1990; Bonabeau *et al.*, 1996; Deneaubourg and Goss, 1989; Seeley *et al.*, 1991) as introduced and discussed in Chapter 9.

It should be recalled that the attribute **computational**, is related to the usage of computers, to computer-based approaches and strategies, i.e. algorithms, neural networks, genetic algorithms (Koza, 1992a; 1992b) and so on. Among the various fields of application we may mention: (1) *Computational Biology* and *Bioinformatics* dedicated to advancing the scientific understanding of living systems through computation, with emphasis on the role of computing and informatics in advancing molecular biology. Computational Biology is dedicated to the discovery and implementation of algorithms which facilitate the understanding of biological processes by using applications of many technologies such as machine learning. (2) *Agent-based Computational Economics* (ACE) is the computational study of economies modelled as evolutionary systems of autonomous interacting agents. ACE is the application to economics of the basic complex adaptive systems paradigm. (3) *Computational Physics* relates to the computational aspects of physical problems, dealing with techniques for the numerical solution of mathematical equations arising in all areas of physics. (4) *Computational neurosciences, computational neurobiology*, focus on compartmental modeling and simulations of biological neural systems (Hecht-Nielsen, 1989). (5) *Computational linguistics* uses computers as models for computable processes thought to underlie the understanding and the production of languages. This requires theories about human linguistic capabilities. *"In looking at language as a cognitive process, we deal with issues that have been the focus of linguistic study for many years ... . But we look at language from a different perspective. The emphasis is less on the structure of the sounds or sentences, and more on the structure of the processes by which they are produced and understood ... . In the forty years since digital computers were first developed, people have programmed them to perform many activities that we think of as requiring some form of intelligence. In doing this, they have developed new ways of talking about knowledge - what it is, and how it can be stored, modified, and used."* (Winograd, 1983), page 1. (6) Another example is given by *computational emergence*, based on computing through the collective behavior of single computational agents (Cariani, 1992). There are even (7) Computational Chemistry (Arkin and Ross, 1994); (8) Computational Complexity; (9) Computational Ecology, and so on.

With reference to decisions coming from emergent computation, in a Collective Being a situation may never occur where two agents *contemporarily* take on an **incompatible** behavior which *disaggregates* the system. In a flock there is room for different agent directions, different speeds, but while keeping the *identity* of the flock as perceived by the observer.

TheBehavior of Collective Beings emerges from continuous balancing and interactions between individual behavior, in such a way as to keep the advantages of acting collectively, that is, to keep the Collective Being's *functional identity*.

Let us now turn to a fundamental question: what are the differences, if any, between *Systems* and *Collective Beings*? The answer is that a system emerges from the interaction between component elements, whether these are systems themselves or not. As introduced above, the concept of Collective Being refers to a collective behavior which emerges from autonomous agents. This collective behavior is related to the sharing of the *same* cognitive system and of cognitive models intended as set of common micro-behavioral rules, as in the case introduced below based upon the Cellular Automata schema (Duparc, 1985). Collective Beings can be suitably managed through a dynamic usage of models, although such a technique may be used to manage and model autonomous system behavior in general. According to what has been said so far, to identify a Collective Being requires more conditions than those required to identify a System. The expression *Collective Beings* refers to all contexts where emergent properties are dynamically manifested. Particular reference is made to the problem of modelling emergence arising from collective behavior in social systems.

The ways in which emergent phenomena affect the very components from which they emerge will now be considered.

Figure 3.1 represents the case in which interacting components give rise to a process of emergence. As mentioned above the interaction among components is a *necessary* condition for such an emergence. Depending upon the kind of interactions, that which is emerging may have non-trivial properties which single individual components do not possess. The process of the birth of global properties which are different from local individual component properties may occur in at least two ways: (1) due to a pre-existing, superimposed, organization or structure among the components, as in the case in which electronic or mechanical components interact becoming either a *working* apparatus with functional properties or just a useless energy consuming device; (2) because interactions among components allow self-organization as in phenomena such as superconductivity, ferromagnetism, Bènard rollers, etc. In both cases the phenomena are related to an observer, to the cognitive model involved.

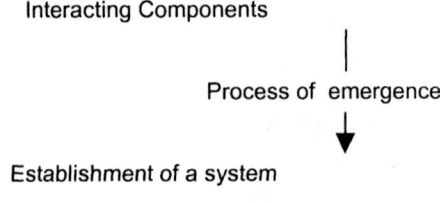

Figure 3-1. Emergence of a system

The emergent system affects, in turn, its interacting components (Fig. 3.2), for instance, because they are used by the emergent system itself in different ways (e. g., the wear and tear of the components of a machine, traffic affecting the driving style of individuals, school having impact on individual students, and so on).

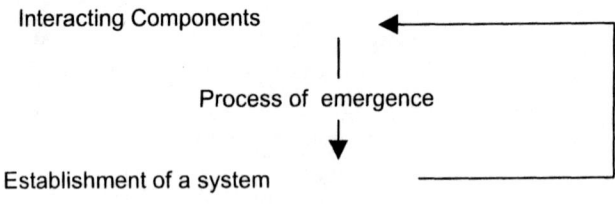

Figure 3-2. The emergent system affects components

As shown in Fig. 3.3, when the interacting components are autonomous systems provided with the same cognitive system and behave using the same cognitive model they may give rise to: (1) simply a set of systems related by an organization, a structure influencing the components, such as in assembly lines, or in rule-based social systems such as armies; (2) Collective Beings emerging from interactions among autonomous systems allowing self-organization such as in phenomena governing flocks, swarms, herds, fish schools (Radakov, 1972; Badalamenti *et al.*, 2002). The emergence of *mono-dimensional* self-organized communities (the attribute refers to the fact that they are characterized by the use of a single model and not multiple models, as in Collective Beings) has been studied from different points of view and by adopting different approaches (Ashby, 1947; Camazine, 1991; Camazine and Sneyd 1991; Collins and Jefferson, 1991; Darling, 1938; Deneubourg *et al.*, 1987; Theraulaz and Gervet, 1992; Allen, 1994). The emergence of a Collective Being depends upon the kind of cognitive model used at that particular moment by the autonomous agents (all agents are provided with the same cognitive system). A set of flying birds, for instance, does not always give rise to the emergence of a flock: it depends on the cognitive model used at that particular moment in time.

Figure 3-3. Emergence of a Collective Being

As in Fig. 3.4 the emergent Collective Being affects the behavior of individual components. This is related to the collective application of the same cognitive model. For instance, the emergence of a flock affects the behavior of single birds making collective the activity of looking for food, of trying to escape a predator, of following a common path.

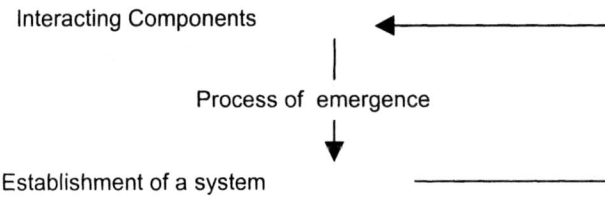

Figure 3.4. The emergent Collective Being affects single systems.

Scheme 3.1 shows a scheme illustrating the concept of Collective Being, characterized by the fact that the agents may simultaneously give rise to different systems by assuming different roles at the same time (simultaneously belonging to different systems) or dynamically give rise to different systems by assuming the same roles at different times (dynamically belonging to different systems). As introduced in Chapter 5 this is an ergodic behavior.

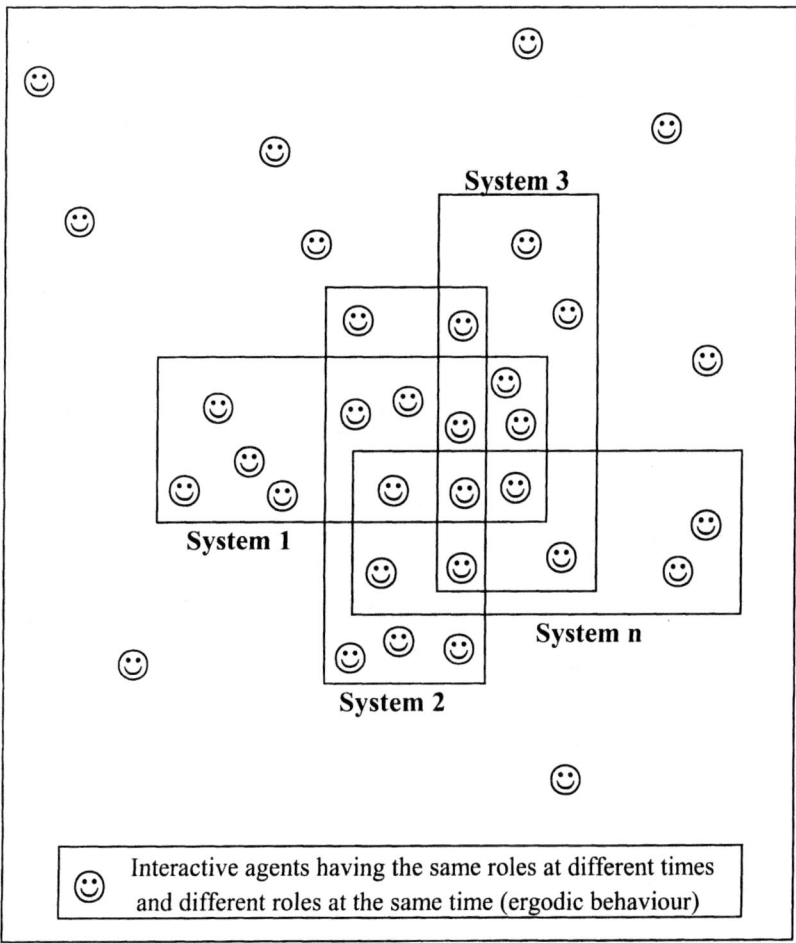

Scheme 3.1. Illustration of the concept of Collective Being

3.2.8 Modelling Collective Beings

Obviously, the modelling of Collective Beings, is a very difficult issue, which can be approached in a virtually unlimited number of different ways. Here we will show, through an example, how it would possible to build a model of a Collective Being within the spirit of the original von Bertalanffy approach (cfr. Von Bertalanffy, 1968, Chapter 3), relying, in turn, on the framework of dynamical systems theory. This proposal constitutes a generalization of the original Bertalanffy analysis, based on autonomous systems of differential equations.

Consider three systems, S, S^2, and S^3 each defined by a suitable set of macroscopic state variables; some of which are common to more than one system. A list of the associations between the systems and their state variables is

$$S^1: (q_1, q_2, q_3, q_5, q_6, q_7)$$

$$S^2: (q_4, q_5, q_6, q_8) \qquad\qquad (3.1)$$

$$S^3: (q_6, q_7, q_9)$$

Here, the symbol q_i is used to denote the i-th state variable, which is assumed to be a function only of time t. A graphical representation of the reciprocal interrelationship between these systems, due to the fact that some state variables are in common, is shown below.

It is assumed that the dynamical evolution of these systems can be described by three different systems of autonomous differential equations, holding simultaneously, such as:

$$S^1: \begin{cases} dq_1 / dt = f_1 (q_1, q_2, q_3, q_5, q_6, q_7) \\[2mm] dq_2 / dt = f_2 (q_1, q_2, q_3, q_5, q_6, q_7) \\[2mm] dq_3 / dt = f_3 (q_1, q_2, q_3, q_5, q_6, q_7) \\[2mm] dq_5 / dt = f_5 (q_1, q_2, q_3, q_5, q_6, q_7) \\[2mm] dq_6 / dt = f_6 (q_1, q_2, q_3, q_5, q_6, q_7) \\[2mm] dq_7 / dt = f_7 (q_1, q_2, q_3, q_5, q_6, q_7) \end{cases} \qquad (3.2)$$

$$S^2: \begin{cases} dq_4 \, /dt = f_4 \, (q_4 \, , \, q_5 \, , \, q_6 \, , \, q_8) \\[2em] dq_5 \, /dt = f'_5 \, (q_4 \, , \, q_5 \, , \, q_6 \, , \, q_8) \\[2em] dq_6 \, /dt = f'_6 \, (q_4 \, , \, q_5 \, , \, q_6 \, , \, q_8) \\[2em] dq_8 \, /dt = f_8 \, (q_4 \, , \, q_5 \, , \, q_6 \, , \, q_8) \end{cases} \qquad (3.3)$$

$$S^3: \begin{cases} dq_6 \, /dt = f''_6 \, (q_6 \, , \, q_7 \, , \, q_9) \\[2em] dq_7 \, /dt = f'_7 \, (q_6 \, , \, q_7 \, , \, q_9) \\[2em] dq_9 \, /dt = f_9 \, (q_6 \, , \, q_7 \, , \, q_9) \end{cases} \qquad (3.4)$$

It should be noted that the common state variables, because of their simultaneous occurrence in different systems, *must simultaneously* behave as components of different systems; the relevant equations are listed separately below:

$$\begin{cases} dq_6 \, /dt = f_6 \, (q_1, q_2, q_3 \, , q_5 \, , q_6 \, , q_7) \\[2em] dq_6 \, /dt = f'_6 \, (q_4 \, , q_5 \, , q_6 \, , q_8) \\[2em] dq_6 \, /dt = f''_6 \, (q_6 \, , q_7 \, , q_9) \end{cases} \qquad (3.5)$$

$$\begin{cases} dq_5 \, /dt = f_5 \, (q_1, q_2, q_3 \, , q_5 \, , q_6 \, , q_7) \\[2em] dq_5 \, /dt = f'_5 \, (q_4 \, , q_5 \, , q_6 \, , q_8) \end{cases} \qquad (3.6)$$

$$\begin{cases} dq_7 \,/dt = f_7 \,(q_1, \, q_2, \, q_3, \, q_5, \, q_6, \, q_7) \\[3em] dq_7 \,/dt = f'_7 \,(q_6, \, q_7, \, q_9) \end{cases} \tag{3.7}$$

A set of functional constraints can be directly derived from these equations, and expressed through a set of functional equations, which can be considered as the constitutive equations of the Collective Being described by the simultaneous validity of equations (3.5), (3.6), and (3.7); The functional equations are:

$$\begin{cases} f_6 \,(q_1, \, q_2, \, q_3, \, q_5, \, q_6, \, q_7) - f'_6 \,(q_4, \, q_5, \, q_6, \, q_8) = f'_6 \,(q_4, \, q_5, \, q_6, \, q_8) - f''_6 \,(q_6, \, q_7, \, q_9) \\[2em] f_5 \,(q_1, \, q_2, \, q_3, \, q_5, \, q_6, \, q_7) = f'_5 \,(q_4, \, q_5, \, q_6, \, q_8) \\[2em] f_7 \,(q_1, \, q_2, \, q_3, \, q_5, \, q_6, \, q_7) = f'_7 \,(q_6, \, q_7, \, q_9) \end{cases} \tag{3.8}$$

From this description some consequences already become clear. The first one stems from the fact that the very existence of the functional equations (3.8) introduces a set of constraints which lowers the number of degrees of freedom of the original description based upon the systems of equations (3.2), (3.3), (3.4). Thus, while the set of these simultaneous equations required a 9-dimensional phase space, the presence of the three constraints expressed by (3.8) allows for an evolution only in a 6-dimensional phase space (nine minus three). The reduction of the number of degrees of freedom has far-reaching implications regarding stability of the motions of the whole system. Namely, it is easy to understand that, on increasing the dimensionality of the phase space, the number of ways in which an equilibrium state can become unstable also increases. For instance, a spiral motion towards an equilibrium point in two-dimensional space could be simply the two-dimensional shadow of helicoidal motion wandering away from that point along the third dimension. Thus, the equilibrium point, apparently stable when observed in two dimensions, could really be unstable when observed in three dimensions. A reverse reasoning shows that, on the contrary, a reduction in the number of degrees of freedom increases stability.

For instance, by constraining such a spiral motion to lie only on a two-dimensional plane, escape along the third dimension (and whence the loss of stability) would be precluded. It can thus be concluded that, in general, a Collective Being is an entity which, in principle, is more stable than its local constituent parts, and this stability is, in turn, granted only by the constraints defining it.

This explains, using a further example, why some collective behaviors, such as those of two-dimensional flocks, seem to violate well-known theorems of Physics, such as the Mermin-Wagner theorem which states that a stable two-dimensional configuration cannot exist (Mermin and Wagner, 1966). This occurs because a flock exists and survives as a consequence of suitable constraints between the motions of individual birds belonging to it and the presence of these constraints lowers the dimensionality of the available phase space, in turn increasing the stability of the whole system and rendering untenable the thermodynamic arguments upon which the Mermin-Wagner theorem itself is based.

It would be technically possible, of course, to shift from the description of the Collective Being based on the systems of equations (3.2), (3.3), (3.4) to a new description, based on suitable *collective variables*, in which we could describe only the collective evolution (in this case a collective motion) resulting from the interactions of these systems. Such a description would be easy to implement by resorting to linearizations of system equations around some equilibrium state and to diagonalization procedures such as the well-known ones used in linear algebra. Without entering into mathematical details, it can be recalled that the introduction of collective variables is a widely used tool in theoretical physics, where it allows a shift from a representation of a system based, for example, upon a set of isolated atoms, mutually interacting in a very complicated way, to a new collective representation (physically equivalent to the previous one) based on isolated atoms interacting in a simple way only with suitable collective excitations (so-called *quasi-particles*).

It should be stressed here that the modelling framework introduced is, of course, subject to strong limitations, essentially due to the fact that it is only of a macroscopic nature, and thus unsuited to dealing with the problem of the dynamical description of the *appearance* of a Collective Being. Namely, the transient phenomena occurring *during* the phase transition giving rise to a Collective Being cannot be studied by resorting only to a macroscopic framework. We need to go down at least one level and introduce a *microscopic* description of the local behavior of the individual agents. Only in this way can advantage be taken of the technical tools of the theory of phase transitions and of statistical mechanics in order to develop and evaluate a more consistent theory of the birth and evolution of a Collective Being.

A framework similar to the one proposed here was adopted, for instance within the context of ecological modelling, by authors such as Yokozawa and Hara (see, for example, Yokozawa and Hara, 1999). The difference with respect to our approach lies mainly in the fact that they described each system through a single equation, introducing a measure within the system space, in order to define neighbour relationships among different systems. Through computer simulations they obtained two interesting results: (1) the behavior of the whole system, when allowing only nearest-neighbour interactions between systems, can be mimicked by an equivalent model in which long-range interactions are allowed; (2) parameter values exist which allow for the occurrence of stable multi-layered hierarchical structures. Results of this kind can be obtained even within our modelling approach, being simply a consequence of the fact that different systems share the same components. The emergence of coherent behavior is just a by-product of the form of sharing introduced within the models adopted.

3.2.9 Kinds of Collective Behaviors and Collective Beings

This section introduces a suitable classification scheme which relates Collective Behaviors (and whence Collective Beings) to the properties of the agents involved. The properties taken into consideration are:
1. Possession of a cognitive system;
2. Possession of a cognitive system identical to that of all other agents;
3. Dynamic using, at any given point in time, of the *same* cognitive model, among the various ones available, used by any other agent;
4. Ability to use *simultaneously* different cognitive models;
5. Association with a specific role (or to a number of different specific roles).

Different types of systems relating to different levels or organizations and processes of emergence are considered here. The following systems are considered organized:
-   Hetero-organized systems, established by applying organization, structure, between elements;
-   Self-organized systems;
-   Self-organized systems showing intrinsically emergent properties;

−   Self-organized systems showing emergent collective behavior;
−   Self-organized systems behaving as Collective Beings.

For the establishment of a process of emergence the agent properties mentioned may be:

- (Y) necessary;
- (N) incompatible;
- (-) non-necessary and thus irrelevant.

Table 3.1 presents a summary of agent properties against types of systems indicating how agent properties are necessary, incompatible or simply irrelevant for various types of systems. Table 3.2 shows the same summary presented through specific examples.

Table 3-3. Systemic levels and agent properties.

Agent properties	1. Hetero-organized systems, established by applying organization, structure, between elements	2. Self-organized Systems	3. Self-organized systems showing intrinsically emergent properties	4. Self-organized systems showing emergent collective behavior	5. Self-organized systems behaving as Collective Beings
1. Possession of a cognitive system	-	-	-	Y	Y
2. Possession of a cognitive system identical to that of all other agents	-	-	-	Y	Y
3. Dynamic use at any given poin in time of the *same* cognitive model, among the various ones available, used by any other agent	-	-	-	Y	Y
4. Ability to use *simultaneously* different cognitive models	-	-	-	Y	Y
5. Association with a specific role (or a number of different specific roles)	Y	N	N	Y	Y

(Y) necessary;   (N) incompatible;   (-) non-necessary and thus irrelevant

Table 3-4. Examples of systemic levels and agents properties

Agent properties	1. Hetero-organized systems, established by applying organization, structure, between elements	2. Self-organized Systems	3. Self-organized systems showing intrinsically emergent properties	4. Self-organized systems showing emergent collective behavior	5. Self-organized systems behaving as Collective Beings
	Automatic assembly line (robots); electronic, mechanical device	Belousov-Zhabotinsky reaction, Bénard's rollers, Coherence in light emission typical of the laser	Collective behavior in condensed matter, in phase transitions, in magnetism, superconductivity, and biophysics, such as in DNA replication.	Swarms, flocks, herds, fish schools, ant-hills, bee-hives, computational ecologies, ecosystems	Families, workers, consumers, students, traffic, crowds
1. Possession of a cognitive system	-	-	-	Y	Y
2. Possession of a cognitive system identical to that of all other agents	-	-	-	Y	Y
3. Dynamic use at any given point in time of the *same* cognitive model, among the various ones available, used by any other agent	-	-	-	Y	Y
4. Ability to use *simultaneously* different cognitive models	-	-	-	N	Y
5. Association with a specific role (or a number of different specific roles)	Y	N	N	Y	Y

(Y) necessary; (N) incompatible; (-) non-necessary and thus irrelevant.

3.2.10 Affecting the Behavior of Collective Beings

The possibility of using, inducing, destroying, controlling, and predicting processes of emergence and, more generally, the behavior of a Collective Being is a crucial issue. It is related, in turn, to the fact that the agents of a Collective Being can behave at different systemic levels both giving arise to multiple-systems and preserving the memory of past behavior, while changing cognitive model.

The emergence of a Collective Being involves and requires different simultaneous interdependent processes related to the management of cognitive models, interactions between agents, multiple-systems and sub-systems. The complexity of a Collective Being, therefore, allows many different approaches to influence it.

Before considering a list of these approaches reference will be made to the concept of *emergent software*. This is endowed with a well-defined meaning within the so-called *computational ecosystems* (Gustavsson and Fredericksson, 2001; Huberman and Hogg, 1993). As is well known, in these systems *computational processes*, consisting of active programs within a suitable hardware-software environment, may *generate* others through competition or co-operation in a context of networked resources. This situation is typical for systems of networked computational resources such as file servers, telecommunication resources, printers and other systems whose operation contrasts with the traditional methods of organized scheduling based on *serial processing*. In such a situation the system is run by a software emerging from interacting active tasks performing traditional computing processes.

Within this context it is possible to take into account the *collective behaviors* emerging from interacting autonomous agents, each of them acting as a *computational device*. This implies that the collective behaviors may be considered as being governed by rules *emerging* from the behavior of interacting agents. The latter do nothing but apply, in different ways (e. g., asynchronously, with different parameters), the *same* behavioral rules specified by the common adopted cognitive model (in this case corresponding to that of the designer of the programming language). Such emergent rules are the *emergent software* running special kinds of Collective Beings acting as computational devices.

Regarding the main problem, that is, the ways of influencing a Collective Being, it should be noted that external influences may, in principle, act at different levels, dealing with (a) single agents, (b) interactions between agents, (c) interaction between agents and multiple-systems and (d) between multiple-systems themselves.

In the first two cases a way of influencing a Collective Being is to act on the cognitive model adopted by the agents. Social systems designers have a long experience about how to influence cognitive models through life styles, religions, languages, social constraints and superimposed values. This takes place through individual and social interactions and their regulation.

In the other two cases the designer must take advantage of the interaction with or between emergent systems.

Systems and multiple-systems may interact at different levels by using *self*-designed rules superimposed upon those of the environment, such as those introduced by ethics and laws. As mentioned in Section 3.2.4 and discussed below in Section 6.4, this new, superimposed, environment may be metaphorically named *social software*. **The *social software* is emerging from the *software* running interaction processes, based on the dynamical use in time of the same cognitive models and social rules, occurring within human agents and the emergent Collective Beings.** The concept of *social software* will be also discussed in Section 6.2 which deals with *professed and practiced* ethics; in it the *practiced ethics* will be considered as an *emergent social software*. Ethics will be introduced as the most important source of cognitive models for agents of social systems.

Influencing the emergent social software of Collective Beings is another possible approach towards their control. If it is not possible or not convenient to directly influence the cognitive system of the interacting agents, then it is possible to consider how *to perturb or induce* the establishment of social software, through learning processes by acting upon:

- environmental conditions;
- communication processes;
- hardware, physiological conditions;
- assumed standards;
- linguistic usages.

Another way of exerting an influence is based on the fact that the interacting agents of a Collective Being "belong" simultaneously to different systems. It is therefore possible to act upon agents and multiple-systems involved in different contemporary systemic processes by influencing single, networked, related variables. By acting upon one variable in one system it is possible to influence it and those related to it in another. In such a way it is possible to influence reaction times, availability and reliability of resources.

Other approaches to control a Collective Being may be:

- The induction of a conflict between agents simultaneously belonging to different systems;
- Taking into account that a Collective Being is not autonomous from other systems (multiple belonging of components);

- Constraining the interactions between Collective Beings of the same kind, i.e. grouping together agents provided with the same cognitive systems and interacting by using the same cognitive models. A Collective Being may interact, for instance, with an apparently similar *simulated Collective Being* having unexpected, destabilizing behavior;
- By influencing the propagation of information, by modifying, confusing, applying *long-range interactions* (on this point, see, for instance, Barrè *et al.,* 2001);
- By acting upon critical conditions leading to the emergence of the Collective Beings;
- By confusing the learning processes occurring within a Collective Being through the simulation of distorted reactions.

All the possible approaches introduced above should be conceptually based on *DYSAM*. In the contrary case (often occurring within designers of social systems) unexpected (and very dangerous) effects can derive from rough attempts to influence a Collective Being.

3.2.11 Further remarks

With reference to the concept of Collective Being introduced in 3.4.5 it is important to underline how the requirement that interacting agents equipped with the *same cognitive system* may

simultaneously belong to multiple-systems or

dynamically belong to different systems over time

does not always refer to **all** components. In case (a) the components are not required to *simultaneously* belong to *all* the multiple-systems made emergent by the collective behavior whereas in case (b) the components are *all* expected to belong to a specific system at a given time.

The *consistency* of a Collective Being is given by *local* rules, depending upon the properties of agents and upon the level of description and observation. Components of a society may be detected as cooperative in the short term and highly competitive in the long term or vice versa. The *dynamical belonging* may refer to competitive systems such as in economics. The *incompatible* belonging refers only to *formal* belonging, because incompatibilities and contradictions assumed at certain levels of description may be managed by cognitive systems adopting multi-modeling according to the particular context. This is the typical case of multi-ethical societies, as discussed in Chapter 6.

Another remark should be made regarding case (a), where it is not required that all components simultaneously belong to all multiple-systems which become emergent. Two extreme circumstances are possible:

- all components simultaneously belong to all the multiple-systems which become emergent;
- all multiple-systems which emerge from the collective behavior of component agents have only a single common element.

In the latter case, the multiple-systems are still a Collective Being because even through the one common element each system may influence the other systems.

On rising to an even higher level of complexity, societies of *multi-Collective Beings* can emerge from the interactions between Collective Beings. The latter are observed at a level such that they can be assumed as being distinct one from another and from the surrounding environment. On the other hand, they may be considered as a single Collective Being at another level of observation. For instance, families may be assumed to be different interactive Collective Beings when considered as the families of students belonging to a school, while they may be considered as single Collective Beings from a marketing point of view.

What advantages derive from the introduction of the concept of Collective Being?

A first example of the application of this concept can be found within the domain of computational communities (such as, for example, computer networks) where a node may be part of *different networks* playing different roles depending upon the software used to interact with other nodes. The same node may be affected by learning processes and data accumulation related to its simultaneous use by different network communities. By using the concept of Collective Being it is possible to exert an influence on another networked community without following the approach based on interacting with it as if it were another system, but by influencing only the nodes which also form part of the other network.

Within social systems one of the main applications of the concept of Collective Being relates to the possibility of influencing systems by acting upon the common elements of different systems. This can occur when social systems are influenced or manipulated in a way corresponding to the *manipulation of consent in western, post-democratic societies* where real consensus is no longer necessary while it is now possible to *buy* it (Barns, 1995; Minati 2004b).

The classical approach allowing manipulation has been based on techniques such as trying to convince, to produce a sharing of common ideas and points of view, discussing, communicating, making the talkers, the leaders, and the projects appealing, as in traditional democratic social systems, whose political operation is based on elections, on voting.

In a multi-dimensional multiple-system, such as a Collective Being, it is possible to set control and manipulative procedures by acting upon one dimension (for some reason considered to be the most accessible one) in order to influence another. This has been done for centuries by controlling social pathologies (violence, rebellions, crimes, refusal to respect rules, etc.) and sometimes, unfortunately, even through the use of games, sport, religion, drugs, pornography, etc.

The concept of Collective Beings allows the discovery of **how and why** the classical democratic approach is becoming more and more *misused, but not side-stepped or destroyed*, making traditional democratic societies only formally democratic, exhibiting a sort of cosmetic democracy.

The approach invented for managing, deciding and orientating formally democratic societies in a non-democratic way is based upon some of the principles introduced with reference to Collective Beings. In the traditional approach people are supposed to be convinced of a certain point of view, to support candidates, to vote in some way. That was the age of political parties, churches and ideologies.

Change began to take place when winners began to be elected by smaller and smaller percentages of voters, with people becoming less and less interested in taking an active part in the elections. One way to manipulate consensus is to progressively reduce the percentage of the electorate allowing candidates to be elected by a relatively small number of voters who are convinced, induced or forced, by various means, to vote for a candidate. **The focus has shifted to the percentage of votes obtained by the candidates with little attention being paid to the percentage of the electorate actually voting.** The first step in devitalizing democracies has been made: it is no longer the majority of the population or even of a significantly large part of it which decides but the majority of a smaller and smaller portion of the population.

The second and most important step which took place recently, based on the intensive use of new technologies and the control of information: TV, radio, newspapers, magazines, standardizing behaviors, fashions, foods and language. Some marketing technologies (advertising with its psychological effects) have been applied directly to the process of obtaining political consensus. In western democracies **marketing technologies have been used for years to obtain (i.e., to buy) political consensus, by considering the decision to vote for a candidate equivalent to buying a product** (Evans, 1995; Herman and Chomsky, 1988; Minati, 2004b).

References

Allen, P. M., 1994, Evolutionary Complex Systems: The Self-Organization of Communities. In: *Complexity and Self Organization in Social and Economic Systems* (F. Fang and M. Sanglier, eds.), Springer, Bejing, pp. 109–134.

Anderson, J. J., 1980, A stochastic model for the size of fish schools, *Fish Bulletin* **79**:315-323.

Anderson, J. R., 1983, *The Architecture of Cognition*. Harvard University Press, Cambridge, MA.

Anderson, J. R., and Lebière, C., 1999, *The atomic components of thought*. Erlbaum, Hillsdale, NJ.

Anderson, J., R., 1995, *Cognitive Psychology and its Implication* (4th edition). Freeman, New York.

Anderson, P. W, and Stein, D. L., 1985, Broken Symmetry, Emergent Properties, Dissipative Structures, Life. Are they related? In: *Self-Organizing Systems: The Emergence of Order* (F. E. Yates, ed.), Plenum, New York, pp. 445-457.

Anderson, P. W.,1981, Can broken symmetry occur in driven systems? In: *Equilibrium and Nonequilibrium Statistical Mechanics* (G. Nicolis, G. Dewel and P. Turner, eds.), Wiley, New York, pp. 289-297.

Aoki, I., 1984, Internal dynamics of fish schools in relation to inter-fish distance, *Bulletin of the Japanese Society of Scientific Fisheries* **48**:1081-1088.

Arkin, A., and Ross, J., 1994, Computational Functions in Biochemical Reaction Networks, *Biophysical Journal* **67**:560-578.

Aron, S., Deneubourg, J. L., Goss, S., and Pasteels, J. M., 1990, Functional self-organization illustrated by inter-nest traffic in ants: the case of the Argentine ants. In: *Biological Motion* (W. Alt and G. Hoffmann, eds.), Springer, Berlin, pp. 533-547.

Ashby, W. R., 1947, Principles of the Self-Organizing Dynamic System, *Journal of General Psychology* **37**:125-128.

Baas, N. A., Emmeche, C., 1997, On Emergence and Explanation, *Intellectica* **25**:67-83.

Baas, N. A., 1994, Emergence, hierarchies and hyperstructures. In: *Alife III, Santa Fe Studies in the Science of Complexity, Proc. Volume XVII*, (C. G. Langton, ed.), Addison-Wesley, Redwood City, CA, pp. 515-537.

Badalamenti, F., D'anna, G., Di Gregorio, S., Pipitone, C., and Trunfio, G. A., 2002, A First Cellular Automata Model of Red Mullet Behaviour. In: *Emergence in Complex Cognitive, Social and Biological Systems*, (G. Minati and E. Pessa, eds.), Kluwer, New York, pp. 17-30.

Barns, I., 1995, Manufacturing consensus? Reflections on the U. K. National Consensus Conference on Plant Biotechnology, *Science as Culture* **5**:200-217.

Barrè, J., Dauxois, T., and Ruffo, S., 2001, Clustering in a model with repulsive long-range interactions, *Physica A* **295**: 254-260.

Beckermann, A., Flohr, H., and Kim, J., (eds.), 1992, *Emergence or reduction?*, de Gruyter, Berlin.

Beckers, R., Deneubourg, J. L., Goss, S., and Pasteels, J. M., 1990, Collective decision making through food recruitment, *Insectes Sociaux* **37**:258-267.

Belic, M. R., Skarka, V., Deneubourg, J. L., and Lax, M., 1986, Mathematical model of honeycomb construction, *Journal of mathematical Biology* **24**:437-449.

Benjafield, J. G., 1992, *Cognition*. Prentice-Hall, Englewood Cliffs, NJ.

Bieberich, E., 2000, Probing quantum coherence in a biological system by means of DNA amplification, *BioSystems* **57**:109-124.

Bonabeau, E., 1997, From classical models of morphogenesis to agent-based models of pattern formation, *Artificial Life* **3**:191-211.

Bonabeau, E., and Dessalles, J.-L., 1997, Detection and emergence, *Intellectica* 2:85-94.

Bonabeau, E., and Theraulaz, G., 1994, Why Do We Need Artificial Life? *Artificial Life* 1:303-325.

Bonabeau, E., Dorigo, M., and Theraulaz, G., 2000, Inspiration for optimization from social insect behaviour *Nature* **406**:39-42.

Bonabeau, E., Dorigo, M., and Theraulaz, G., 1999, *Swarm Intelligence: from natural to artificial systems*. Oxford University Press, UK.

Bonabeau, E., Theraulaz, G., and Deneubourg, J. L., 1996, Quantitative study of the fixed threshold model for the regulation of division of labour in insect societies, *Proceedings of the Royal Society of London* **B 263**:1565-1569.

Breder, C. M., 1954, Equations descriptive of fish schools and other animal aggregations, *Ecology* **35**:361-370.

Brooks, R. A., 1992, Artificial Life and Real Robots. In: *Toward a Practice of Autonomous Systems: Proceedings of First European Conference on Artificial Life* (F. J. Varela and P. Bourgine, eds.), MIT Press, Cambridge, MA, pp. 3-10.

Bushev, M., 1994, *Synergetics: Chaos, order, self-organization*. World Scientific, London.

Camazine, S., 1991, Self-organizing pattern formation on the combs of honey bee colonies, *Behavioral Ecology and Sociobiology* **28**:61-76.

Camazine, S., and Sneyd, J., 1991, A model of collective nectar source selection by honey bees: self-organization through simple rules, *Journal of Theoretical Biology* **149**:547-551.

Cao, Y. U., Fukunaga, A. S., and Kahng, A. B., 1977, Cooperative Mobile Robotics: Antecedent and Directions, *Autonomous Robots* 4:1-23.

Cariani, P., 1992, Emergence and artificial life. In: *Artificial Life II* (C. Langton, D. Farmer and S. Rasmussen, eds.), Addison-Wesley, Redwood City, CA, pp.775-797.

Cliff, D. T., Harvey, I., Husbands, P., 1993, Explorations in Evolutionary Robotics, *Adaptive Behavior* **2**:73-110.

Collins, D., and Jefferson, D., 1991, Ant Farm: Towards Simulated Evolution. In: *Artificial Life III* (C. Langton, C. Taylor, J. Farmer, and S. Rasmussen, eds.), Addison Wesley, pp. 603-629.

Corning, P., 2002, The Re-emergence of "Emergence":A Venerable Concept in Search of a Theory, *Complexity* 7:18-30.

Couclelis, H., and Gale, N., 1986, Space and Spaces, *Geografiske Annaler* **68 B**:1-12.

Cruchtfield, J. P., 1994, The Calculi of Emergence: Computation, Dynamics and Induction, *Physica D* **75**:11-54.

Darling, F. F., 1938, *Bird flocks and the breeding cycle*. Cambridge University Press, Cambridge, UK.

Delgado, J., and Solé, R. V., 1998, Mean Field Theory of Fluid Neural Networks. *Physical Review E* **57**:2204-2211.

Deneubourg, J. L., and Goss, S., 1989, Collective patterns and decision-making, *Ethology Ecology & Evolution* 1:295-331.

Deneubourg, J. L., Aron, S., Goss, S., and Pasteels, J. M., 1990, Self-organizing exploratory pattern of the Argentine Ant, *Journal of Insect Behavior* **32**:159-168.

Deneubourg, J. L., Goss, S., Franks, N., and Pasteels, J. M., 1989, The blind leading the blind: Modeling chemically mediated army ant raid patterns, *Journal of Insect Behavior* **23**:719-725.

Deneubourg, J. L., Goss, S., Franks, N., Sendova-Franks, A., Detrain, C., and Chretien, L., 1991, The Dynamics of Collective Sorting: Robot-Like Ant and Ant-Like Robot. In:

Proceedings of SAB90-First Conference on Simulation of Adaptive Behavior: From Animals to Animats (J. A. Meyer and S. W. Wilson, eds.), MIT Press/Bradford Books, Cambridge, MA, pp. 356-365.

Deneubourg, J. L., Goss, S., Pasteels, J. M., Fresneau, D., and Lachaud, J. P., 1987, Self-organization mechanisms in ant societies (II): Learning in foraging and division of labor. In: *From Individual to Collective Behavior in Social Insects*, (J. M. Pasteels and J. L. Deneubourg, eds.), Birkhausser, Basel, pp. 177-196.

Deneubourg, J. L., Pasteels, J. M., and Verhaeghe, J. C., 1983, Probabilistic behavior in Ants: A strategy of errors, *Journal of Mathematical Biology* 105:259-271.

Dorigo, M., Bonabeau, E., Theraulauz, G., 2000, Ant Algorithms and Stigmergy, *Future Generation Computer Systems* 16:851-871.

Duparc, J. H., 1985, Generalization of 'life'. In: *Dynamical Systems and Cellular Automata* (J. Demongeot, E. Golès and M. Tchuente, eds.), Academic Press, London, pp. 57-60.

Echelle, A. A., and Kornfield, I., (eds.), 1984, *Evolution of Fish Species Flocks*. University of Maine Press, Orono, ME.

Evans, R. G., 1995, Manufacturing consensus, marketing truth: guidelines for economic evaluation, *Annals of Internal Medicine* 122:59-60.

Forrest, S., 1990, *Emergent computation*. MIT Press, Cambridge, MA.

Franks, N. R., Gomez, N., Goss, S., and Deneubourg, J. L., 1991, The blind leading the blind: Testing a model of self-organization (Hymenoptera: Formicidae), *Journal of Insect Behavior* 4:583-607.

Freundschuh, S., and Egenhofer, M., 1997, Human Conception of spaces: Implications for geographic information systems *Transactions in GIS* 2:361-375.

Gärdenfors, P., 2000, *Conceptual Spaces: The Geometry of Thought*, MIT Press, Cambridge, MA.

Genakopolos, G., and Gray, L., 1991, When seeing farther is not seeing better, *The Bulletin of the Santa Fe Institute* 62:4-7.

Gierer, A., and Meinhardt, H., 1972, A theory of Biological Pattern formation, *Kybernetik* 12:30-39.

Goldstein, J., 1999, Emergence as a construct: History and issues. *Emergence* 1:49-72.

Golley, F. B., 1993, *An History of the ecosystem concept in ecology: more than the sum of its parts*. Yale University Press, New Haven and London.

Goss, S., Beckers, R., Deneubourg, J.-L., Aron, S., and Pasteels, J. M., 1990, How trail laying and trail following can solve foraging problems for ant colonies. In: *Behavioral Mechanisms of Food Selection*, (R. N. Hughes, ed.), Springer, Berlin-Heidelberg, pp. 661-678.

Gustavsson, R, and Fredericksson, M., 2001, Coordination and control of computational ecosystems. A vision of the Future. In: *Coordination of Internet Agents: Models, Technologies, and Applications*, (A. Omnicini, M. Klusch, F. Zambonelli, and R. Tolkschorf, eds.), Springer, Berlin, pp. 443-469.

Haken, H., 1979, Pattern formation and pattern recognition - an attempt at a synthesis. In: *Pattern formation by dynamical systems and pattern recognition*, (H. Haken, ed.), Springer, Heidelberg.

Haken, H., 1981, Erfolgsgeheimnisse der Natur, Deutsche Verlags-Anstalt, Stuttgart.

Haken, H., 1983, *Advanced Synergetics*. Springer, Berlin-Heidelberg-New York.

Haken, H., 1987, Synergetics: An Approach to Self-Organization. In: *Self-organizing systems: The emergence of order* (F. E. Yates, ed.), Plenum, New York, pp. 417-434.

Hecht-Nielsen, R., 1989, *Neurocomputing*. Addison Wesley, Reading, MA

Hemerlijjk, C. K., 2001, Computer Models of Social Emergent Behavior. In: *International Encyclopaedia of the Social & Behavioral Sciences* (N. J. Smelser and P. B. Baltes, eds.), Pergamon Press, Oxford, UK, pp. 14212-14216.

Herman, E. S., and Chomsky, N., 1988, *Manufacturing Consent*. Pantheon Books, New York.

Hilborn, R., 1991, Modeling the stability of fish schools: exchange of individual fish between schools of skipjack tuna (Katsuwonus pelamis), *Canadian Journal of Fisheries and Aquatic Sciences* **48**:1081-1091.

Hino, T., 2002, Breeding bird community and mixed-species flocking in a deciduous broad-leaved forest in western Madagascar, *Ornithological Science* **1**:111-116.

Holland, J. H., 1998, *Emergence from Chaos to Order*. Perseus Books, Cambridge, MA.

Holldobler, B., and Wilson, E. O., 1990, *The Ants*. Springer, Berlin.

Huberman, B. A., and Hogg, T., 1993, The Emergence of Computational Ecologies. In: *Lectures in Complex Systems*, (L. Nadel and D. Stein, eds.), Addison-Wesley, Reading, MA, pp.163-205.

Iberall, A. S., and Soodak, H., 1978, Physical basis for complex systems-Some propositions relating levels of organization, *Collective Phenomena* **3**:9-24.

Johnson, S., 2001, *Emergence: The connected lives of ants, brains, cities, and software*. Simon & Schuster, New York

Jullien, M., and Thiollay, J. M., 1998, Multi-species territoriality and dynamics of neotropical forest understory bird flocks, *Journal of Animal Ecology* **67**:227-252.

Kauffman, S. A., 1993, *The Origins of Order: Self-organization and Selection in Evolution.*. Oxford University Press, New York.

Koza, J. R., 1992a, Genetic Evolution and Co-evolution of Computer Programs. In: *Artificial Life II* (C. Langton, C. Taylor, J. Farmer, and S. Rasmussen, eds.), Addison-Wesley, Redwood City, CA, pp. 603-629.

Koza, J. R., 1992b, *Genetic Programming*. MIT Press, Cambridge, MA.

Lewes, G. H., 1877, *The Physical Basis of Mind. With Illustrations. Being the Second Series of Problems of Life and Mind*. Trübner, London.

Lindauer, M., 1961, *Communication among social bees*. Harvard University Press, Cambridge, MA.

Lundh, D., Olsson B., and Narayanan, A., (eds.), 1997, *Proceedings of BCEC97, BioComputing and Emergent Computation*. World Scientific, Singapore.

Manrubia, S. C, Delgado, J., and Luque, B., 2001, Small-world behaviour in a system of mobile elements, *Europhysics Letters* **53**:693-699.

Maturana, H. R. and Varela, F., 1992, *The Tree of Knowledge: The Biological Roots of Human Understanding*. Shambhala, Boston, MA.

Maturana, H. R., and Varela, F., 1980, *Autopoiesis and Cognition: The Realization of the Living*. Reidel, Dordrecht.

Mayer, B., and Rasmussen, S., 1998, Self-reproduction of a dynamical hierarchy in a chemical system. In: *Artificial Life VI*, (C. Adami, R. Belew, H. Kitano and C. Taylor, eds.), MIT Press, Cambridge, MA, pp. 123-129.

Mermin, N. D., and Wagner, H., 1966, Absence of ferromagnetism or antiferromagnetism in one- or two-dimensional isotropic Heisenberg models *Physical Rewiew Letters* **17**:1133-1136.

Mikhailov, A. S., and Calenbuhr, V., 2002, *From Cells to Societies. Models of Complex Coherent Action*. Springer, Berlin-Heidelberg-New York

Miller, J. G., 1978, *Living Systems*. McGraw Hill Books, New York.

Millonas, M. M., 1992, A Connectionist Type Model of Self-Organized Foraging and Emergent Behaviour in Ant Swarms, *Journal of Theoretical Biology* **159**:529-542.

Millonas, M. M., 1994, Swarms, phase transitions, and collective intelligence. In: *Artificial Life III*, (C. G. Langton, ed.), Addison-Welsey, Reading, MA, pp. 417-445.

Minati, G., 2001a, *Esseri Collettivi*. Apogeo, Milano, Italy.

Minati, G., 2001b, The Concept of Emergence in Systemics. *General Systems Bulletin* 30:15-19.

Minati, G., 2001c, Experimenting with the DYnamic uSAge of Models (DYSAM) approach: the cases of corporate communication and education. In: *Proceedings or the 45th Conference of the International Society for the Systems Sciences* (J. Wilby and J. K. Allend, eds.), Asilomar, CA, 01-94, pp. 1-15.

Minati, G., 2004a, Towards a Systemics of Emergence, In *Proceedings or the 48th Conference of the International Society for the Systems Sciences*, (J. K. Allen, J. Wilby, eds.), Asilomar, CA, 04-11, pp. 1-14.

Minati, G., 2004b, Buying consent in the "free markets": the end of democracy?, *World Futures* 60:29-37.

Minati, G., Penna, M. P., and Pessa, E., 1997, A conceptual framework for self-organization and merging processes in social systems. In: *Systems For Sustainability: People, Organisations and Environment*, (F. A. Stowell, R. Ison, R., Armson, J. Holloway, S. Jackson and S., McRobb, eds.), Plenum, New York.

Minati, G., Penna, M. P., and Pessa, E., 1998, Thermodynamic and Logical Openness in General Systems, *Systems Research and Behavioral Science* 15:131-145.

Minati, G., and Brahms, S., 2002, The DYnamic uSAge of Models (DYSAM). In: *Emergence in Complex Cognitive, Social and Biological Systems* (G. Minati and E. Pessa, eds.), Kluwer, New York, pp. 41-52.

Minsky, M., 1985, *The society of mind*. Simon & Schuster, New York.

Montoya, J. M, and Solé, R. V, 2002, Small world patterns in food webs, *Journal of Theoretical Biology* 214:405-412.

Morgan, C. L., 1923, *Emergent Evolution*. Williams & Norgate, London.

Munn, C. A., and Terborgh, J. W., 1979, Multi-species territoriality in neotropical foraging flocks, *Condor* 81:338-347.

Neisser, U., 1967, *Cognitive Psychology*, Prentice-Hall, Englewood Cliffs, N.J.

Newell, A., and Simon, H. A., 1976, Computer Science as Empirical Inquiry: Symbols and Search, *Communications of the ACM* 19:113-126.

Nicholls, J. L., and Baldassarre, G. A., 1990, Habitat selection and interspecific associations of piping plovers along the Atlantic and Gulf Coasts of the United States, *Wilson Bulletin* 102:581-590.

Nicolis, G., and Prigogine, I., 1977, *Self-Organization in Nonequilibrium Systems: From Dissipative Structures to Order through Fluctuations*. Wiley, New York.

Parrish, J. K., and Edelstein-Keshet, L., 1999, Complexity, Pattern, and Evolutionary Trade-Offs in animal aggregation, *Science* 284:99-101.

Parrish, J. K., and Hamner, W. M., (eds.), 1998, *Animal Groups in Three Dimensions*. Cambridge Univ. Press, Cambridge, UK.

Pasteels, J. M., Deneubourg, J. L., and Goss, S., 1987, Transmission and amplification of information in a changing environment: The case of insects societies. In: *Laws of nature and Human Conduct*, (I. Prigogine and M. Sanglier, eds.), Gordes, Brussels, pp. 129-156.

Payne, T. L., 1980, Life history and habits. In: *The Southern Pine Beetle, USDA – Forrest Service Technical Bulletin* (R. C. Tatcher, J. L. Searcy, J. E. Coster, and G. D. Hertel, eds.), No 1631, pp. 7 28.

Pessa, E., 1998, Emergence, Self-Organization, and Quantum Theory. In: *Proceedings of the First Italian Conference on Systemics* (G. Minati, ed.), Apogeo scientifica, Milano, Italy, pp. 59-79.

Pessa, E., 2000, Cognitive Modelling and Dynamical Systems Theory, *La Nuova Critica* **35**:53-93.

Pessa, E., 2002, What is emergence? In: *Emergence in Complex Cognitive, Social and Biological Systems* (G. Minati and E. Pessa, eds.), Kluwer, New York, pp. 379-382.

Pessa, E., Penna, M. P., and Minati, G., 2004, Collective Phenomena in Living Systems and in Social organization, *Chaos & Complexity Letters*, pp.173-184.

Pitcher, T. J., and Parrish, J. K., 1993, Functions of shoaling behavior in teleost. In: *Behavior of Teleost Fishes* (T. J. Pitcher, ed.), Chapman and Hall, London, pp. 363-439.

Prigogine, I., 1967, Dissipative Structures in Chemical Systems. In: *Fast Reactions and Primary Processes in Chemical Kinetics*, (S. Claesson, ed.), Interscience, New York., pp. 371-382.

Prigogine, I., and Glansdorff, P., 1971, *Thermodynamic Theory of Structure, Stability and Fluctuations*. Wiley, New York.

Radakov, D. V., 1972, *Schooling in the Ecology of Fish*. Wiley, New York.

Rauch, E. M., Millonas, M. M., and Chialvo, D. R., 1995, Pattern Formation and Functionality in Swarm Models, *Physics Letters A* **207**:185-193.

Ray, T. S., 1991, An approach to the synthesis of life. In: *Artificial Life II* (C. G. Langton, C. Taylor, J. D. Farmer, and S. Rasmussen, eds.), Addison-Wesley, Redwood City, CA, pp. 371-408.

Ray, T. S., 1994, An evolutionary approach to synthetic biology: Zen and the art of creating life. *Artificial Life* **1**:179-209.

Reynolds, C., 1987, Flocks, Herds, and Schools: A distributed Behavioral Model, *Computer Graphics* **21**:25-34.

Sakaguchi, H., 2003, Self-organization of hierarchical structures in non-locally coupled replicator models, *Physics Letters A* **313**:188-191.

Seeley, T. D., Camazine, S., and Sneyd, J., 1991, Collective decision-making in honey bees: how colonies choose among nectar sources, *Behavioral Ecology and Sociobiology* **28**:277-290.

Sewell, G. L., 1986, *Quantum Theory of Collective Phenomena*. Oxford University Press, Oxford, UK.

Skinner, B. F., 1938, *The behavior of organisms: an experimental analysis*. Appleton-Century-Crofts, New York.

Skinner, B. F., 1953, *Science and Human Behavior*, The MacMillan Company, New York

Solé, R. V, and Montoya, J. M., 2001, Complexity and fragility in ecological networks, In: *Proceedings Of The Royal Society Of London Series B-Biological Sciences* **268**: 2039-2045.

Solé, R. V., and Miramontes, O., 1995, Information at the edge of chaos in Fluid Neural Networks, *Physica D* **80**:171-180.

Solé, R. V., Miramontes, O., and Goodwin, B. C., 1993, Oscillations and Chaos in ant societies *Journal of Theoretical Biology* **161**:343-357.

Theraulaz, G., and Bonabeau, E., 1995a, Coordination in distributed building, *Science* **269**:686-688.

Theraulaz, G., and Bonabeau, E., 1995b, Modelling the collective building of complex architectures in social insects with lattice swarms, *Journal of Theoretical Biology* **177**:381-400.

Theraulaz, G., and Deneubourg, J. L., 1994, Swarm Intelligence in Social Insects and Emergence of Cultural Swarm pattern. In: *The Ethological Roots of Culture. Proceedings of NATO Advanced Study Institute* (A. Gardner, B. Gardner, B. Chiarelli, and F. Plooij, eds.) Cortona, Italy. Series D: Behavioural and Social Sciences - Vol. 78, Kluwer, Dordrecht, pp. 107-123.

Theraulaz, G., and Gervet, J., 1992, Les performance collectives des sociétés d'insectes, *Psychologie Francaise* **37**:7-14.

Theraulaz, G., Goss, S., Gervet, J., and Deneubourg, J. L., 1990, Swarm Intelligence in Wasps Colonies: an example of task assignment in multi-agents systems. In: *Proceedings of the 1990 IEEE International Symposium on Intelligent Control* (A. Meystel , J. Herath and S. Gray, eds.), IEEE Computer Society Press, Los Alamitos, CA, pp. 135-143.

Toner, J., and Tu, Y., 1995, Long-Range Order in a Two-Dimensional Dynamical XY Model: How Birds Fly Together, *Physical Review Letters* **75**:4326-4329.

Tulving, E., 1985, How many memory systems are there? *American Psychologist* **40**:385-398.

Turchin, V. A., 1989, Population consequences of aggregative movement, *Journal of Animal Ecology* **58**:75-100.

Turchin, V. A., 1998, *Quantitative analysis of movement: Measuring and modeling population redistribution in animals and plants*. Sinaurer Associates, Sunderland, MA.

Varela, F., Thompson, E., and Rosch, E., 1991, *The Embodied Mind: Cognitive Science and Human Experience*. MIT Press, Cambridge, MA.

Vitiello, G., 2001, *My double Unveiled: the dissipative quantum model of brain*. Benjamins, Amsterdam.

Von Bertalanffy, L., 1968, *General System Theory. Development, Applications*. George Braziller, New York.

Von Foerster, H, 1979, Cybernetics of Cybernetics. In: *Communication and Control in Society* (K. Krippendorff , ed.), Gordon and Breach, New York, pp. 5-8.

Von Glasersfeld, E., 1996, *Radical Constructivism: A Way of Learning*. Falmer Press, London.

Watzlawick, P., (ed.), 1983, *The Invented Reality*. Norton, New York.

Wilson, E. O., 1962, Chemical Communication among workers of the fire ant *Selenopsis saevissima*, *Animal Behavior* **10**:134-164

Wilson, E. O., 1971, *The Insect Societies*. The Belknam Press of Harvard University Press, Cambridge, MA and London, UK.

Wilson, E. O., 1975, *Sociobiology*. The Belknam Press of Harvard University Press, Cambridge, MA and London, UK.

Winograd, T., 1983, *Language as a Cognitive Process: Syntax*. Addison-Wesley, Reading, MA.

Yokozawa, M., and Hara, T., 1999, Global versus local coupling models and theoretical stability analysis of size-structure dynamics in plant populations, *Ecological Modelling* **118**·61- 72.

Chapter 4

HOW TO MODEL EMERGENCE: TRADITIONAL METHODS

4.1 A general classification of models of emergence
4.2 Dynamical Systems Theory for modelling emergence: the basic concepts
4.3 Dynamical Systems Theory for modelling emergence: bifurcation phenomena
4.4 Emergence phenomena in spatio-temporal systems and Dissipative Structures
4.5 The intrinsic limitations of traditional methods

4.1 A general classification of models of emergence

The previous Chapter illustrated just how complex is the concept of 'emergence' and also presented a possible classification of the different kinds of emergence, inspired by the proposal of Crutchfield in 1994 (Crutchfield, 1994; for different but somewhat related classifications, see Bedau, 1997; Ronald *et al.*, 1999; Rueger, 2000). Before briefly reviewing the complex world of the methods designed to describe, and sometimes manage, emergence within both the models and the analysis of experimental data, the reader should be aware that the attribute 'emergent' is often accompanied by another attribute: 'self-organizing'. Conceptually, self-organization is different from emergence, although it is sometimes confused with the latter. Strictly speaking, a system is defined as self-organizing when it is able to change its internal structure and its function in response to external circumstances (Banzhaf, 2002). This means that a system showing emergence must, of course, be self-organizing. However, the contrary is not necessarily true: we can build models of self-organizing systems, which do not show the most interesting forms of emergence. Despite such a conceptual difference, the study of self-organizing systems is very useful as it provides the main theoretical tools for dealing with emergence.

Before dealing with these, it should be said that the choice of a particular tool, or a set of them, depends, in turn, upon the general framework adopted by the researcher when studying emergence. The possible frameworks can be roughly classified along two main dimensions: that relating to the role played within them by *general principles*, and the other to the role of *homogeneity* of individual components of the system under study. Regarding the former, a rough approximation will be introduced, which allows a distinction only between *ideal* and *non-ideal* models of emergence (Pessa, 2000). For the latter, however, the approximation consists only of distinguishing between *homogeneity-based* and *heterogeneity-based* models of emergence.

Let us now briefly describe the meaning of the attributes introduced above, starting from 'ideal'. A model of emergence can be qualified in this way if it is characterized by a top-down structure, based on general principles assumed to be universally (or at least largely) valid, covering the widest possible spectrum of phenomena. This feature allows the deduction of particular consequences and forecasting only if suitable mathematical tools are available. In turn, this implies that the search for these tools becomes the main concern for the researcher trying to build an ideal model of emergence. Usually such a search is difficult, requiring a high level of mathematical competence; in most cases the tools required by the researcher simply do not exist and one is aware, from the beginning, that the model of emergence taken in consideration will be only a very rough approximation, with respect to the initial requirements. However, when an ideal model of emergence produces a result, one can be sure that this is not a consequence of some *ad hoc* assumption or of some mathematical trick, but derives in a logical way from 'first principles'. This allows for the *control* of emergence phenomena foreseen by the model, as it is always possible to understand, in mathematical terms, how to act upon the system described by the model to produce or to eliminate such phenomena. For this reason, many researchers (particularly physicists) think that ideal models of emergence are the most reliable. Such reliability, however, carries a high cost: currently, only a restricted number of scientists have the abilities needed for building and/or understanding such models of emergence. In addition, ideal models of emergence, owing to their very nature, tend to neglect the description and the role of the environment in which a system is embedded. Indeed, a model which pretends to be universal should, in principle, be applied to the entire Universe, that is to a system which, by definition, has no external boundary.

Such an absence of boundaries, however, in most practical cases is a shortcoming, as a description of most systems cannot be made without, at the same time, taking into account the nature of their boundaries and of their external environment.

Moving on now to 'non-ideal' models of emergence, they can be characterized essentially by the *difficulty in controlling* the emergence itself. In other words, when these models exhibit emergent behaviors (most frequently by resorting to very simple algorithms), they cannot be forecast, nor can the mechanisms for producing or eliminating a given behavior be identified. This occurs because most models of this kind are a mixture of general principles and of specific choices, which give rise to a mathematical structure so complicated as to make it very difficult to foresee its operational features. In these cases, often the only way to obtain information about model behavior is by using suitable computer simulations. Amongst the specific choices mentioned above, the most popular ones concern the form of the model equations and/or suitable boundary conditions. An obvious advantage of these models is their simplicity: even a person with a modest mathematical competence can understand the model laws, run a computer simulation of its behavior, and interpret its outcome. For this reason these models are very popular within the scientific community and used in many different domains. These include neural networks, cellular automata, artificial life, dissipative structures, and so on. The next Chapter will consider some of them in more detail.

Let us turn, at this point, to the distinction between 'homogeneity-based' and 'heterogeneity-based' models of emergence. As already mentioned in Section 3.1 in the previous Chapter, emergence is connected, amongst other things, to the existence of different *observational* or *descriptional* levels, the 'lowest' of which is generally assumed to be the 'base' level, in which the system under study is described as an assembly of *components*. The behaviors emergent at the 'higher' levels are produced by interactions between these components. But which framework should be adopted to describe the components? According to homogeneity-based models of emergence we should neglect any differences between the components and treat them all being equivalent to one another. For centuries this was the approach followed by physicists and mathematicians, as it led to simpler and more tractable models. On the other hand, it has the flaw of being unable to account for emergence in biological systems (and *a fortiori* in cognitive, social and economic ones), which derives precisely from the differences between individual components. Thus biologically-oriented researchers tend to adopt heterogeneity-based models of emergence, in which each component is endowed with a particular 'individuality' and the features of resulting emergent behaviors rely heavily upon the interactions between the various individualities. Of course, heterogeneity-based models of emergence are far and away more difficult to deal with than homogeneity-based ones.

Using the two dimensions introduced above, together with their associated bipolar distinctions, one can propose a classification scheme for possible models of emergence, useful both for fitting existing models (discussed later in this Chapter and in the next one) and for suggesting new possibilities. This scheme is shown in Table 4.1.

Table 4-1. A general classification scheme for models of emergence

	Ideal models	Non-ideal models
Homogeneity-based Models	− Spontaneous Symmetry Breaking in Quantum Field Theory − Noise-induced phase transitions	− Dissipative structures − Cellular Automata
Heterogeneity-based Models	− Spin glasses − Collective Beings (?) − Third-Quantization Models (?)	− Neural Networks − Immune Networks − Artificial Life − Agent-based models

The scheme contains a number of different models of emergence. The question marks denote models which are not fully developed and currently under study.

For practical reasons, existing models of emergence can be divided into two main categories: models based only upon *traditional methods*, and those which can also be based upon *non-traditional methods*. The former category includes models relying on mathematical formalisms designed to obtain a *full and exact knowledge* of the phenomena to be modeled. For these models, therefore, any reference to probability, uncertainty, noise, fluctuations, is excluded from the start. In the history of science, especially of physics, such models were very popular and the mechanistic view claims that all scientific models should belong to this category. However, even the founding fathers of General Systems Theory, in their first attempts, relied upon these models (see the framework adopted by Von Bertalanffy and by Mesarovic and Takahara, 1975), and the systemics field still includes supporters of their use even to describe and manage social and economic systems (starting from Forrester, 1968). For this reason our discussion begins with models of emergence based upon traditional methods, and this Chapter will be devoted only to these. Models based also upon non-traditional methods will be dealt with in the next Chapter.

Before discussing particular models of emergence it must be stressed that most of them share a common conceptual (and mathematical) basis. This has occurred because in most cases emergence is a temporal or a spatio-temporal phenomenon and, as a consequence, its modeling uses suitable *evolution equations*, whether they be relative to time as the only independent variable or include other independent variables, such as spatial ones. In turn, as these equations are of the utmost importance in most domains of science and technology, their theory has been extensively developed over a long period and is currently known as *Dynamical Systems Theory*. Some of the main concepts of this theory have already been introduced in Chapter 1 and others are described in Appendix 1. The next section presents some aspects of this theory only inasmuch as their use in building models of emergence. Amongst all the available tools for managing emergence, Dynamical Systems Theory is still the most important. Despite its severe limitations, no-one seriously interested in emergence can ignore it.

4.2 Dynamical Systems Theory for modelling emergence: the basic concepts

Over the last fifty years, Dynamical Systems Theory (hereafter abbreviated to DST) has developed with such impetus that the number of mathematicians and physicists involved has grown enormously. However, this did not produce a general consensus about the subject matter of this theory. In fact, when DST was born, it was simply a theory about the general behaviors of solutions of autonomous systems of ordinary differential equations. It was within this context that the main concepts of the theory (equilibrium points, attractors, stability, bifurcations) were originally developed, and this very same context was used by Von Bertalanffy when founding General System Theory, as described in Chapter 1.

In principle, such a situation may seem somewhat strange, as ordinary differential equations are not the only possible tool for describing dynamical evolution. However, other kinds of evolutionary equations such as, for example, *difference equations*, in which the independent variable is discretized, presented far greater difficulties than ordinary differential equations (for a review see, for instance, Saaty, 1981). Such a situation persisted up to the Sixties when the pioneering work of mathematicians such as E.Lorenz, led to the discovery of *chaotic deterministic* behaviors, opening up an entirely new and fascinating research domain for DST. It soon became clear that such behaviors were a consequence not only of complicated systems of differential equations, such as the Lorenz system cited in Chapter 1, but even of simpler *recurrence relations*, such as the celebrated *logistic map*:

$$x_{n+1} = 4 \, \lambda \, x_n \, (1 - x_n)$$

where λ, whose value must lie between 0 and 1, is a suitable parameter. This produced intensive research activity which attempted to generalize the concepts and the apparatus of DST to recurrence maps and difference equations. This effort was successful (see, for example, Collet and Eckmann, 1980; Gumowski and Mira, 1980; for more recent reviews see Mikhailov and Loskutov, 1996; Alligood *et al.*, 1997) and led to a considerable extension of the domain covered by this theory.

Further enlargement came from the study of chaotic attractors which requires a profound analysis of the statistical features of the behavior of solutions of evolution equations, thus favouring the inclusion of some statistical mechanics and ergodic theory within the domain of DST. In turn, this pushed researchers towards a further generalization of the concept of evolution equations, to include more general dynamical flows on topological varieties, no longer amenable to traditional difference or differential equations (for a good review of these developments see Eckmann and Ruelle, 1985; see also Sinai, 1989; Ruelle, 1995). At the same time, most researchers interested in DST began to include within this domain the study of all phenomena related to Partial Differential Equations (PDE), traditionally considered to be more difficult to deal with than ordinary differential equations.

As a consequence of all these developments, it is not easy to rigorously delimit the subject of DST. Here we adopt a very conservative view, excluding PDE, focusing mainly upon examples from ordinary differential equations. Although this choice forces us to add new sections for the topics lying outside this classification, it has the advantage of maintaining a close link with the tradition of good old classical mechanics and General Systems Theory.

Our discussion on the contribution of DST to models of emergence starts by looking at a generic autonomous system of ordinary differential equations having the form:

$$dQ_i / dt = F_i (Q_1, \dots, Q_n) \quad, \quad i = 1, \dots, n$$

Here time, t, is the only independent variable, and Q_i the i-th dependent variable. The total number n of the latter gives the number of *degrees of freedom* of the system. The symbols F_i denote given functions of their arguments.

Let us now briefly recall some fundamental concepts about systems of this kind (for further details see standard textbooks such as Davis, 1962; Jordan and Smith, 1977; Guckenheimer and Holmes, 1983; Haken, 1983; Tabor, 1989; Glendinning, 1994; Mikhailov and Loskutov, 1996). Some of these concepts have already been introduced in Chapter 1 and are further discussed in the next Chapter as well as in both Appendices, so here we will be brief. First, we define *phase space* as the space spanned by coordinates Q_i. Within it every *state* of the system, defined by a particular set of values of these variables, will be represented by a point. Moreover, every solution of the system itself, being described by n functions of time $Q_1 = Q_1(t)$, ..., $Q_n = Q_n(t)$, will correspond to the displacement of the point representative of the state of the system along a particular trajectory within the phase space, named *phase trajectory*. A simple example of phase trajectories can be obtained by looking at the popular equation of *harmonic oscillator* :

$$d^2x/dt^2 + \omega^2 x = 0$$

where ω denotes the frequency. By introducing an auxiliary dependent variable defined by:

$$y = dx/dt$$

the previous equation can be written in the form of a system of two first-order equations:

$$dx/dt = y, \quad dy/dt = - \omega^2 x$$

Therefore the phase space is 2-dimensional, being spanned by the two variables x and y. If now we eliminate the independent variable t from these two equations, we obtain a single equation describing the phase trajectories, having the form:

$$dy/dx = - \omega^2 (x/y)$$

The latter can, in this case, be easily solved by quadratures, giving the relationship:

$$y^2 + \omega^2 x^2 = C$$

where C denotes a constant whose value depends upon the initial values of x and y. This algebraic relationship describes, in the plane of coordinates x and y, a family of ellipses concentric with the origin which coincide with the phase trajectories of a harmonic oscillator.

Given a generic system with n degrees of freedom, we can now consider a subset S of its n-dimensional phase space with a finite n-dimensional volume. On examining each point belonging to S when the system is allowed to evolve for a given finite time Δt, we see that each point will be translated, along a given phase trajectory, to another location within phase space. Considering this new subset of phase space, denoted by S', consisting of all the new locations deriving from the evolution of each point belonging to S, two cases are possible: either the n-dimensional volumes of S and S' are equal or they are different. In the former case we have a *conservative* system, whereas in the latter the system is described as *dissipative*.

The most well-known and widely studied conservative systems are the *Hamiltonian* ones. They are characterized by the existence of a *Hamiltonian* (or *energy) function*, whose value is kept constant during the evolution of the system, provided it is an autonomous system. The state variables of these systems, in the case of n degrees of freedom, are given by n *generalized coordinates* q_i and by the associated n *generalized momenta* p_i. The generic Hamiltonian is a function of q_i and p_i (and sometimes even an explicit function of t) often denoted by $H = H(q_i, p_i)$. The dynamical evolution equations can be written under the form of *Hamilton's equations* :

$$dq_i/dt = \partial H/\partial p_i \quad , \quad dp_i/dt = -\partial H/\partial q_i$$

These are simply a generalization of motion equations of classical mechanics. Often, instead of starting from the Hamiltonian function H, some prefer to derive the whole description of the dynamical evolution of a Hamiltonian system from a general principle, that of *least action*, which states that, amongst all possible motions connecting a given configuration of generalized coordinates present at time $t1$, to another configuration of generalized coordinates present at time $t2$, that which physically occurs will minimize the value of the *action* :

$$I = \int_{t1}^{t2} dt\, L(q_i, dq_i/dt)$$

where the symbol L denotes another function, this time of generalized coordinates and of their time derivatives, usually called a *Lagrangian function*. It allows for a rigorous definition of generalized momenta through the formula:

$$p_i = \partial L/\partial (dq_i/dt)$$

Moreover, the Lagrangian and the Hamiltonian function are connected through the *Legendre transformation*:

$$H = \Sigma_i \, p_i \, (dq_i/dt) - L \, .$$

It is, of course, possible to write the dynamical evolution equations using only the Lagrangian function L . These are called *Lagrange's equations* and have the form:

$$\partial L/\partial q_i - (d/dt) \, \partial L/\partial (dq_i/dt) = 0 \, .$$

Before discussing the usefulness of Hamiltonian systems for modeling emergence, a simple example may be useful. Consider once again a harmonic oscillator of unit mass. Elementary physics shows that, during its motion, we have the conservation of total energy E, given by :

$$E = T + U$$

where T is the *kinetic energy* and U the *potential energy*. Their explicit expressions in this case have the well-known form:

$$T = (1/2)(dx/dt)^2 \quad , \quad U = (1/2) \, \omega^2 x^2$$

where x denotes, as usual, the displacement from the equilibrium position. If we identify this variable with the generalized coordinate, by putting $q = x$ (recalling that this system has only one degree of freedom), elementary computations show that, when we choose a Lagrangian function given by:

$$L = T - U = (1/2)(dq/dt)^2 - (1/2) \, \omega^2 q^2$$

the Lagrange equation written above is exactly equivalent to the harmonic oscillator equation of motion:

$$d^2q/dt^2 + \omega^2 q = 0$$

Once the Lagrangian function is known, one can derive from it the generalized momentum:

$$p = \partial L/\partial (dq/dt) = dq/dt$$

Now, by resorting to the Legendre transformation, an explicit form of the Hamiltonian function can be found, which turns out to be:

$$H = p\,(dq/dt) - L = (1/2)\,p^2 + (1/2)\,\omega^2\,q^2 = T + U$$

It coincides with the expression of total energy, as expected.

After this elementary example, we can focus our attention on the general features of the dynamical evolution of Hamiltonian systems. First of all, as these systems are the most natural generalization of systems studied by classical mechanics, they are, and have been for a long time, the most well-known and widely studied. For this reason, they are still preferred for modeling natural phenomena. It can be shown that the dynamical evolution equations contained within all the major physical theories of the twentieth century, such as General Relativity, Quantum Mechanics, Quantum Field Theory, are always formulated in terms of Hamiltonian systems or of suitable generalizations of them. One could even say that all reliable models of physical phenomena are tailored to the scheme described above, including those related to emergence. This raises a formidable problem, as we are now aware that the majority of physical, as well as biological, social and cognitive phenomena, cannot, in principle, fit within this framework. And the reason is simple: in most phenomena there is no conservation of a quantity resembling physical energy (or a Hamiltonian function). As described below, attempts have been made to modify the structure of physical theories in order to keep the advantages of Hamiltonian systems without resorting to some form of 'energy conservation' but the current situation can be summarised as follows: *Hamiltonian systems are the best candidates for building ideal models, but unfortunately they can produce only unrealistic 'toy models'.*

A second point regards a general behavioral feature of every Hamiltonian system: its evolution *can never stop*. In other words, any system of this kind will never reach an equilibrium state, even after an infinite time. This is because its evolution keeps the *n*-dimensional volume of subsets of phase space constant. Namely, an equilibrium state, within phase space, corresponds to a point, characterized by zero volume. Now, evolution towards an equilibrium state should correspond to the shrinking of a finite volume subset of phase space (containing, for instance, a given subset of the possible initial states of the system) so as to reduce it to a single point (the final equilibrium state). But in this way we would have non-conservation of volume in phase space, which is forbidden in Hamiltonian systems. This leads to the conclusion that a Hamiltonian system does not allow for the existence of equilibrium states. This does not mean, however, that the evolution of systems of this kind is devoid of interest. It can be shown that, besides standard behaviors, such as *periodic* ones (the typical cases are the harmonic oscillator or a planet orbiting the sun), these systems can exhibit *deterministic chaotic behaviors* (see, besides the references cited above, Lichtenberg and Lieberman, 1983). Should the transition from chaotic motion to a periodic one be considered as a form of emergence, then Hamiltonian systems can help in building models of this kind. Strangely enough, most researchers dealing with this transition (or with the reverse one) have made little use of the mathematics of systems of this kind.

Notwithstanding these unfavourable features of Hamiltonian systems, physicists continue to attribute a special meaning to the states of these systems which correspond to *minima of the potential energy function*. More precisely, those corresponding to *global minima* (that is, characterized by the minimum possible value of potential energy within the set of all allowable values of this energy) are called *ground states*, whereas those corresponding to *local minima* are called *metastable equilibrium states*. It is possible to show that all these states are *locally stable states*, in the sense that, by putting the system in a state close enough to one of these states, it will never go away from that state. Of course, a Hamiltonian system will *never fall* into one of these states, as it cannot lose energy. However, most model builders assume that a suitable effect from the external environment, such as a friction, would, after some time (possibly infinite), lead to one of these equilibrium states being reached. They neglect, however, to describe this action, as it would force them to go outside the framework of Hamiltonian systems. These states, therefore, hold a very strange theoretical status: they can be identified but in principle one can not explain, within an ideal theory, how they can be reached, as such an explanation would go beyond the possibilities of the theory itself. This situation renders unsatisfactory any attempt to build a reliable model of emergent phenomena by resorting only to Hamiltonian systems. Unfortunately, the latter comprise almost all the tools used so far for building reliable models!

The conclusions drawn in the above discussion about the usefulness of Hamiltonian systems force us to shift our attention towards dissipative systems, in the hope that they may exhibit a greater variety of behaviors. Namely, nothing within them prevents systems from reaching equilibrium states. The first important concept needed to deal with dissipative systems is that of *attractor*. The latter is a region of phase space where all behaviors (i.e., phase trajectories) starting within a given subset of the phase space itself, the so-called *attraction basin*, will be confined for $t \to +\infty$. An attractor may consist of a single point, in which case it is called an *equilibrium point*, or a curve within phase space, corresponding to an asymptotic dynamical behavior. If this curve is closed, the asymptotic behavior is *periodic* and the curve is then called a *limit cycle*. It is even possible to have attractors consisting in a whole subset of phase space (of course having a dimensionality, say m, lesser than that of the phase space itself, here denoted by n, and whence characterized by a zero n-dimensional volume). The latter correspond typically to *chaotic* behaviors in which the dynamical evolution of the system tends toward an unpredictable wandering within the domain of phase space corresponding to the attractor. These qualitative notions can be reformulated in a more rigorous (and abstract) way (see Lanford, 1981) by defining an attractor as a bounded subset of zero volume of phase space satisfying the following conditions:

a.1) the subset as a whole cannot vary through the effects of dynamical evolution of the system;

a.2) there exists a suitable domain of phase space which shrinks to the attractor as an effect of dynamical evolution of the system;

a.3) the subset corresponding to the attractor cannot be further sub-divided into two non-intersecting sub-subsets, in order for each of them to be, in turn, invariant under the effects of dynamical system evolution.

Again, simple examples can elucidate the concept of attractor. Let us start with the equation of a *damped harmonic oscillator* of unit mass:

$$d^2x/dt^2 + \gamma(dx/dt) + \omega^2 x = 0$$

where γ denotes the damping coefficient. As the energy of this system is no longer conserved, it is a dissipative system. It can be formulated as a first-order autonomous system by introducing an auxiliary variable $y = dx/dt$. We thus obtain:

$$dx/dt = y \ , \ dy/dt = -\omega^2 x - \gamma y$$

The equilibrium points can be found by putting equal to zero all time derivatives or, equivalently, by solving the algebraic system obtained by putting equal to zero the right-hand members of both system equations, that is:

$$y = 0 \ , \ -\omega^2 x - \gamma y = 0$$

This system has only one solution:

$x = 0$, $y = 0$

which constitutes the only equilibrium point of the damped harmonic oscillator. In intuitive terms, the damping term will counteract the dynamical motion of the oscillator so as to bring it, after a (theoretically) infinite time to its equilibrium position.

Another, more complicated, example is given by the *Van der Pol equation*:

$$d^2x/dt^2 - \gamma(1 - x^2)(dx/dt) + \omega^2 x = 0$$

which can be obtained from that of the damped harmonic oscillator simply by assuming that the damping coefficient is a suitable function of x. By reducing it to a system of two first-order equations using the previous technique we easily obtain:

$$dx/dt = y \quad , \quad dy/dt = -\omega^2 x + \gamma(1 - x^2)y$$

By proceeding as in the previous example, we again obtain only one equilibrium point given by $x = 0$, $y = 0$. However, the nonlinear nature of these equations leads to a suspicion that some other kind of attractor could be present. To investigate, let us introduce new *polar* dependent variables $\rho = \rho(t)$ and $\varphi = \varphi(t)$, related to the ones above by the standard relationships:

$$x = \rho\cos\varphi \quad , \quad y = \rho\sin\varphi$$

If these two relationships are substituted into the two system equations, multiplying the first equation by $\cos\varphi$ and the second by $\sin\varphi$, and finally summing both members of the two equations thus obtained, we arrive, after straightforward procedures, to the following equation for the time evolution of $\rho(t)$:

$$d\rho/dt = (1 - \omega^2)\rho\sin\varphi\cos\varphi + \gamma\rho\sin^2\varphi(1 - \rho^2\cos^2\varphi)$$

Recalling that, by definition, ρ represents, within the phase plane, the distance of the instantaneous state of the system from the equilibrium state $x = 0$, $y = 0$, its evolution thus tells us whether the system will asymptotically approach this equilibrium state or be captured by some other kind of attractor. Unfortunately, the equation giving $d\rho/dt$ is too complicated to be amenable to an exact solution. We therefore resort to a crude approximation, based on the fact that all functions of φ appearing within this equation are periodic (with period 2π) and bounded. This enables us to substitute each of these functions with its average value over the interval between zero and 2π. In practice, all we are doing is neglecting bounded fluctuations around the average evolution of $\rho(t)$. From a computational point of view, elementary applications of integral calculus will give, for the averages lying between zero and 2π of the three functions of φ entering into the equation for $\rho(t)$, the following results (for brevity we denote the average of a generic function $f(\varphi)$ within the latter interval by writing $<f(\varphi)>$):

$$<\sin \varphi \cos \varphi> = 0 \ , \ <\sin^2 \varphi> = \frac{1}{2} \ , \ <\sin^2 \varphi \ \cos^2 \varphi> = 1/8$$

Therefore the equation for $d\rho/dt$ can be approximated by:

$$d\rho/dt = (\gamma\rho/2)\,[1-(\rho^2/4)]$$

Elementary mathematics now tells us that, for $0 < \rho < 2$, the right-hand member of this equation is positive, so that the distance of the instantaneous state of the system from the equilibrium point increases with time. On the contrary, when $\rho > 2$, the right-hand member becomes negative, so that the distance decreases with time. These two circumstances, combined with the fact that $d\rho/dt$ tends to vanish when $\rho = 2$, show that the circumference in phase space defined by $\rho = 2$ is an attractor for the dynamical evolution of a system obeying the Van der Pol equation. Namely, if the system starts with an initial value of ρ less than 2, this value will increase with time, owing to the positive nature of $d\rho/dt$, but with a progressively smaller rate as ρ increases towards a value of 2. However, if the system starts with an initial value of ρ greater than 2, this value will decrease with time, owing to the negative nature of $d\rho/dt$, but with a progressively smaller rate of decrease as ρ decreases towards 2. Being given by a closed trajectory, the attractor thus found consists of a periodic behavior with fixed amplitude, that is a *limit cycle*. Systems of this kind are sometimes described as *self-oscillating*, because their fall into a limit cycle situation, with the associated periodic behavior, is absolutely independent of the initial state as well as the initial behavior (be it periodic or not) of the system itself. Although the above analysis is based upon a crude approximation, a more refined analysis (which will not be reported here) confirms the essential features of our findings.

Here we spare the reader from a discussion of a chaotic attractor, as this would require a number of new mathematical concepts. On the other hand, chaotic deterministic behaviors are introduced in Chapter 1 and further discussed in Appendices I and II. The only properties of these behaviors which are relevant for our further discussions are:

    c.1) chaotic deterministic behaviors are unpredictable over the long term;

c.2) they are *sensitive* to initial conditions.

The latter property means that, if we have a system exhibiting chaotic deterministic behavior, on introducing two points in the phase space, the difference between the solutions originating from these points will progressively increase with time, however close the initial points. In formal terms, this can be briefly denoted by:

$$dx/dt = F(x)$$

an *n*-dimensional autonomous system, with *x* and *F* representing, respectively, an *n*-dimensional vector of dependent variables and an *n*-dimensional vector function. On introducing two different initial states (or points in the phase space), denoted by $x_1(0)$ and $x_2(0)$, their difference is given by:

$$\xi(0) = x_2(0) - x_1(0)$$

Using $x_1(t)$ and $x_2(t)$ as the two solutions of our system having, respectively, initial states $x_1(0)$ and $x_2(0)$, the property of sensitivity to initial conditions means that the difference:

$$\xi(t) = x_2(t) - x_1(t)$$

increases with *t*. In general, the growth of $\xi(t)$ will follow a law of the form:

$$\xi(t) = \sum_j C_j \, e_j \exp(\lambda_j \, t)$$

where all subscripts range from 1 to *n*, C_j are suitable constants and e_j suitable constant vectors. The constant coefficients λ_j are called *Lyapunov exponents* of the system.

The practical computation of Lyapunov exponents would be very useful, as they would provide direct information about the features of the system dynamics (for instance, whether its motion is regular or chaotic). Unfortunately, this is very difficult and a number of approximate methods have been proposed (see the pioneering, and still topical, paper by Benettin *et al.*, 1976; amongst the most popular methods is that proposed by Wolf *et al.*, 1985; a readable review is to be found in De Souza-Machado *et al.*, 1990). Neglecting this topic which, over the past few years, has taken on the dimension of a true autonomous research domain, below some useful properties of Lyapunov exponents are listed:

L.1) an *n*-dimensional system has *n* Lyapunov exponents;

L.2) if at least one of the Lyapunov exponents is positive, then the system
exhibits chaotic behavior; if all Lyapunov exponents are zero or
negative, then the system only allows regular motion;

L.3) the sum of all Lyapunov exponents is zero for a Hamiltonian system,
whereas it is less than zero for a dissipative system;

L.4) if there is a phase trajectory not ending in an equilibrium point, then one
of the Lyapunov exponents is zero;

L.5) in every Hamiltonian system at least two Lyapunov exponents are zero.

These properties suggest that, generally speaking, the variation in behaviors
of dissipative systems is far greater than in the case of Hamiltonian systems
as decades of ongoing research have confirmed. There is practically no
behavior, however complex, which is not a dynamical attractor of a suitable
dissipative system. This fact has led most researchers to resort to this class of
systems to describe phenomena of emergence. The next section explores this
possibility.

4.3 Dynamical Systems Theory for modelling emergence: bifurcation phenomena

With such a large variety of possible attractors of dynamical systems, the
most useful concept is probably that of *stability*. The history of stability is a
long one, starting from Poincaré and Lyapunov, during which many different
definitions of the concept have been proposed. However, excluding the
technical details, for which the reader is referred to the literature already
cited (a useful overview is given by Lefschetz, 1977) we recall here that the
stability of a given attractor is determined through the behavior of the system
when it is pushed away from that attractor by a suitable perturbation and
then allowed to freely evolve. When this behavior consists in a spontaneous
re-approach towards the original attractor, the latter is *stable*. The various
definitions of stability differ on the basis of: the modality of this re-approach
(some requiring a final 'fall' onto the attractor, others are satisfied when the
system does not wander away from the attractor); the time involved (infinite
or finite); amplitude of the admissible perturbations (small or of any size). In
any case, whatever the definition of stability adopted, a determination of the
stability of a given attractor is, in general, a very difficult problem.

For this reason stability theory is still incomplete and only for particular
kinds of attractors are effective techniques available. The most well-known
case, of course, is that of equilibrium points. The stability of these can be

studied using a number of different techniques, the most popular of which is that based on *linearization*. In the case of autonomous systems of ordinary differential equations having the general form introduced in the previous section:

$$dQ_i/dt = F_i(Q_1, \ldots, Q_n), \quad i = 1, \ldots, n$$

this method requires the following steps:

S.1) Identification of the *fixed points* of the system, given by the solutions of the systems of equations:
$$F_i(Q_1, \ldots, Q_n) = 0$$

S.2) Introduction of small deviations $\xi_i(t)$ around each fixed point Q_0 through:

$$\xi_i(t) = Q_i(t) - Q_0$$

S.3) Development in power series of the right-hand members of system equations around Q_0 and retention only of terms of first order in $Q_i(t) - Q_0$, so as to transform the original system equations into a system of first order linear differential equations with constant coefficients for the small deviations $\xi_i(t)$. This system, known as the *linearized system*, has the general form:

$$d\xi_i(t)/dt = \sum_k A_{ik}\, \xi_k(t)$$

where A_{ik} denotes the elements of matrix A of constant coefficients thus obtained;

S.4) Determination of the stability features of the fixed point under consideration by looking at the eigenvalues of the matrix A, sometimes called the *Lyapunov matrix*; in particular, if all real parts of these eigenvalues are negative, the fixed point is *stable*.

This method, also known as the *Lyapunov first method*, is very easy to implement in the case of systems of two or three differential equations, where the computations are within the grasp of a first-year undergraduate student. It is, however, plagued by a number of shortcomings, such as:

d.1) When a system contains a large number of equations, the method runs into serious computational difficulties,

d.2) The structure of phase trajectories of the linearized system in the neighbourhood of a fixed point is topologically equivalent to that of the original system, except in the case where the Lyapunov matrix has a pair of imaginary conjugated eigenvalues with a zero-valued real part; thus in the latter case the linearization method cannot be applied, as it could give wrong answers (a nice example is given in Tabor, 1989, pp. 29-30);

d.3) The method is unable to detect the presence of other kinds of attractors (such as limit cycles).

These shortcomings can, however, be avoided by resorting to other methods (whose description will not be given here) so that, in principle, the problem of the determination of the stability features of fixed points can be considered as solved. Unfortunately, the situation is more serious for other kinds of attractors. For periodic behaviors, limit cycles and more complex dynamical attractors very few theories are available, and their application requires a high degree of mathematical competence, and considerable ingenuity. Often, resorting to computer simulations is the only way to obtain information about models which can not be studied using traditional analytical methods. Anyway, with one method or the other, over the last century our knowledge about the phase space structure of dynamical systems has increased enormously and this accumulated experience allows us to deal with a more difficult problem: what happens if the *actual structure* of a dynamical system is changed by an action coming from the external environment?

Recalling the concept of self-organizing system introduced at the beginning of this Chapter, one sees that this change of structure is merely a self-organization process, so that, to solve this problem, a theory of self-organization is needed. And clearly, such a theory is the first step towards a theory of emergence. The simplest approach for tackling the question is to take advantage of the concept of *parameter*. This term denotes a quantity entering into the evolution equations of a dynamical system, without it being confused with its dependent (or state) variables. What is important is that a parameter is a variable quantity whose value, however, is not fixed by those of dependent variables, but determined by the external environment (comprising the choices of model builder, the context and so on). For instance, in the equation of a damped harmonic oscillator:

$$d^2x/dt^2 + \gamma(dx/dt) + \omega^2 x = 0$$

the symbols γ and ω^2 denote suitable parameters, whose values are chosen on the basis of external criteria. It is important here to understand that the changes inparameter values correspond to the simplest form of change in the *structure* of a dynamical system. For example, when $\gamma = 22$, $\omega^2 = 21$, the system described by the equation:

$d^2x/dt^2 + 22 \, (dx/dt) + 21 \, x = 0$

is very different from the system corresponding to the choice $\gamma = 2$, $\omega^2 =$ 101:

$d^2x/dt^2 + 2 \, (dx/dt) + 101 \, x = 0$

Namely, as seen in standard textbooks on differential equations (see, for example, Boyce and DiPrima, 1977), if we choose initial conditions given by $x(0) = 0$, $dx/dt(0) = 20$, the former has the solution:

$x(t) = \exp(-t) - \exp(-21 \, t)$

describing an exponential and regular decay towards zero, whereas the latter, under the same initial conditions, has the solution:

$x(t) = 2 \exp(-t) \sin(10 \, t)$

describing an oscillatory decay towards zero. In other words, by varying parameter values, we obtain two very different behaviors.

Within this context the previous question can now be reformulated as: how does the behavior of a given dynamical system change when we change the values of its parameters? In attempting to answer this question, the most widely studied phenomenon is *bifurcation*. This term denotes a change in the number or type of attractors as a consequence of changes in parameter values (for a general treatment of this subject, besides the literature on dynamical systems cited above, see also Iooss and Joseph, 1981; a very useful review of the main recipes in bifurcation theory, particularly suited to the needs of those involved in practical computation, can be found in Rand and Armbruster, 1987). In most simple cases (i.e., those dealing with a single parameter), a bifurcation takes place when the value of a parameter, the so-called *bifurcation* (or *critical*) *parameter*, crosses a *critical value*, which thus appears as a separator between two structurally different states of affairs: one with values of the bifurcation parameter less than the critical value, and the other with values greater than the critical value. This suggests not only that models admitting bifurcation phenomena are the best suited to describe self-organizing systems, but induces one to postulate a close analogy between bifurcation phenomena and *phase transitions* in physical systems. Namely, the two different states of affairs, before and after critical value, can be considered as being analogous to different *phases* of matter, the critical value itself being viewed as the *critical point* of a phase transition. We shall see later, however, that such an analogy breaks down when we take into account the fact that the values of dependent variables undergo unavoidable fluctuations, due both to the limited sensitivity of our measuring instruments and to the coupling between the system and a noisy environment. Despite this, most researchers, from Prigogine onwards (Nicolis & Prigogine, 1977), upheld the validity of such an analogy. It is based on the fact that, in general, a bifurcation is associated with another phenomenon known as *symmetry breaking*. From a mathematical point of view, the solutions of dynamical evolution equations are invariant in form with respect to symmetry transformations which are only a proper subset of symmetry transformations leaving invariant the form of the evolution equations themselves. Later in the following Chapter we will enter into greater detail about this important topic. Here we merely recall that the connection between symmetry breaking and bifurcation phenomena is a well-known and widely studied subject (see, for example, Sattinger, 1978; 1980; Cicogna, 1981; Vanderbauwhede, 1982; Sattinger, 1983; Golubitsky and Schaeffer, 1985).

Bifurcation phenomenology is, in fact, so rich that it can offer a potential description of most *pattern formation* processes and associated emergence phenomena. Mathematicians usually distinguish between three different classes of bifurcations:

b.1) *subtle bifurcations*, in which the number of attractors remains constant, but their types undergochanges; the most celebrated example of such a class is a *Hopf bifurcation*, in which an attractor, when a parameter crosses a critical value, changes its structure from that of an equilibrium point to that of a limit cycle, i.e., a periodic motion. Such a bifurcation plays a major role in describing the birth of *self-oscillatory* phenomena. An example of a Hopf bifurcation is given by the Van der Pol equation cited in the previous section. If we assume that the values of the two parameters γ and ω^2 entering into this equation satisfy the constraint $\gamma^2 < 4\ \omega^2$, then it is possible to show, by choosing γ as a critical parameter, that a Hopf bifurcation occurs at the critical value $\gamma = 0$. More precisely, the limit cycle will be present when $\gamma > 0$ and absent when $\gamma < 0$. The theory of Hopf bifurcation is very complex and the determination of the stability of the associated limit cycle requires a considerable computational effort (standard and very good textbooks on this topic are Marsden and McCracken, 1976; Hassard *et al.*, 1981);

b.2) *catastrophic bifurcations*, in which the number of attractors changes. An elementary example of such a class of bifurcations is given by the unidimensional motion of a mass point in a force field characterized by the potential:

$$U(x) = k(x^2/2) + \lambda x^4 + \alpha \qquad\qquad (1)$$

where $\lambda > 0$, $\alpha > 0$ are parameters whose value is kept constant, whereas k represents the bifurcation parameter, and it is assumed that its value can change as a consequence of external influences. Straightforward mathematical considerations show that when $k > 0$, the potential $U(x)$ will have *only one* minimum, corresponding to a stable equilibrium point, located at $x = 0$, and having the value $U(0) = \alpha$. On the contrary, when $k < 0$, the potential $U(x)$ will have *two* equivalent minima, located at $x_1 = (|k|/4\lambda)^{1/2}$ and $x_2 = -(|k|/4\lambda)^{1/2}$, and corresponding to two stable equilibrium points, in which the value of $U(x)$ is given by:

$$U(x_1) = U(x_2) = \alpha + (5k^2/16\lambda)$$

Thus the *critical value* $k = 0$ marks the transition from a situation characterized by only one attractor to a situation characterized by two attractors. It is important to stress here that this system provides the simplest possible example of *symmetry breaking* associated with a bifurcation. To illustrate this, let us write in an explicit way the dynamical equation associated with the potential $U(x)$. It can be seen that, as the motion of the point (which, to simplify the expressions, we assume to be of unit mass) obeys the Second Principle of Newtonian dynamics, that is:

$$d^2x/dt^2 = -\partial U(x)/\partial x$$

the dynamical equation describing the motion itself will be:

$$d^2x/dt^2 + kx + 4\lambda x^3 = 0$$

Through direct substitution it can be verified that the form of this equation remains unchanged on changing the dependent variable:

$$x' = -x$$

It is also said that such a transformation specifies the *symmetry* of this equation. Clearly, when $k > 0$, even the equilibrium solution of the equation itself fulfils such a symmetry. Namely, as in this case we have the only equilibrium point given by $x = 0$, it obviously does not change its value in the above transformation. However, the situation changes when $k < 0$. This time the dynamical equation continues to be invariant in form with respect to the symmetry transformation, but this is no longer true for its equilibrium solutions. It can, in fact, be immediately verified that the transformation produces a shift of x_1 into x_2 and of x_2 into x_1 : a profound change in the nature of each equilibrium solution. We can thus say that the bifurcation occurring in correspondence to $k = 0$ produced a symmetry breaking with respect to the situation occurring for $k > 0$. This example can be considered as prototypical of many other kinds of symmetry breaking phenomena.

b.3) *plosive bifurcations*, in which the attractor size undergoes a sudden, discontinuous change as the bifurcation parameter crosses the critical value; a very simple example of a plosive bifurcation is the system represented by the nonlinear differential equation:

$$dx/dt = \alpha x \{x^2 - (\alpha + 1)x + [(\alpha^2 + 1)/4]\} \tag{2}$$

where α is the bifurcation parameter; the equilibrium points (attractors) of (2) are defined by :

$$dx/dt = 0 \tag{3}$$

which implies:

$$\alpha x \{x^2 - (\alpha + 1)x + [(\alpha^2 + 1)/4]\} = 0 \tag{4}$$

Straightforward algebra shows that when $\alpha < 0$, equation (4) has one and only one real solution, given by $x = 0$; a linearization of (2) around such a solution gives the following equation which governs the behavior of a small perturbation ξ of zero solution:

$$d\xi/dt = \alpha [(\alpha^2 + 1)/4] \xi \tag{5}$$

As the coefficient of ξ in the right-hand member of (5) is negative for $\alpha < 0$, the equilibrium point corresponding to $x = 0$ is *stable* for $\alpha < 0$. It can now be seen that, when $\alpha = 0$ (the critical value), equation (4) has two real roots: again $x = 0$, and the further root $x = \frac{1}{2}$ (with double multiplicity).

Moreover, when $\alpha > 0$, (4) has three real roots, i.e., the previous root $x = 0$ and the other two roots:

$$x = [\alpha + 1 \pm (2\alpha)^{1/2}]/2 \tag{6}$$

A stability analysis performed using the linearization method around each of these three roots, shows that when $\alpha > 0$, both the roots $x=0$ and $x = [\alpha + 1 + (2\alpha)^{1/2}]/2$ correspond to *unstable* equilibrium points, whereas the third root $x = [\alpha + 1 - (2\alpha)^{1/2}]/2$ represents the only *stable* equilibrium point. Thus, the bifurcation which occurs for $\alpha = 0$ gives rise to a sudden change from a stable equilibrium state, corresponding to $x = 0$, to a new stable equilibrium state, corresponding to $x = \frac{1}{2}$, or lying near to this value for positive and small enough values of α. Such a sudden change in size of the attractor is an example of a plosive bifurcation.

Mathematicians also introduced another classification of bifurcations into two categories: *local* and *global* (*cfr.* Ott, 1993). A local bifurcation gives rise to changes in attractor structure only within a small neighbourhood of phase space of the system under study. On the contrary, a global bifurcation results from a connection between distant attractors, and gives rise to sudden structural changes over large domains of phase space. The different categories of bifurcations sketched above are the object of intense study by mathematicians. A general theory of bifurcation, however, covering all possible phenomenologies is still lacking. Although for particular kinds of bifurcation well-known algorithms exist (implemented even through computer programs for symbolic manipulation) which allow us to forecast their phenomenology in a detailed way, in other cases we are still forced to resort to numerical simulations. Mathematicians and theoretical physicists (see the literature cited above) have, however, shown how nonlinear systems can always be described by suitable *canonical forms*, which are valid near bifurcation points.

Despite the wide variety of behaviors associated with bifurcation phenomena, the models described in this section are little more than elementary models of emergence. That is, they appear to lack the formal tools necessary for representing two main features of emergent phenomena (as introduced in the previous Chapter): the coherence of 'emergent' entities and the existence of a hierarchy of levels of description. To improve modelling tools in this area two further ingredients are required: new independent variables (typically of a spatial nature), besides the temporal one, and the presence of noise or fluctuations. In the following sections we discuss some models of emergence taking advantage of these new features. Before embarking upon such a discussion, however, we recall that, within the framework of traditional dynamical systems theory illustrated above, there are still some possibilities of describing the formation of hierarchies of levels of description.

The need for such developments arose, typically, within mathematical models of biological population dynamics (excellent textbooks include Okubo, 1980; Svirezhev and Logofet, 1983; Murray, 1989; see also Chapter 7 in Mikhailov, 1990; biologically oriented readers will profit from Doucet and Sloep, 1992). Namely, experimental observations of populations of species of living beings showed, besides the well-known interrelationships between different species, the presence of hierarchies, such as those connected with trophic chains, which require suitable modelling tools. A first proposal was made by Auger (for further details, see Auger 1980; 1983; 1985; 1989; Auger and Poggiale, 1995; 1996). The starting point of his analysis is an autonomous system of ordinary differential equations having the usual form:

$$dQ_i/dt = F_i(Q_1, \ldots, Q_n) \quad , \quad i = 1, \ldots, n$$

Here, the variable $Q_i(t)$ can be interpreted as the number (or the proportion) of elements in the system under study lying in the i-th state at time t. For instance, it could be the population density at time t of the i-th species within an ecological system. It must thus be stressed that in most cases, the whole set of the n states can be subdivided into a number of disjointed subsets in such a way that interactions between the variables associated with states lying within the same subset are strong, whereas interactions between variables associated with states lying in different subsets are weak. There are a number of different methods for doing this, all based upon one general principle: in the absence of interactions with other subsets the equations governing interactions within the same subset must be characterized by the *conservation of a global quantity* (a so-called *constant of motion*). It is to be stressed that, despite the fact that models of population dynamics do not generally belong to the class of Hamiltonian systems, most of them do allow for the presence of a constant of motion. The most celebrated case are the *Lotka-Volterra equations*, describing prey-predator interactions. They have the form:

$$dx/dt = a x - c x y \quad , \quad dy/dt = -b y + c x y$$

where x and y denote, respectively, the proportion of prey and the proportion of predators. The symbols a, b and c denote suitable positive parameters. This system has a family of periodic solutions, both for x and y, whose amplitudes are univocally determined by the chosen initial conditions. Within the phase plane spanned by x and y it can be shown that these solutions consist of concentric closed phase trajectories, all of which surround the only fixed point of this system, given by:

$$x_0 = b/c \quad , \quad y_0 = a/c$$

It is possible to prove, with suitable mathematical procedures, that this system allows for a constant of motion, given by:

$$\{[\exp(X)]/X\}^b \ \{[\exp(Y)]/Y\}^a = \text{constant} \ , \ X = x/x_0 \ , \ Y = y/y_0$$

As most models of interacting populations are merely suitable generalizations of the Lotka-Volterra model, they are expected to be often characterized by the presence of (exact or approximate) constants of motion.

In order to place the arguments of Auger within a more formal setting, let A be the total number of different subsets and $N\alpha$ the total number of different states belonging to the α-th subset. Moreover, let $(Q^\alpha)_i$ be the proportion of elements of the α-th subset lying in the i-th state, index i ranging, of course, from 1 to $N\alpha$. Now, if interactions within the same subset are much stronger than interactions between different subsets, it can be seen that the original system can be rewritten as:

$$\varepsilon \, d(Q^\alpha)_i \, /dt = f^\alpha{}_i[(Q^\alpha)_1, \dots ,(Q^\alpha)_{N\alpha}] + \varepsilon \Sigma_\beta f^{\alpha\beta}{}_i[(Q^\alpha)_1, \dots ,(Q^\alpha)_{N\alpha} \ (Q^\beta)_1, \dots ,(Q^\beta)_{N\beta}]$$

where ε denotes a small parameter, functions $f^\alpha{}_i$ the intra-group interactions (within the same subset) and functions $f^{\alpha\beta}{}_i$ the inter-group interactions (between different subsets).

At this point Auger introduces suitable *collective* (or *global*) dependent variables, identified with the constants of motion for each subset, in the absence of inter-group interactions. If V^α is the constant of motion relative to the α-th subset, it can be shown that these new collective variables obey the evolution equations:

$$dV^\alpha \, /dt = \varepsilon \, \Sigma_\beta \ \Sigma_i \, (R^\alpha)_i f^{\alpha\beta}{}_i[(Q^\alpha)_1, \dots ,(Q^\alpha)_{N\alpha} ,(Q^\beta)_1, \dots ,(Q^\beta)_{N\beta}]$$

where:

$$(R^\alpha)_i = \partial V^\alpha \, /\partial (Q^\alpha)_i$$

These equations govern the evolution of *slow* variables, as it can be shown that the rates of change of the variables V^α with time is much smaller than those of the original variables $(Q^\alpha)_i$. Thus the conceptual hierarchy produced by the introduction of the new variables is equivalent to a time-scale hierarchy. Of course, the new variables V^α describe a new, higher level of description with respect to that of the original variables $(Q^\alpha)_i$. The above equations simply describe how the dynamics of this new level depend upon those of the lower level. Besides, if one introduces further approximations, such as the replacement of the lower-level variables $(Q^\alpha)_i$ with their time averages, it is possible to rewrite these equations only in terms of the new collective variables V^α . The latter should describe the evolutionary laws holding at the new higher level. Auger claims that, if their right-hand members differ in form from those appearing in the right-hand members of the equations describing the evolution of $(Q^\alpha)_i$, then we can speak of a *functional emergence*. This means that the model with global variables shows a functional dependence upon them which is different from that appearing in the model using the original variables. Auger's proposal shows that the usual apparatus of evolutionary equations allows for a feasible description of hierarchical levels. We therefore believe that it should be embodied within any future theory of emergent phenomena.

A number of computer simulations of models of population dynamics seem to support the arguments put forward by Auger. For instance, Yokozawa and Hara (*cfr.* Yokozawa and Hara, 1999) studied the so-called *replicator models*, which are merely generalizations of the Lotka-Volterra model described by equations of the form:

$$dx_i /dt = x_i (a_{i0} + \Sigma_k w_{ik} x_k)$$

which hold for a system of N different interacting species, x_i denoting the population density of the i-th species. However, in this case the coefficients w_{ik} are no longer constants, but represented by functions of the ratio $r = x_k / x_i$. More precisely, these authors assumed $w_{ik}(r)$ to be given by:

$$w_{ik}(r) = r^\beta \text{ for } 0 < r < R_C , \quad w_{ik}(r) = R_C{}^\beta \text{ for } r > R_C$$

The results of computer simulations show that, for suitable choices of parameter values, the distribution of population sizes is characterized by a *layered structure*, in which a small number of species are associated with large population sizes, whereas a high number of species are associated with small population sizes. This can be interpreted as an *emergence of hierarchical levels*, with the species associated with large population sizes lying at the higher hierarchical levels. In a sense, their sizes, in the overall system dynamics, play a role approximately similar to that of the global variables introduced by Auger, as the dynamical evolution of species with small size populations is constrained by that of the species with larger population sizes. These findings were confirmed by further simulations (*cfr.* Sakaguchi, 2003) using different choices of the functions $w_{ik}(r)$.

Finally, another possible way to introduce hierarchical levels would be to resort to dynamical evolution equations in which the dependent variables are, in turn, directly influenced by global, or collective variables such as, for example, the global average value of the dependent variables themselves or their average correlation. Such an approach has been fostered by a number of authors (see, for instance, Nicolis, 1986). It is, however, not very popular, as it requires the introduction of dynamical evolution equations, typically in the form of integro-differential equations, which are far more complicated than ordinary differential equations. However, there is a small number of contributions to models of this kind which include a model of neural field dynamics (Ermentrout and Cowan, 1979), other dynamical models based on integro-differential equations (see, for example, Mikhailov, 1990, Chapter 5) and the underlying bifurcation theory (for an example see Olmstead *et al.*, 1986).

4.4 Emergence phenomena in spatio-temporal systems and Dissipative Structures

As shown in the previous section, the introduction, besides the temporal one, of other independent variables (typically of a spatial nature) is mandatory for reaching a more realistic description of emergence phenomena. Of course, this forces us to resort to the mathematics of PDEs. This is more complicated than that of ordinary differential equations (an introductory textbook is Duchateau and Zachmann, 1986; a good treatise on the applications of PDEs to mathematical physics is Bitsadze, 1980; popular advanced textbooks include Sneddon, 1957; Copson, 1975). Notwithstanding this, the theory of bifurcation phenomena associated with PDEs has been developed over a long period (see Sattinger, 1973, and the references on bifurcation theory cited in the previous section).

In general terms, a dynamical model describing self-organization and emergence within a spatio-temporal context comprises the following components:

E.1) a *dynamical evolution law* for its dependent variable, often expressed through a PDE or a system of PDEs;

E.2) suitable *initial* and *boundary conditions*;

E.3) a number of *parameters*, some of which are considered as *critical parameters* (also called *bifurcation parameters* or *control parameters*).

In most cases, the dependence of model state variables upon spatial coordinates is expressed through the action of suitable *operators*, the most widely-used ones, in the case of n spatial coordinates x_1 , ..., x_n , being the well known:

$$grad \equiv (\partial/\partial x_1 , ..., \partial/\partial x_n) \quad , \quad \Delta \equiv (\partial^2/\partial x_1^2 + ... + \partial^2/\partial x_n^2)$$

The operator Δ is also called *Laplace operator* or *Laplacian*. Therefore, the general form of the evolution equations for a model of this kind, involving N dependent (or state) variables s_1 , ..., s_N , will be:

$$L_t s_i = g_i (\{s_j \}) + h_i (\{grad\ s_j \}, \{\Delta s_j \})$$

where the braces denote a generic dependence on all variables s_1 , ..., s_N, and L_t a generic time operator.

A first difference with the situation occurring in ordinary differential equations consists of the fact that here we have different kinds of equilibrium states, that is:

a)   *stationary homogeneous equilibrium states*, defined through the vanishing both of spatial and time derivatives, and whence found by solving the system of equations:

$$g_i (\{s_j \}) = 0$$

b)   *stationary non-homogeneous equilibrium states*, defined through the vanishing only of time derivatives, and whence found by solving the system of differential equations:

$$g_i (\{s_j \}) + h_i (\{grad\ s_j \}, \{\Delta s_j \}) = 0$$

Possible kinds of constraints usually imposed on these equations allow a classification of spatio-temporal models of self-organization into three different categories:

A) *models of bistable media*; in this case the evolution equation for each single isolated variable, obtained from the general evolution equations for the same variable by neglecting all terms containing spatial operators, as well as any interaction with the other dependent variables, leads to two different homogeneous stationary states, which are both stable in the presence of small perturbations;

B) *models of excitable media*; in this case the evolution equation for each single isolated variable, obtained as before, leads to a single homogeneous stationary state, stable under small perturbations;

C) *models of oscillatory media*; in this case the evolution equation for each single isolated variable leads to a stable limit cycle solution.

The number of models so far built, belonging to each of these three categories, is so vast high that even a simple list of them would fill a book. Therefore, only a few representative cases are presented below, with the aim of understanding just how useful these models could be for improving the modeling of emergence phenomena.

Let us start with a model typical of category A), proposed by F. Schlögl at the beginning of the 1970s (good references on this model are Albano *et al.*, 1984; Schöll, 1986). The model is based upon a hypothetical scheme of chemical reactions involving three substances A , B and X, given by:

$$A + 2 X \leftrightarrow 3 X \quad , \quad X \leftrightarrow B$$

Within this model it is postulated that the concentrations of A and B be kept constant at nonequilibrium values which can, however, be changed by an external action. If we further assume that the reactions take place within a 1-dimensional vessel and that in it substance X can diffuse, we easily obtain that the *Schlögl model* reduces to the single PDE for the concentration ρ of X:

$$\partial \rho / \partial t = D \, \partial^2 \rho / \partial x^2 - k_2 \, \rho^3 + v \rho^2 - k_3 \, \rho + \mu$$

where $v = k_1 \, \rho_A$, $\mu = k_4 \, \rho_B$, and k_1 , k_2 , k_3 and k_4 denote suitable reaction rates.

The stationary homogeneous equilibrium states of the model are solutions of:

$$- k_2 \, \rho^3 + v \rho^2 - k_3 \, \rho + \mu = 0$$

A little algebra shows that, if we take v as a critical parameter, there is a critical value v_c such that, when $v > v_c$, this equation has three real solutions, denoted here by ρ_1 , ρ_2 and ρ_3 , ordered in such a way that $\rho_1 < \rho_2 < \rho_3$ provided the value of μ be within a given interval. To be more precise, the critical value is given by $v_c = (3 \, k_2 \, k_3)^{1/2}$.

A linearized stability analysis around these three equilibrium points shows that both ρ_1 and ρ_3 are stable, whereas ρ_2 is unstable. This confirms that we are in the presence of a bistable medium. The main question now, is whether this model allows for other stable non-homogeneous stationary equilibrium states. A more sophisticated analysis, however, showed (see the literature cited above, and Mikhailov, 1990, Chapter 2) that such a circumstance can never occur, whatever the choice of boundary conditions. Thus, the model behavior could seem trivial, were it not for the existence of solutions which describe *propagating wave fronts*. These are also known as *trigger waves*. Their role is that of changing the values of ρ in a bounded region of the space within which the system is constrained (the 1-dimensional vessel) so as to transform values close to those of stable homogeneous stationary solutions into values close to the other stable homogeneous stationary solution. In other words, these trigger waves support the transition from one stationary solution to the other. While their velocity is fixed by the boundary conditions and the values of system parameters, the form of their front changes with time. It is possible to have more than one wave front at the same time, depending upon the initial conditions. However, when two wave fronts collide, they undergo mutual annihilation.

It is possible to study non-homogeneous stationary solutions of this model as if the latter were describing the motion of a point particle within a given force 'potential'. Namely, the equation representing these solutions has the form:

$$D \, \partial^2 \rho / \partial x^2 - k_2 \, \rho^3 + v \rho^2 - k_3 \, \rho + \mu = 0$$

and is therefore similar to the equations of motion of classical mechanics (the only difference being that, in the latter the independent variable is given by time, whereas here it is given by spatial coordinate x). This allows for the introduction of a 'potential' $U(\rho)$ defined by:

$$U(\rho) (= \int_0^\rho (- k_2 u^3 + vu^2 - k_3 u + \mu) \, du$$

It can be shown that the stable homogeneous stationary solutions ρ_1 and ρ_3 are *minima* of $- U(\rho)$. This allows us to draw a parallel between this model and first-order phase transitions occurring within physical systems. That is, we can identify the minima of $- U(\rho)$ with the different *phases* of matter and call a phase *metastable* if it corresponds to a minimum of $- U(\rho)$ which is not the lowest one. A metastable phase is unstable with respect to sufficiently strong perturbations. This means that, starting from an initial condition in which there is already a nucleus of the stable phase above a critical dimension, then we will have the birth of a trigger wave which will support a transition of the whole medium from the metastable phase to the stable one. Leaving aside further quantitative details and despite such an interesting analogy, the behavior of bistable media (as confirmed by the analyses of other models in this category, as well as from general theorems such as those proved by Chafee, 1975) does not qualify them as good candidates for modeling emergent phenomena.

Let us now focus our attention on models belonging to class *B*): *excitable media*. These currently enjoy wide popularity, which has brought them to a large audience, including not only mathematicians and physicists, but even biologists, philosophers, psychologists, economists and social scientists. According to widespread opinion they provide the best possible models of emergent phenomena. Owing to such great interest, the number of different models of this kind so far proposed is, as to be expected, very high. Within them, however, we focus our attention on a particular subclass, sometimes called *activator-inhibitor* models, in order to classify them into categories and identify prototypical forms.

The basic idea underlying an activator-inhibitor model is that all interesting dynamics should derive from suitable competition between two opposing influences: that exerted over short spatial ranges by an *activator*, and the other exerted over long spatial ranges by an *inhibitor*. The role of the activator is to give rise to local deviations from the homogeneous stationary state, whereas the inhibitor prevents these local inhomogeneities from reaching macroscopic dimensions. Of course, there are different cases:

-    the influence of short-range activation on the dynamics is *less* than that of long-range inhibition; in this case every deviation from the homogeneous stationary state will, sooner or later, be destroyed and the system will fall back into that state;

-    the influence of short-range activation on the dynamics is *greater* than that of long-range inhibition; in this case every deviation from the homogeneous stationary state, however small, will grow in an unlimited way giving rise to chaotic and unforeseeable dynamics;

-    the influence of short-range activation on the dynamics is *exactly counterbalanced* by the long-range inhibition; in this case stable complex structures can form, consisting of spatially inhomogeneous patterns or even in spatio-temporal inhomogeneous patterns.

Thus, activator-inhibitor models are potentially able to describe the emergence of any complex pattern from a homogeneous state, merely as a consequence of variations in suitable parameter values. And this is precisely what is meant by the term 'self-organization'!

To transform this general idea into workable models one must, however, introduce precise choices in the description both of the activator and the inhibitor. For the moment, these choices can be subdivided into two main categories: those associated with *reaction-diffusion systems* and those associated with *non-linear dispersive wave equations*. Each of these categories is briefly discussed below.

Reaction-diffusion systems. This term is applied to models described using systems of PDEs, in which *non-linearity* plays the role of local activator and *diffusion* is the long-range inhibitor. If we introduce suitable dependent variables c_i $(i = 1, ..., n)$, each of which can be interpreted, for instance, as the (space and time dependent) concentration of a given chemical species, the general form of a reaction-diffusion system, in a mono-dimensional spatial domain, is:

$$\partial c_i / \partial t = \partial / \partial x \left(\sum_k D_{ik} \, \partial c_k / \partial x \right) + F_i \left(c_1, ..., c_n \right)$$

Here the first term in the right-hand side of the equation describes the effect of diffusion, whereas the second describes the activation by the non-linear functions F_i $(c_1, ..., c_n)$. Coefficients D_{ii} are called *diffusion coefficients*, whereas coefficients D_{ik} $(i \neq k)$ are the *mutual diffusion coefficients*. In principle, these coefficients could be functions of spatial coordinates, even though they are usually taken as constant., Besides suitable initial conditions, these equations are normally supplemented with a set of *boundary conditions*. In the case of a finite spatial domain of length L, the most popular are the *Dirichlet conditions*, fixing the values of dependent variables at the two extremes of this domain (typically with coordinates 0 and L), and the *Neumann conditions*, fixing the values of spatial derivatives of dependent variables at the same points.

There is a long tradition of such models, dating from the early proposals of Rashevsky (for a comparison with modern treatments, see Rashevsky, 1960, Vol. I, Chaps. VIII-XI) and Von Bertalanffy (*cfr.* Von Bertalanffy, 1940; 1950; 1968, Chap. 5). However, the first explicit construction of a model of this kind came from the mathematical genius of A. M. Turing (*cfr.* Turing, 1952). He clearly showed how the emergence of a stationary non-homogeneous equilibrium solution from a pre-existing stationary homogeneous state was due to a phenomenon of bifurcation, generally of a catastrophic type, occurring as a consequence of the loss of stability of the homogeneous state when a suitable model parameter crossed a critical value. His paper opened up a new area of research which, however, owes its development and popularity to the work of Prigogine and co-workers (see Lefever and Prigogine, 1968; Glansdorff and Prigogine, 1971; Nicolis and Prigogine, 1977). They showed that models of systems where non-homogeneous structures could occur were to be interpreted as models of *open systems* characterized by an exchange, not only of energy, but also of matter with the external environment. This interpretation was supported with a suitable generalization of the standard thermodynamics of closed systems in equilibrium states, the so-called *thermodynamics of irreversible processes*, which led to a generalization of the standard Second Principle of Thermodynamics. In this way, it was possible to interpret the bifurcations introduced by Turing as processes of transition from a disordered (homogeneous) state to an ordered (non-homogeneous) state, made possible by a decrease in the internal entropy of the system, allowed by the fact that we are dealing with an open system. As the non-homogeneous structures arising from such transitions are kept stable owing to the existence of a suitable *dissipation* (here occurring under the form of diffusion) which counterbalances the activation produced by non-linearity, Prigogine coined the term *dissipative structures*.

One of the first and most celebrated models of a dissipative structure was proposed in 1968 by Lefever and Prigogine (*cfr.* Lefever and Prigogine, 1968). It is now universally known under the name of 'Brusselator' (after the city of Brussels, which was home to Prigogine's school). The Brusselator was built to describe a system composed of four chemical substances, whose concentrations A, B, X and Y, interact through the chemical reactions:

$$r.1) \ A \rightarrow X \quad , \quad r.2) \ B + X \rightarrow Y + D \quad , \quad r.3) \ 2X + Y \rightarrow 3X \quad ,$$
$$r.4) \ X \rightarrow E$$

It is assumed that the reaction products, of concentrations D and E are removed from the space domain in which the reactions occur. It can be shown that, in the presence of a mono-dimensional spatial domain and diffusion, system evolution can be described by the following system of PDEs:

$$\partial X / \partial t = D_1 \, \partial^2 X / \partial x^2 + A - (B + 1) X + X^2 Y$$

$$\partial Y / \partial t = D_2 \, \partial^2 Y / \partial x^2 + B X - X^2 Y$$

supplemented by the Neumann boundary conditions:

$$\partial X / \partial x(0) = \partial X / \partial x(L) = \partial Y / \partial x(0) = \partial Y / \partial x(L) = 0$$

where L denotes the length of the reaction domain.

It can be immediately seen that this system has only one stationary homogeneous equilibrium solution:

$$X_0 = A \quad , \quad Y_0 = B / A$$

A tedious but straightforward linear stability analysis shows that this homogeneous state can, however, become unstable. More precisely, if the diffusion coefficients satisfy the inequality:

$$D_2 - D_1 > (L / \pi)^2$$

then the two eigenvalues of the Lyapunov matrix, introduced in the previous section, are both real. In this case, if B is the bifurcation parameter, it can be shown that there exists a *critical curve* $B(m)$, where m is a positive integer ($m = 1, 2, ...$), defined by:

$$B(m) = 1 + (D_1/D_2)A^2 + (1/D_2)(L \, A \,/m \, \pi)^2 + D_1(m \, \pi/L)^2$$

such that, if, for some value of m, the value of B is greater than that of the corresponding $B(m)$, then the homogeneous state is *unstable* and we now have the occurrence of two possible different *stable* non-homogeneous states which, to a first approximation, are spatially periodic with a wavenumber given by m. It can also be seen that when m is real, instead of being an integer, the curve $B(m)$ would be characterized by a single global minimum q given by:

$$q = (A)^{1/2} (L /\pi) (D_1 D_2)^{-1/4} \quad , \quad B(q) = [1 + (D_1/D_2)^{1/2} A]^2$$

Thus the loss of stability of the homogeneous state occurs in correspondence with the integer closest to q and the value of B closest to $B(q)$. If, for instance, we choose $L = 2 \pi$, $A = 1$, $D_1 = 1$, $D_2 = 16$, we find that $q = 1$ and that $B(1) = 25/16$. Thus, on increasing the value of B, when we cross the *critical value* $B(1)$, we observe a bifurcation phenomenon, with the loss of stability of the homogeneous state and the birth of two new stable non-homogeneous structures which, to a first approximation, can be described by a function of the spatial coordinate directly proportional to $cos(\pi x /L)$.

We will not pursue further such analysis, for which the reader is referred to the literature (see, for example, Auchmuty and Nicolis, 1975; Nicolis and Prigogine, 1977). What matters here is that the following points be stressed:

a) The recipe used in this kind of model for obtaining emergence is *stability exchange* between different stationary solutions (for instance, homogeneous and non-homogeneous ones); this implies that the evolution equations must have a form allowing, from the start, all the various solutions; this explains why PDEs, with their unlimited variety of solutions, are more suitable than ordinary differential equations;

b) Despite the fact that the emergence of a particular dissipative structure is simply the acquisition of stability by a particular non-homogeneous solution of the evolution equations which, in principle, should be known in advance to the mathematically sophisticated model builder, this phenomenon seems to be one of *intrinsic emergence,* ias defined in Section 3.1 . That is, the bifurcation allowing for stability exchange is a catastrophic bifurcation giving rise not to one, but to *two* different non-homogeneous solutions. The mathematical formalism, however, in principle does not allow one to know *which* of the two solutions will be chosen by the system under study;

c) A bifurcation giving rise to the occurrence of a non-homogeneous solution is associated with a *symmetry breaking*, just as in the case of phase transitions occurring in condensed matter physics: both the Brusselator equation and its stationary homogeneous solution are invariant with respect to space translations, i.e., changes in the spatial coordinate of the form $x' = x + h$, where h denotes an arbitrary spatial shift. On the contrary, the non-homogeneous solution corresponding to a dissipative structure is no longer invariant with respect to these transformations, because, for instance, $cos(\pi x /L)$ is generally different from $cos[\pi (x + h) /L]$.

All these features seem to support the idea that reaction-diffusion systems or, more in general, activator-inhibitor systems allowing for bifurcation phenomena associated with the occurrence of dissipative structures are the best candidates for modeling intrinsic emergence, at least within a macroscopic framework. As this opinion was shared by many researchers, it gave rise to a veritable explosion in the number of investigations on these topics (for reviews see, besides the literature cited above, Vasilev *et al.,* 1979; Belintsev, 1983; Krinsky, 1984; Vasilev *et al.,* 1986; Garrido, 1988; Tirapegui and Zeller, 1993; 1996; Golubitsky *et al.,* 1999; Mori and Kuramoto, 2001). This vast amount of research produced a number of interesting findings, including an impressive variety of different structures and behaviors arising, through the mechanism of bifurcation, from models based on reaction-diffusion equations. Besides the spatially quasi-periodic structures cited in the case of the Brusselator (now considered to be no more than a 'toy model'), in the literature there are examples of *contrasting dissipative structures*, localized within a particular spatial sub-domain and associated with sharp boundaries (the first examples were given by Kerner and Osipov; 1978), *standing waves, travelling pulses, self-oscillating structures, travelling structures, chaotic spatio-temporal behaviors*. Such a wide variety of forms of emergence was paralleled by the variety of different kinds of bifurcation phenomena. This led to a number of biologists resorting to models allowing for dissipative structures in order to explain *morphogenesis*, that is the formation of ordered structures in biological organisms, as observed, for instance, in embryonic development (see, for example, Beloussov, 1998; a short review of the shortcomings of this approach can be found in Tsikolia, 2003). In fact, application of the theory of dissipative structures to biology had, as its starting point, the pioneering paper by Gierer and Meinhardt (Gierer and Meinhardt, 1972) in which they proposed, for the first time, that ordered structures arose from a balance between short-range activation and long-range inhibition.

Despite this success, the reliability of reaction-diffusion systems capable of forming dissipative structures, as models of intrinsic emergence, still has to be demonstrated. This, in turn, requires one to prove in a rigorous way that reaction-diffusion models meet the following requirements:

- the form of dissipative structures emerging from a bifurcation should depend only on parameter values and never on initial conditions; not too great an influence of the kind of boundary conditions would also be desirable;

- the bifurcation phenomena characterizing these models should be *structurally stable*, that is small perturbations from the environment (for instance through the variation of another parameter, apart from the critical one, or through the explicit introduction of a new term in the right-hand side of the evolution equations) should not alter the qualitative nature of those bifurcation phenomena. In particular, a bifurcation giving rise to *two* different dissipative structures should still give rise to *two* different dissipative structures (eventually slightly modified) in the presence of any kind of perturbation, provided it be small enough;

- the reaction-diffusion equations should coincide with the macroscopic limit of a *microscopic theory*, demonstrating, in concomitance with the bifurcation point, the occurrence of a stable long-range correlation between the individual local fluctuations, in a way which should be independent of the size of the system being considered. In other words, the bifurcation phenomenon should be the macroscopic counterpart of a *collective behavior* occurring simultaneously within the whole system and driving the behavioral features of the single fluctuations.

Much work has been devoted to verifying whether these requirements are satisfied or not. Here we limit ourselves to summarizing the main results. Regarding requirement 1) it soon became clear that this is very difficult to satisfy, even in the simplest models. For instance, it was shown that, to prevent the dependence of the emerging structures upon the initial conditions, one must adopt particular and unrealistic laws of variation with time of equation parameters (Chernavskii and Ruijgrok, 1982). With regard to the influence of boundary conditions, the situation is even worse as the extreme sensitivity of solutions of PDEs to the nature of these conditions is well known. It is thus evident that dissipative structures cannot belong to *ideal* models of emergence, in the sense defined in the first section of this Chapter.

Requirement 2) has been the subject of intensive research by mathematicians, owing to the development of the theory of *imperfect bifurcations* (Matkowsky and Reiss, 1977; Golubitsky and Schaeffer, 1979). These occur when the bifurcation phenomenon, leading to the emergence of a *multiplicity* of different structures (for instance *two*, as in the case of the Brusselator) where a control parameter crosses a critical value, is destroyed in the presence of even very small variations of a second control parameter, leading to *only one* emergent structure. Clearly, if small perturbations transform a normal bifurcation into an imperfect bifurcation, the emergence described by the model can no be longer considered as *intrinsic*. Namely, as perturbations are ubiquitous in nature and the mathematical theory of imperfect bifurcations allows us to foresee the precise form of the only emergent structure, we can say that models of dissipative structures can, at the most, describe *pattern formation* but not intrinsic emergence. In practice, a number of researchers showed the presence of imperfect bifurcations in most models of this kind, obviously including the Brusselator itself (see,for example, Erneux and Cohen, 1983). Moreover, in 1985 Fernández proved in a rigorous way that *all reaction-diffusion models* have only imperfect bifurcations (Fernández, 1985). This result destroyed any hope of solving the problem of describing intrinsic emergence by resorting to dissipative structure models.

After such a list of shortcomings, we might expect that requirement 3) would also be far from being fulfilled. This is, of course, the case. To understand why, we must remember that in the physical world, at the critical point of a phase transition involving a profound structural change (a typical case being the transition from the paramagnetic to the ferromagnetic state), there is both a progressive slowing down of the fluctuations and an increase in the distance within which they are reciprocally correlated (the so-called *correlation radius*). Both circumstances allow for the emergence of a collective behavior only at the critical point. As a matter of fact, any microscopic theory of fluctuations which includes these circumstances gives rise, at the macroscopic limit, to forecasts in accordance with experimental observations on this kind of phase transition. This is often spoken of as a *classical behavior* of fluctuations. As far back as1980, Nitzan and Ortoleva showed that when noisy fluctuations are added to reaction-diffusion equations, their behavior is *non-classical*, when the wavenumber of the first approximation of the emergent structure is greater than zero (as long as the dimensionality of the spatial domain in which the system is embedded is less than 5,), and also when the wavenumber of first approximation to the emergent structure is exactly zero (with a spatial domain dimensionality of less than 7) (Nitzan and Ortoleva, 1980; see also Stein, 1980). As the dimensionality of physical space is no greater than 3, this result means that a physically plausible reaction-diffusion model cannot be considered as the macroscopic limit of a microscopic theory describing the emergence of collective behaviors. In other words, every dissipative structure emerging from a bifurcation phenomenon tends to be destroyed by the fluctuations which, precisely at the critical point, behave in the wrong way. To summarize, the bifurcation phenomena associated with reaction-diffusion models *are not the analogues* of phase transitions occurring in the physical world, at least not of those involving structural changes.

The question now is: as none of requirements 1), 2) or 3) has been satisfied, are dissipative structure models still useful? The answer is yes, if we limit ourselves to the macroscopic description of the phenomenology of pattern formation. But the answer is no, if we are looking for models of intrinsic emergence. At this point one further question arises: what is missing in reaction-diffusion models which could render them suitable for intrinsic emergence? The preceding arguments lead to only one answer: the inclusion of fluctuations and a correct formalism for dealing with them. Later in the book, we will see how it is possible to remedy this shortcoming. Here, however, we will spend a few words on the results obtained when dealing with the other category of activator-inhibitor models, that of *non-linear dispersive wave equations*.

Non-linear dispersive wave equations. Before starting, it should be recalled that the term *dispersion* arose within the context of waves, described in mono-dimensional space spanned by a single coordinate x, by generic functions of the quantity $k\,x - \omega\,t$. Here k is a *wavenumber*, and ω a *frequency*. A trivial example is given by a linear monochromatic stationary plane wave described by a function such as $sin(k\,x - \omega\,t)$. Obviously in 3-dimensional space this quantity is generalized by $\Sigma_i\,k_i\,x_i - \omega\,t$, where the 3-dimensional vector of components k_i is the *wave vector*. For the sake of simplicity in what follows, we refer always to the mono-dimensional case.

In general, the frequency ω is not independent of the wavenumber k, but is related to it by a *dispersion relation*. The latter characterizes the properties of the medium supporting the propagation of the waves themselves. A very well-known case is that of *linear* waves, in which the dispersion relation has the simple form:

$$\omega = c\,k$$

where c is the (constant) *propagation velocity*. In many other cases, however, we have *non-linear* dispersion relations (a classical book on non-linear waves is Whitham, 1974). Generally speaking, once given a dispersion relation, it is possible to extract from it a PDE governing the dynamics of wave propagation by replacing the frequency and the wavenumber in the relation itself with suitable differential operators according to the simple prescriptions:

$$\omega \to -j\,\partial/\partial t \;, \quad k \to j\,\partial/\partial x$$

where $j = \sqrt{-1}$ is the imaginary unit. For instance, in the case of monochromatic linear waves which can travel in both spatial directions, we have the dispersion relation:

$$\omega^2 = c^2\,k^2$$

owing to the fact that the possible propagation velocities are both c and $-c$. Now, substituting as described above and denoting with $\psi\,(x\,,\,t)$ a quantity which is a suitable property of wave motion, one immediately obtains the PDE governing ψ under the form:

$$\partial^2\psi/\partial t^2 = c^2\,\partial^2\psi/\partial x^2$$

This is the well-known *D'Alembert equation* describing a number of important physical phenomena such as the vibrating string, the propagation of sound or electromagnetic waves and so on.

One now has to take into account that this simple example represents an extreme idealization (even in the case of a laser source), because in the real physical world two circumstances modify the previous description: 1) the dispersion relations are far from linear and, in the simplest cases, have a form such as:

$$\omega^2 = c^2\,k^2 + D(k)$$

$D(k)$ often being called the *dispersion term*; 2) the real waves never coincide with single monochromatic plane waves but generally consist in *wave packets*, arising from the superposition of a number (possibly an infinite one) of different monochromatic plane waves, each associated with a different value of k. Let us now introduce *phase velocity* $v(k)$ defined as:

$$v(k) = \partial \omega / \partial k$$

It can be immediately seen that when $D(k) = 0$, the phase velocity is given by c (or $-c$) and is independent of k, whereas when $D(k)$ is not zero, even the phase velocity depends upon k. This produces a *spreading over time* of wave packets: namely different components associated with different values of k propagate with different velocities (the latter depending upon k) and go out of phase with one another. In short, dispersion produces spreading.

At this point, we must take into account that, besides dispersion, one can introduce another contribution from *nonlinearity*. The latter consists of the fact that the coefficients of the dispersion relation (essentially the coefficient c described above) can be made to depend, in turn, upon the quantity describing the main property of wave motion, previously denoted by ψ. For instance, c could be given by an expressionsuch as:

$$c = c_0 \left(1 + \beta_1 \, \psi + \beta_2 \, \psi^2 \right)$$

where c_0, β_1 and β_2 are suitable parameters. A number of studies have shown how the introduction of nonlinearity can produce an effect exactly opposite to that of dispersion, by *sharpening*, instead of spreading, the wave packet distribution. This is due to the fact that the parts of the wave characterized by higher values of ψ have a propagation velocity greater than those corresponding to lower values of ψ (for a simple analysis of the role of dispersion and nonlinearity Korpel and Banerjee, 1984 is highly recommended).

The above considerations suggest that, if an exact balance between the effect of spreading due to dispersion, and the effect of sharpening, due to nonlinearity, could be introduced, then we could obtain, as a solution of a suitable PDE, an equilibrium structure, localized in space but propagating as a whole while keeping its profile unchanged. In fact, such structures have been observed experimentally, being known today as *solitary waves* or *solitons* . The first scientific observation of a soliton in a water channel was reported by the Scottish engineer John Scott Russell in 1845 (for a historical account, see Bullough, 1988) but only over the last forty years has the study of solitons become one of the most important research topics in physics and mathematics. The importance of this subject stems from its intrinsic interdisciplinarity, as it crosses and links together very different domains of physics, mathematics, engineering, biology and the earth sciences. In turn, this given rise to huge number of books, papers, conferences, scientific journals devoted to it (the most popular textbooks on this subject include Ablowitz and Segur, 1981; Eilenberger, 1981; Dodd *et al.*, 1982; Newell,

1985; Rajaraman, 1987; Lakshmanan, 1988; Tabor, 1989; Toda, 1990; Makhankov, 1991; Infeld and Rowlands, 2000; Nekorkin and Velarde, 2002; Nettel, 2003; Scott, 2003).While it is impossible even to summarize the conspicuous body of results so far obtained in this field, here we limit ourselves to some elementary considerations, having only the aim of assessing the relevance of solitons for models of emergence.

We start by stressing the fact that, in this domain, research activity has been focused upon some prototypical PDEs, each of which was shown to describe the evolution of phenomena in many different fields of science. The most popular (and historically the first) of these equations is the *Korteweg - de Vries* (KdV) *equation*, first proposed in 1895 to describe the motion of small-amplitude waves in shallow water. Its general form, where $\psi(x, t)$ is the wave amplitude, is:

$$\partial \psi / \partial t + c_0 (1 + \beta_1 \psi) \partial \psi / \partial x + \gamma \partial^3 \psi / \partial x^3 = 0$$

From this the associated dispersion relation is of the kind:

$$\omega = c k + \gamma k^3$$

whereas the nonlinearity is given by:

$$c = c_0 (1 + \beta_1 \psi)$$

Suitable scaling transformations acting upon the independent variables, as well as a suitable shift of the origin for the original dependent variable ψ, lead to a simpler form of the KdV equation where u is the new dependent variable, and is given by:

$$\partial u / \partial t + 6 u \, \partial u / \partial x + \partial^3 u / \partial x^3 = 0$$

It is possible to show that, if u is assumed to be a function only of the quantity:

$$z = x - v t \quad (v = \text{propagation velocity})$$

and if we impose that, for $x \to \pm \infty$, u, as well as its first and second derivative with respect to z, will all be vanishingly small, then the previous equation has a solution of the form:

$$u = (v/2) \, sech^2 \, [(\sqrt{v}/2)(x - v t)]$$

This solution describes a bell-like profile (resembling a Gaussian curve) which travels along the x axis with velocity v, without changing its form. Here is a concrete example of a soliton!

Without pursuing further the discussion of the KdV equation, here we list only two other widely used PDEs which can provide solitonic solutions. The first is the *Nonlinear Schrödinger equation*, having the form:

$$j \, \partial u / \partial t = \partial^2 u / \partial x^2 + g \, |u|^2 u$$

where g denotes a suitable parameter and the variable u is a complex quantity. This equation generalizes the well-known Schrödinger equation, holding in Quantum Mechanics as well as being used to describe, amongst other things, the evolution of gravity waves in deep water, pulse propagation along optical fibers and superfluid dynamics. The other example is the *Sine-Gordon equation*, which, in its simplest form, is written as:

$$\partial^2 u / \partial t^2 - \partial^2 u / \partial x^2 + \sin u = 0$$

It describes, for instance, the self-induced transparency in nonlinear optical materials. However, its popularity is mainly due to the fact that its solitonic solutions are considered as the classical counterparts of the 'quanta' occurring in quantum theories of fundamental interactions (see, for instance, Manton *et al.*, 2004).

In general, the above equations as well as other PDEs studied in soliton theory provide solutions not only describing the propagation of single solitons, but also *multisolitonic* solutions which describe a number of different solitons propagating simultaneously in different directions. This gives rise to an extremely rich phenomenology of observed behaviors: colliding solitons (in many cases they emerge from the collision completely unaltered), annihilations of soliton-antisoliton pairs, soliton splitting, bound states of interacting solitons, oscillating solitons, chaotic solitons, vortices (in two-dimensional cases) and more, with a wide variety of phenomena which parallels that of the phenomena observed in the physical world. This circumstance obviously leads to the question of whether models allowing for solitonic solutions are better than reaction-diffusion models for describing emergence. To be more precise, such a question subdivides into two classes of different questions:

1. Are solitons stable? Do they arise from bifurcation phenomena? Can the effects of these phenomena be forecast in advance?

2. What happens to solitons in the presence of fluctuations? Can the latter destroy the picture of soliton behavior existing in the absence of fluctuations?

Regarding the questions in class 1) it must be recalled that, even though the problem of the stability of solitonic solutions has been considered from the very beginning of soliton theory as a very important one, for many years stability analyses were carried out only for particular cases and without the support of a general theory. And even after such a theory was built (see Grillakis *et al.*, 1987; 1990), only in recent years has it been applied and further generalized (see, for instance, Kishvar and Pelinovsky, 2000; Pelinovsky and Kivshar, 2000; Kevrekidis *et al.*, 2003; Crasovan *et al.*, 2003; Campbell *et al.*, 2004). This growth of interest was due to rapid developments in the technology of high-speed pulse communications based on the propagation of solitons along optical fibers. As this propagation depends upon the effects of inter-soliton interactions, usually modelled using systems of coupled nonlinear Schrödinger equations, a theory of the stability of propagating solitons could have immediate technologicalimpact.

The most relevant findings of these theoretical analyses can be thus summarized:

a. There are precise conditions placed upon the equation parameters which provide stability for solitonic solutions; such(remarkable) stability has also been observed experimentally in solitonic propagations occurring in the physical world; in general, such stability appears to be more common than in the case of dissipative structures arising from reaction-diffusion equations;

b. In a number of cases solitons occur as a consequence of bifurcation phenomena; this typically happens for multisolitonic solutions;

c. Contrary to reaction-diffusion systems, the effect of a bifurcation upon solitonic solutions can, in principle, always be predicted, as there is no problem of choosing a particular one out of the various solutions arising from the bifurcation; namely different solitonic solutions can simultaneously share the same spatial domain.

Let us now turn to the questions of class 2). Most frequently, fluctuations are introduced by resorting to a disordered medium (which is the case of interest in technological applications of solitons). In this case, computer simulations demonstrated that, in the presence of suitable conditions, a very strange and unexpected phenomenon occurs: when a solitary wave propagates in such a medium, it initially splits into a number of smaller solitons which mutually collide. Over time, however, the larger solitons, after each collision, continue to grow in amplitude, whereas the smaller ones continue to decrease. After a sufficient length of time, only one large soliton survives. In other words, a soliton propagating in a disordered medium tends to *react in an active way*, such as to restate its coherence, despite the random influences tending to destroy it. Such behavior has been confirmed by theoretical analyses (see, for example, Clausen *et al.*, 1999; Chertkov *et al.*, 2001) which formulated the conditions under which this circumstance can occur.

One could ask why solitons behave in such a different way from dissipative structures, where the effects of fluctuations can be dramatic. Without introducing too much mathematics, the simple answer is that all PDEs allowing solitonic solutions are examples of conservative systems. Now, the existence of conservation laws (and soliton equations have an infinite set of them) entails the existence of global constraints on system dynamics which, under suitable conditions, allow it to maintain its coherence and prevent it from evolving by being fully driven by dissipation. We thus conclude that, even though solitonic solutions of non-linear dispersive wave equations cannot be considered as models of intrinsic emergence, they do appear to be able to model, quite adequately, the processes of *keeping coherence* after a structure has been produced by an emergence phenomenon.

4.5 The intrinsic limitations of traditional methods

All the arguments presented so far in this Chapter lead to a unique conclusion: models based only upon traditional methods are unsuitable for describing intrinsic emergence. In other words, intrinsic emergence appears as the result of a suitable combination of both dynamical rules and fluctuations (whether they be produced by noise, quantum effects, impurities or anything else). Therefore traditional methods must necessarily be complemented by non-traditional methods. The latter, however, are difficult to accept in a scientific environment which still believes in the myth of objectivity, of exact results, of mathematical guarantees. On the contrary, non-traditional methods not only introduce probability, noise, uncertainty, fuzziness, but often require one to completely abandon the idea of the existence of a single model and even of a "best" model. Of course, theorems are welcomed, but in most cases the only available strategies are based upon computer simulations. The existence of a formal model does not necessarily imply that we can derive all its possible consequences.

On the other hand, such a conclusion was to be expected. The very definition of intrinsic emergence implies that, in principle, we cannot design a mathematical model for predicting it. This implies that we cannot *control* emergence completely, but only try to *manage* its effects, through a mutual interaction with them. It is within this management, however, that models based upon traditional methods turn out to be useful. Namely, from a macroscopic point of view, every process of emergence gives rise to a particular pattern-formation phenomenology, whose features can be captured in an excellent way by most models discussed in this Chapter. In a sense, these models should act as a *macroscopic limit* of a microscopic theory of emergence (based, of course, on non-traditional methods). Therefore, the main shortcoming of these models is not the fact that they are only macroscopic models of pattern formation, but rather that they lack any grounding on microscopic features. In this regard, the situation changes from one model to another. For instance, this lack of grounding is particularly strong in reaction-diffusion models, whereas this is not the case for nonlinear dispersive wave equations.

One very difficult question concerns the (eventual) limitations of traditional models in the kinds of pattern-formation processes they are able to describe. Here, there are two differing opinions: the first holds that these models can describe, in principle, *any kind* of pattern formation, whereas the second insists that they can be used only to describe the simplest cases of pattern formation occurring at a simple physical level. Currently, the former seems to be the most widespread amongst physicists and (surprisingly) economists, social scientists and population biologists. Most models of this kind have, in fact, been employed to describe the evolution of economic macrosystems (see, amongst others, Barnett *et al.*, 1996; Brian Arthur *et al.*, 1997; Punzo, 2001), industrial organisations, human and animal populations and urban development. On the contrary, the latter appears to be typical of philosophers and of most biologists. They effectively claim that pattern formation, as observed at the physical level (such as, for instance, in phase transitions) is intrinsically different from that characterizing living beings and, *a fortiori*, cognitive or social systems. According to them, the homogeneity inherent in traditional models cannot account for the prominent role of *individuality* in all biological processes.

Thus the problem is still unsolved and the debate between the two points of view will probably continue for the foreseeable future. Here, we stress the fact that the whole question becomes meaningless when adopting non-traditional models and approaches based upon the non-uniqueness of models. This provides the basis for the next Chapter where we deal with a (necessarily incomplete) discussion of non-traditional models. Clearly, such a discussion is possible only after gaining a knowledge of the results already obtained using traditional models which has been the subject of the present Chapter.

References

Ablowitz, M. J., and Segur, H., (eds.), 1981, *Solitons and Inverse Scattering Transform*. Society for Industrial and Applied Mathematics, Philadelphia, PA.

Albano, A. M., Abraham, N.B., Chyba, D.E., and Martelli, M., 1984, Bifurcations, propagating solutions, and phase transitions in a nonlinear chemical reaction with diffusion, *American Journal of Physics*, **52**:161-167.

Alligood, K., Sauer, T., and Yorke, J. A., 1997, *Chaos: An introduction to Dynamical Systems*. Springer, New York.

Auchmuty, J. F. G., and Nicolis, G., 1975, Bifurcation analysis of nonlinear reaction-diffusion equations. I: Evolution equations and the steady state solutions. *Bulletin of Mathematical Biology*, **37**: 325-365.

Auger, P., 1980, Coupling between N levels of observation of a system (biological or physical) resulting in creation of structure, *International Journal of General Systems* **8**:82-100.

Auger, P., 1983, Hierarchically organized populations: Interactions between individual, population and ecosystem levels, *Mathematical Biosciences* **65**:269-289.

Auger, P., 1985, Dynamics in hierarchically organized systems. In: *Dynamics of macrosystems*, (J.P.Aubin, D.Saari, and K.Sigmund, eds.), Springer, Berlin, pp. 203-212.

Auger, P., 1989, *Dynamics and Thermodynamics in hierarchically organized systems: Applications in Physics, Biology and Economics*. Pergamon Press, Oxford, UK.

Auger, P., and Poggiale, J. C., 1995, Emerging properties in population dynamics with different time scales, *Journal of Biological Systems* **3**:591-602.

Auger, P., and Poggiale, J. C., 1996, Aggregation, emergence and immergence in hierarchically organized systems. In: *Third European Congress on Systems Science*, (E. Pessa, M. P. Penna and A. Montesanto, eds.), Kappa, Rome, pp. 43-45.

Barnett, W. A., Kirman, A. P., and Salmon, M., (eds.), 1996, *Nonlinear dynamics and Economics: Proceedings of the Tenth International Symposium in Economic Theory and Econometrics*. Cambridge University Press, Cambridge, UK.

Bedau, M. A., 1997,Weak emergence, *Philosophical Perspectives* **11**:375-399.

Belintsev, B. N., 1983, Dissipative structures and the problem of biological pattern formation, *Soviet Physics Uspekhi* **26**:775-800.

Beloussov, L. V., 1998, *The dynamic architecture of developing organism*. Kluwer, Dordrecht.

Benettin, G., Galgani, L. and Strelcyn, J.-M., 1976. Kolmogorov entropy and numerical experiments, *Physical Review A* **14**:2338-2345.

Bitsadze, A. V., 1980, *Equations of mathematical physics*. Mir, Moscow.

Boyce, W. E., and DiPrima, R. C., 1977, *Elementary differential equations and boundary value problems*, 3rd edition. Wiley, New York.

Brian Arthur, W., Durlauf, S. N., and Lane, D. A., (eds.), 1997, *The Economy as an evolving complex system II: Proceedings*. Perseus Books, Santa Fe, NM.

Bullough, R. K., 1988, "The Wave par excellence", the solitary, progressive great wave of equilibrium of the fluid – an early history of the solitary wave. In: *Solitone*, (M. Lakshmanan, ed.), Springer, Berlin, pp. 150-281.

Campbell, D. K., Flach, S., and Kivshar, Yu. S., 2004, Localizing energy through nonlinearity and discreteness, *Physics Today* **57**:43-49.

Chafee, N., 1975, Asymptotic behavior for solutions of a one-dimensional parabolic equation with homogeneous Neumann boundary conditions, *Journal of Differential Equations* **18**:111-134.

Chernavskii, D. S., and Ruijgrok, T. W., 1982, On the formation of unique dissipative structures, *BioSystems* **15**:75-81.

Chertkov, M., Gabitov, I., Kolokolov, I., and Lebedev, V., 2001, Shedding and interaction of solitons in imperfect medium, *JETP Letters* **74**:357-361.

Cicogna, G., 1981, Symmetry breakdown from bifurcation, *Lettere al Nuovo Cimento* **31**:600-602.

Clausen, C. B., Kivshar, Yu. S., Bang, O., and Christiansen, P. L., 1999, Quasiperiodic envelope solitons, *Physical Review Letters* **83**:4740-4743.

Collet, P., and Eckmann, J. P., 1980, *Iterated maps of the interval as Dynamical Systems.* Birkhäuser, Boston.

Copson, E. T., 1975, *Partial differential equations.* Cambridge University Press, Cambridge, UK.

Crasovan, L. C., Kartashov, Y. V., Mihalache, D., Torner, L., Kivshar, Yu. S., and Perez-Garcia, V. M., 2003, Soliton "molecules": Robust clusters of spatiotemporal optic solitons, *Physical Review E* **67**:046610-046615.

Cruchtfield, J. P., 1994, The Calculi of Emergence: Computation, Dynamics and Induction, *Physica D* **75**:11-54.

Davis, H. T., 1962, *Introduction to nonlinear differential and integral equations.* Dover, New York.

De Souza-Machado, S., Rollins, R. W., Jacobs, D. T. and Hartman, J. L., 1990, Studying chaotic systems using microcomputer simulations and Lyapunov exponents, *American Journal of Physics* **58**:321-329.

Dodd, R. K., Eilbeck, J. C., Gibbon, J., and Morris, H., 1982, *Solitons and nonlinear wave equations.* Academic Press, New York.

Doucet, P., and Sloep, P. B., 1992, *Mathematical modeling in the life sciences.* Ellis Horwood, Chichester, UK.

Duchateau, P., and Zachmann, D. W., 1986, *Partial differential equations.* McGraw-Hill, New York.

Eckmann, J. P. and Ruelle, D., 1985, Ergodic theory of chaos and strange attractors, *Reviews of Modern Physics* **57**:617-656.

Eilenberger, G., 1981, *Solitons: Mathematical methods for physicists.* Springer, New York.

Ermentrout, G. B., and Cowan, J. D., 1979, Temporal oscillations in neuronal nets, *Journal of Mathematical Biology* **7**:263-280.

Erneux, T., and Cohen, D. S., 1983, Imperfect bifurcation near a double eigenvalue: Transitions between nonsymmetric and symmetric patterns, *SIAM Journal of Applied Mathematics* **43**:1042-1060.

Fernández, A., 1985, Global instability of a monoparametric family of vector fields representing the unfolding of a dissipative structure, *Journal of Mathematical Physics* **26**:2632-2633.

Forrester, J. W., 1968, *Principles of Systems.* Wright-Allen Press, Cambridge, MA..

Garrido, L., (ed.), 1988, *Far from equilibrium phase transitions.* Springer, Berlin.

Gierer, A., and Meinhardt, H., 1972, A theory of Biological Pattern formation, *Kybernetik* **12**:30-39.

Glansdorff, P., and Prigogine, I., 1971, *Thermodynamic theory of structure, stability and fluctuations*. Wiley, New York.

Glendinning, P.,1994, *Stability, Instability and Chaos: An introduction to the theory of Nonlinear Differential Equations*. Cambridge University Press, Cambridge, UK.

Golubitsky, M., and Schaeffer, D. G., 1979, A theory for imperfect bifurcations via singularity theory, *Communications in Pure and Applied Mathematics* **32**:21-98.

Golubitsky, M., and Schaeffer, D. G., 1985, *Singularities and groups in bifurcation theory*, vol. I. Springer, Berlin.

Golubitsky, M., Luss, D., and Strogatz, S. H., (eds.), 1999, *Pattern formation in continuous and coupled systems*. Springer, New York.

Grillakis, M., Shatah, J., and Strauss, W., 1987, Stability theory of solitary waves in the presence of symmetry I, *Journal of Functional Analysis* **74**:160-197.

Grillakis, M., Shatah, J., and Strauss, W., 1990, Stability theory of solitary waves in the presence of symmetry II, *Journal of Functional Analysis* **94**:308-348.

Guckenheimer, J., and Holmes, P.,1983, *Nonlinear oscillations, dynamical systems and bifurcation of vector fields*. Springer, Berlin.

Gumowski, I., and Mira, C., 1980, *Dynamique chaotique. Transformations ponctuelles. Transition Ordre-Désordre*. Cepadues, Toulouse.

Haken, H., 1983, *Advanced Synergetics*. Springer, Berlin-Heidelberg-New York.

Hassard, B. D., Kazarinoff, N. D., and Wan, Y.-H., 1981, *Theory and applications of Hopf bifurcation*. Cambridge University Press, Cambridge, UK.

Infeld, E., and Rowlands, G., 2000, *Nonlinear waves, solitons and chaos*. Cambridge University Press, Cambridge, UK.

Iooss, G., and Joseph, D. D., 1981, *Elementary stability and bifurcation theory*. Springer, New York.

Jordan, D. W., and Smith, P., 1977, *Nonlinear ordinary differential equations*. Clarendon Press, Oxford, UK.

Kerner, B. S., and Osipov, V. V., 1978, Nonlinear theory of stationary strata in dissipative systems. *Soviet Physics JETP* **47**:874-885.

Kevrekidis, P. G., Kivshar, Yu. S., and Kovalev, A. S., 2003, Instabilities and bifurcations of nonlinear impurity modes, *Physical Review E* **67**:046604-046608.

Kivshar, Yu. S., and.Pelinovsky, D. E., 2000, Self-focusing and transverse instabilities of solitary waves, *Physics Reports* **331**:117-195.

Korpel, A., and Banerjee, P. P., 1984, A heuristic guide to nonlinear dispersive wave equations and soliton-type solutions, *Proceedings of the IEEE* **72**:1109-1130.

Krinsky, V. I., (ed.), 1984, *Self-organization: Autowaves and structures far from equilibrium*. Springer, Berlin.

Lakshmanan, M., (ed.), 1988, *Solitons* . Springer, Berlin.

Lanford, O. E., 1981, Strange attractors and turbulence. In: *Hydrodynamic instabilities and transition to turbulence*, (H. L. Swinney and J. P. Gollub, eds.), Springer, Berlin, pp. 7-31.

Lefever, R., and Prigogine, I., 1968, Symmetry-breaking instabilities in dissipative systems, *Journal of Chemical Physics* **48**:1695-1700.

Lefschetz, S., 1977, *Differential equations: geometric theory*. Dover, New York.

Lichtenberg, A. J., and Lieberman, M. A., 1983, *Regular and stochastic motion*. Springer, Berlin.

Makhankov, V.G., 1991, *Soliton phenomenology*. Kluwer, Dordrecht.

Manton, N., Sutcliffe, P., Landshoff, P. V., Nelson, D. R., Sciama, D. W. and Weinberg, S., (eds.), 2004, *Topological solitons*. Cambridge University Press, Cambridge, UK.

Marsden, J. E., and McCracken, M., 1976, *The Hopf bifurcation and its applications.* Springer, New York.

Matkowsky, B. J., and Reiss, E. L., 1977, Singular perturbations of bifurcations, *SIAM Journal of Applied Mathematics* **33**:230-255.

Mesarovic, M. D., and Takahara, Y., 1975, *General Systems Theory: Mathematical foundations.* Academic Press, New York.

Mikhailov, A. S., 1990, *Foundations of Synergetics I. Distributed active systems.* Springer, Berlin.

Mikhailov, A. S., and Loskutov, A.Yu., 1996, *Foundations of Synergetics II. Chaos and Noise*, 2nd revised edition. Springer, Berlin.

Mori, H., and Kuramoto, Y., 2001, *Dissipative structures and chaos.* Springer, Berlin.

Murray, J. D., 1989, *Mathematical Biology.* Springer, Berlin.

Nekorkin, V. I., and Velarde, M.G., 2002, *Synergetic phenomena in active lattices. Patterns, waves, solitons, chaos.* Springer, Berlin.

Nettel, S., 2003, *Wave Physics: Oscillations – Solitons – Chaos*, 3rd ed. Springer, Berlin.

Newell, A. C., 1985, *Solitons in Mathematics and Physics.* Society for Industrial and Applied Mathematics, Philadelphia, PA.

Nicolis, G., and Prigogine, I., 1977, *Self-Organization in Nonequilibrium Systems: From Dissipative Structures to Order through Fluctuations.* Wiley, New York.

Nicolis, J. S., 1986, *Dynamics of hierarchical systems. An evolutionary approach.* Springer, Berlin.

Nitzan, A., and Ortoleva, P., 1980, Scaling and Ginzburg criteria for critical bifurcations in nonequilibrium reacting systems, *Physical Review A* **21**:1735-1755.

Okubo, A., 1980, *Diffusion and ecological problems. Mathematical models.* Springer, Berlin.

Olmstead, W. E., Davis, S. H., Rosenblat, S., and Kath, W. L., 1986, Bifurcation with memory, *SIAM Journal of Applied Mathematics* **46**:171-188.

Ott, E.,1993, *Chaos in Dynamical Systems.* Cambridge University Press, Cambridge, UK.

Pelinovsky, D. E., and Kivshar, Yu. S., 2000, Stability criterion for multicomponent solitary waves, *Physical Review E* **62**:8668-8676.

Pessa, E., 2000, Cognitive Modelling and Dynamical Systems Theory, *La Nuova Critica* **35**:53-93.

Punzo, L. F., (ed.), 2001, *Cycles, growth and structural change. Theories and empirical evidence.* Routledge, London.

Rajaraman, R., 1987, *Solitons and Instantons.* North Holland, Amsterdam.

Rand, R. H., and Armbruster, D., 1987, *Perturbation methods, bifurcation theory and computer algebra.* Springer, New York.

Rashevsky, N., 1960, *Mathematical Biophysics. Physico-mathematical foundations of Biology*, 2 voll., 3rd edition. Dover, New York.

Ronald, E. M. A., Sipper, M., and Capcarrère, M. S., 1999, Design, observation, surprise! A test of emergence, *Artificial Life* **5**:225-239.

Rueger, A., 2000, Physical emergence, diachronic and synchronic, *Synthese* **124**:297-322.

Ruelle, D., 1995, *Turbulence, Stange Attractors and Chaos.*World Scientific, Singapore.

Saaty, T. L., 1981, *Modern nonlinear equations.* Dover, New York.

Sakaguchi, H., 2003, Self-organization of hierarchical structures in non-locally coupled replicator models, *Physics Letters A* **313**:188-191.

Sattinger, D. H., 1978, Group representation theory, bifurcation theory and pattern formation, *Journal of Functional Analysis* **28**:58-101.

Sattinger, D. H., 1978, *Topics in stability and bifurcation theory.* Springer, Berlin.

Sattinger, D. H., 1980, Bifurcation and symmetry breaking in applied mathematics, *Bulletin of the American Mathematical Society* **3**:779-819.

Sattinger, D. H., 1983, *Branching in the presence of symmetry*. Society for Industrial and Applied Mathematics, Philadelphia, PA.

Schöll, E., 1986, Influence of boundaries on dissipative structures in the Schlögl model. *Zeitschrift für Physik B – Condensed Matter* **62**:245-253.

Scott, A., 2003, *Nonlinear science: Emergence and dynamics of coherent structures*. Oxford University Press, Oxford, UK.

Sinai, Ya. G. (ed.), 1989, *Dynamical Systems II*. Springer, Berlin.

Sneddon, I., 1957, *Elements of partial differential equations*. McGraw-Hill, New York.

Stein, D. L., 1980, Dissipative structures, broken symmetry, and the theory of equilibrium phase transitions, *Journal of Chemical Physics* **72**:2869-2874.

Svirezhev, Yu. M., and Logofet, D. O., 1983, *Stability of biological communities*. Mir, Moscow.

Tabor, M., 1989, *Chaos and integrability in nonlinear dynamics*. Wiley, New York.

Tirapegui, E., and Zeller, W., (eds.), 1993, *Instabilities and nonequilibrium structures IV*. Kluwer, Dordrecht.

Tirapegui, E., and Zeller, W., (eds.), 1996, *Instabilities and nonequilibrium structures V*. Kluwer, Dordrecht.

Toda, M., 1990, *Nonlinear waves and solitons*. Kluwer, Dordrecht.

Tsikolia, N., 2003, What is a role of the morphogenetic gradients in development? *Rivista di Biologia – Biology Forum* **96**:293-315.

Turing, A. M., 1952, The chemical basis of morphogenesis, *Philosophical Transactions of the Royal Society of London* **B 237**:37-42.

Vanderbauwhede, A., 1982, *Local bifurcation and symmetry*. Pitman, Boston.

Vasilev, V. A., Romanovskii, Yu. M. and Yakhno, V. G., 1979, Autowave processes in distributed kinetic systems, *Soviet Physics Uspekhi* **22**:615-639.

Vasilev, V. A., Romanovskii, Yu. M., Chernavskii, D. S. and Yakhno, V. G., 1986, *Autowave processes in kinetic systems*. Reidel, Dordrecht.

Von Bertalanffy, L., 1940, Der organismus als physikalisches System betrachtet, *Die Naturwissenschaften* **28**:521-531.

Von Bertalanffy, L., 1950, The theory of open systems in Physics and Biology, *Science* **111**:23-29.

Von Bertalanffy, L., 1968, *General System Theory. Development, Applications*. George Braziller, New York

Whitham, G. B., 1974, *Linear and nonlinear waves*. Wiley, New York.

Wolf, A., Swift, J. B., Swinney, H. L. and Vastano, J. A., 1985, Determining Lyapunov exponents from a time series, *Physica D* **16**:285-317.

Yokozawa, M., and Hara, T., 1999, Global versus local coupling models and theoretical stability analysis of size-structure dynamics in plant populations, *Ecological Modelling* **118**:61-72.

Chapter 5

HOW TO MODEL EMERGENCE: NON-TRADITIONAL METHODS

5.1 Synergetics

The arguments presented in the previous chapter evidenced how reliable models of intrinsic emergence must take into account, along with activation and inhibition, the role played by microscopic fluctuations. This requires the introduction of non-traditional methods, which are discussed in this chapter. Let us begin by recalling that the study of conditions which allow the emergence of macroscopic patterns through suitable cooperation of microscopic fluctuations is the subject of *Synergetics*. This was introduced in the 1970s by the physicist Hermann Haken (the most comprehensive textbooks include Haken, 1978; 1983; 1988; Mikhailov, 1990; Mikhailov and Loskutov, 1996). The main differences between the Synergetics approach to modeling emergence and the traditional approaches described in Chapter 4 can be summarized as follows:

- In Synergetics both macroscopic and microscopic descriptions of a complex system are always taken into account, whereas most traditional models deal only with macroscopic ones;

- Equations describing the macroscopic dynamics in Synergetics are introduced not on the basis of phenomenological arguments but as suitable approximations of dynamical equations deriving from the general principles of Physics;
- In their more general formulation, the macroscopic dynamical equations studied by Synergetics always contain terms describing the influence of microscopic fluctuations.

In order to describe the effects of microscopic fluctuations in a process of macroscopic pattern formation Haken introduced a number of technical tools of which the most important is the so-called *Slaving Principle*. This applies to the cases where pattern formation arises from the loss of stability of a homogeneous equilibrium solution, associated with a suitable control parameter crossing a critical value. It should be stressed that within Synergetics the concept of 'stability' generalizes that holding within traditional Dynamical Systems Theory, in such a way as to include even noise and fluctuations. Roughly, one can say that, in a system with many dependent variables, the loss of stability is due to the fact that, at the critical point, the amplitude of the fluctuations in one of them, the so-called *unstable mode*, grows with time, whereas the amplitudes of fluctuations in the other variables, the so-called *stable modes*, continue to decrease over time. If we now assume that the behavior of the system close to the critical point is analogous to that of a system undergoing a *phase transition*, the system under study can be considered as being characterized by a sort of *critical slowing down*, that is a tendency towards zero of the rate of variation of fluctuation amplitudes of its stable modes. Such a property is observed experimentally in physical phase transitions. Thus, this hypothesis, also known as the *adiabatic approximation*, implies that the amplitudes of fluctuations in all stable modes can be expressed in terms of the amplitude of fluctuation in the only unstable mode which, in a sense, *enslaves* all the stable modes, essentially determining the main features of the new emerging pattern. For this reason the amplitude of fluctuations of the unstable mode is called an *order parameter*, as it drives the dynamics of pattern formation.

Technically, the Slaving Principle consists of suitable formulae which, for any specific kind of dynamical evolution equations, allows one to express the amplitudes of every stable mode in terms of the order parameter, that is of the amplitude of the unstable mode. In order to grasp in an intuitive way how this works, a simple example is presented, in which both noise and dependence on spatial variables have been eliminated from the start. This example is based on the following ordinary differential equation:

$$(d^2x/dt^2) + \gamma\,(dx/dt) + k\,x + \beta x^3 = 0$$

It describes a damped nonlinear oscillator and can be considered as a special case of the so-called *Duffing equation*, containing in the left-hand member an external forcing term. This equation has been extensively studied because it models a number of interesting phenomena in nonlinear mechanics and more generally within nonlinear physics. It can be written in the form of an autonomous system of two first-order differential equations:

$$dx/dt = y , \quad dy/dt = -kx - \beta x^3 - \gamma y$$

It is interesting to recall that, by adding a term directly proportional to x to the right-hand member of the first equation, we would obtain a system of equations formally equivalent to those of the *FitzHugh-Nagumo model*, describing in a very rough way the operation of a biological neuron. These remarks have been made to show the important role played by this example in a number of domains of applied mathematics.

If we choose k as a bifurcation parameter, it can immediately be seen that $k = 0$ is a *critical value*, separating two different states of affairs: when $k > 0$ the system allows only one equilibrium point, given by $x = 0$, $y = 0$, whereas when $k < 0$ there are three different equilibrium points:

$$P_1 : x = 0 , y = 0 ;$$

$$P_2 : x = (|k|/\beta)^{1/2} , y = 0 ;$$

$$P_3 : x = -(|k|/\beta)^{1/2} , y = 0$$

Moreover, when $k > 0$ a straightforward linear stability analysis shows that the unique equilibrium point is *stable*, as both eigenvalues of the Lyapunov matrix have negative real parts. If the values of system parameters also fulfill the inequality:

$$k < \gamma^2 /4$$

both these eigenvalues are also real. When $k < 0$, however, the previous equilibrium point becomes *unstable*, because the eigenvalues of the Lyapunov matrix, while still being real, have opposite signs. A similar analysis shows that the two other equilibrium points P_2 and P_3 are both *stable*.

Summarizing the results obtained so far, we can say that $k = 0$ is a *bifurcation point* (namely, on decreasing the value of k to cross this value two new equilibrium solutions arise) where there is an *exchange of stability* regarding the equilibrium point $x = 0$, $y = 0$: from being stable it becomes unstable. Such a circumstance suggests studying the behavior of the system in the immediate neighbourhood of this equilibrium point, when $k < 0$ but its value is very close to the critical value $k = 0$. Let us introduce two auxiliary variables, denoted by ξ and η, describing, respectively, the small displacements of the variables x and y from the equilibrium values $x = 0$, $y = 0$. In terms of these variables the original system will become:

$$d\xi/dt = \eta, \quad d\eta/dt = |k|\xi - \beta\xi^3 - \gamma\eta$$

Neglecting the powers of ξ and η higher than the first order, the linearized system around the equilibrium point becomes:

$$d\xi/dt = \eta, \quad d\eta/dt = |k|\xi - \gamma\eta$$

so that the eigenvalues of the Lyapunov matrix will be given by:

$$\lambda_1 = [-\gamma + (\gamma^2 + 4|k|)^{1/2}]/2, \quad \lambda_2 = [-\gamma - (\gamma^2 + 4|k|)^{1/2}]/2$$

Clearly, one can see that λ_1 is positive, whereas λ_2 is negative.

In order to highlight what is occurring close to the bifurcation point we introduce a linear transformation of the original variables ξ and η, designed so as to reduce the linearized system to a fully diagonal form. As it is well known from standard linear algebra, this can be done in an infinite number of different ways. Here we choose the one where the new auxiliary variables, often called *normal modes*, denoted by u and s, are related to the old variables ξ and η by the linear relationships:

$$\xi = u + s, \quad \eta = \lambda_1 u + \lambda_2 s$$

Straightforward computations show that, in terms of these new variables, the original linearized system assumes the simple form:

$$du/dt = \lambda_1 u, \quad ds/dt = \lambda_2 s$$

Recalling now that λ_1 is positive and λ_2 is negative, we see immediately that u describes the *unstable mode*, responsible for the loss of stability associated with the crossing of the bifurcation point, whereas s describes the *stable mode*. In order to understand the roles played by these modes, however, we must focus on the whole nonlinear system in the variables ξ and η. If we apply to it the previous linear transformation of variables, we obtain, in terms of normal modes, a system having the form:

$$du/dt = \lambda_1 u, \quad ds/dt = \lambda_2 s - \beta (u + s)^3$$

It is now easier to understand the Slaving Principle. It states that the time evolution of stable mode s (or of stable modes, if their number were greater than one) is enslaved by that of unstable mode u, in the sense that the instantaneous values of s can be fully expressed as functions of only the instantaneous values of u. Moreover, the equation obeyed by u (in our case the first of the two equations constituting the system) describes the time evolution of the system's order parameter. The technical form of the Slaving Principle is based upon suitable mathematical formulae describing such a dependence in an explicit form. These formulae are obtained by resorting to the adiabatic approximation cited above. Here we limit ourselves to rather crude arguments, based upon the assumption of a critical slowing down of stable modes close to the bifurcation point which implies that, to a first approximation, the rate of change of s, that is ds/dt, can be taken as being equal to zero. The second equation of the above system then reduces to a simple algebraic expression connecting the values of s to those of u. If, owing to the fact that λ_2 is negative, we introduce the notation:

$$\lambda_2 = - \alpha \quad (\alpha > 0)$$

where α denotes a suitable *damping constant* which, owing to the existence of a critical slowing down, is assumed to be very large, then this equation can be written as:

$$- \alpha s - \beta (u + s)^3 - 0$$

Now the large value of the damping constant leads to the assumption that the value of s close to the bifurcation point will be very small and can be neglected with respect to the value of u, so that the previous equation can be further simplified to:

$$- \alpha s - \beta u^3 = 0$$

from which we finally obtain an approximate expression of s in terms of u :

$$s = - \beta u^3 / \alpha$$

This expression gives a first rough form of the Slaving Principle. A more refined form could be obtained by dividing both members of the equations for normal modes so as to obtain:

$$ds/du = [- \alpha s - \beta (u + s)^3] / \lambda_1 u$$

If we look for a solution of this differential equation by expressing s through a power series in u, we find a better approximation of the Slaving Principle than the previous one. Interestingly, however, the first non-vanishing term of this power series gives the same result found before in such a crude manner.

Without analyzing further our example, we recall that explicit formulations of the Slaving Principle have been derived not only for differential equations, but even for difference equations, recurrence maps and stochastic differential equations. Its usefulness can be immediately recognized, as it allows for:

1   a tremendous reduction in the complexity associated with a system containing a large number of dependent variables and equations; namely, once the few unstable modes are found (only one in the previous example), all other modes, whatever their number, can be expressed in a simple way as algebraic functions of the unstable modes, thus avoiding the search for a direct solution of these systems which, in most cases, would be practically impossible to find;

2   a deeper understanding of the circumstances taking place after the critical point, as well as the roles of different fluctuations in pattern formation, because the latter depends only upon the dynamics and the interactions between a small number of unstable modes.

Our short description of the ideas and methods of Synergetics, should conclude with a discussion of the techniques used within this framework to account for the roles played by noise and fluctuations. For the time being, however, such a discussion will be postponed until Section 5.6, where stochastic systems are dealt with in a general way. Here we limit ourselves to mentioning that the introduction of Synergetics, establishing a unified framework for a number of different investigations dealing with complex and self-organizing systems, triggered an exponential growth in modeling activities relating to emergence and self-organization. It should be stressed here that the goal of all these models, as well as the whole of Synergetics, is always the same: understanding the phenomena and mechanisms relating to structural changes. It is, however, well known that, within Physics, the latter are grouped within the category of *phase transitions* and that every model of structural change is inspired by the theory of physical phase transitions, on the one hand, while on the other, one must compare the arguments and findings of a model with those of this theory. Thus, every discussion about models of emergence would be impossible without an understanding of the main points of the theory. The next section is devoted to a brief illustration of the latter.

5.2 The theory of phase transitions

The expression "phase transitions" denotes, from an intuitive point of view, physical phenomena associated with macroscopic qualitative changes of structure of the system under consideration. Such a definition is, however, rather vague and to put things onto firmer ground, we must resort to classical macroscopic Thermodynamics, which constitutes the best starting point for a more precise analysis of these phenomena (it is virtually impossible here to list the plethora of textbooks on classical Thermodynamics; traditional and comprehensive treatises include Sears, 1955; Callen, 1960; Rumer and Ryvkin, 1980).

Let us begin by recalling that within Thermodynamics a system is described through suitable variables, some of which can be used as *state variables*, as their values define, in a univocal way, the system state. Within the variables used to describe a system a distinction can be made between *extensive* and *intensive* ones. The values of the former depend upon the whole extension of the system, that is, on its spatial volume, while the values of the latter are defined, for each point belonging to the volume occupied by the system, in a way which is independent of the value of this volume. Examples of extensive variables are given by the mass, the entropy, or the internal energy; on the contrary, the mass density, the specific heat and the entropy density are intensive variables. Such a distinction allows for a more precise definition of the concept of "phase", which in this context refers to a part of a system which is homogeneous in terms of the values of some intensive variables and differing, in this same aspect, from other parts of the same system. Moreover, it is required that different phases be separated by well-defined boundaries. This definition captures the essential macroscopic features of most behaviors of physical matter, both from everyday experience and as observed in physical experiments.

When dealing with a system with the coexistence of different phases (such as, for instance, a liquid and its saturated vapor, or a liquid and its crystalline form) sometimes we observe that these phases are in equilibrium. However, it can be immediately seen that a slight external perturbation can induce the (partial or total) disappearance of one phase in favor of another. Such phenomena are called *phase transitions*. In order to characterize them, a suitable generalization of the First Principle of Thermodynamics is required to account for situations in which the total mass of a system is no longer constant. The usual form of First Principle is:

$$dU = dQ - dL = T \, dS - P \, dV$$

where dU is the increase of internal energy, dQ the amount of heat supply, dL the work done against the external environment, T is the absolute temperature, dS the change in entropy, P the pressure and dV the change in volume. In the case of a variable amount of mass, which here will be characterized by a change in the number of moles N, the First Principle is generalized as follows:

$$dU = T \, dS - P \, dV + \mu \, dN$$

The symbol μ denotes a quantity, a function only of T and P, which is called the *chemical potential* of the system under consideration. It can be easily computed as a function of system state variables by resorting to the *Gibbs thermodynamic potential*, defined by:

$$\Phi = U - TS + PV$$

Namely it can be shown that:

$$\mu = \Phi / N$$

Consider now a two-phase system, in which μ_1, N_1, μ_2, N_2 denote, respectively, the chemical potential and the number of moles of each phase. The variation in its Gibbs thermodynamic potential, in the absence of variations in pressure and temperature, will be:

$$d\Phi = \mu_1 \, dN_1 + \mu_2 \, dN_2$$

When the two phases coexist under equilibrium conditions, it can be shown that this implies the condition:

$$d\Phi = 0$$

which, in turn, entails the relationship:

$$\mu_1 = \mu_2$$

The latter is very important in the thermodynamic theory of phase transitions because, as both chemical potentials depend upon T and P, it is equivalent to a relationship between these two state variables which, within the plane of coordinates T and P, describes a *critical curve* along which, and only along which, the coexistence in equilibrium of the two phases is possible. This curve separates two regions of this plane, each characterized by the stability of only one of the two phases and, to induce a phase transition, we are forced to vary the temperature and the pressure along a path which crosses the critical curve. Only in correspondence to this crossing can the phase transition occur.

It should be noted, however, that the above equality cannot tell us what happens to the derivatives of the quantity $\mu_1 - \mu_2$ at the critical curve. The latter, however, characterizes in a critical way some macroscopic properties of the system under study. We recall, for instance, that:

$$(\partial \mu / \partial T)_P = -s \, , \quad (\partial \mu / \partial P)_T = -v$$

where s denotes the entropy per mole and v is the volume of a mole (as is customary in Thermodynamics, the subscripts denote the state variables kept constant while performing the partial derivative). Thus, in the presence of a discontinuity in first order derivatives of $\mu_1 - \mu_2$ at the critical point, the two phases will be characterized by different values of s and v and, as a consequence, by different densities and by a non-vanishing molar heat released or absorbed in the phase transition itself. Phase transitions of this kind are called *first-order phase transitions*. All changes in states of aggregation, such as boiling, melting or solidification, are examples of this category. The discontinuity of first order derivatives of chemical potential at the critical point where this kind of phase transition occurs entails that the transition requires a finite time (the transition ends only after one of the two phases has disappeared), during which the two different phases coexist. This means that, at the onset of the transition, the new phase will appear as very small nuclei (for instance, small crystals, fine liquid droplets, small vapor bubbles) which, as the transition advances, undergo a dynamical growth process. The study of this latter is a very complicated matter, as it requires a detailed knowledge of intermolecular forces. In any case, it is mostly influenced by the fact that the interface between the two phases carries an additional energy, implying that every increase in its surface requires an additional amount of work. This circumstance plays an important role in the initial stage of the transition, when the growth of a nucleus, implying an increase of the surface of its interface, can be energetically disadvantageous, not being compensated by the advantage produced by the increase in its volume (in very small nuclei, surface effects are more important than volume effects). Thus, it could occur that, under suitable conditions (for instance, the absence of external disturbances, very slow heat liberation or absorption, absence of impurities), the phase transition cannot take place and the initial phase continues to survive, even after crossing the critical curve, in a situation which is no longer thermodynamically stable, but *metastable*. This means that, in the presence of suitable disturbances, the metastable initial phase disappears and suddenly a phase transition produces the new stable phase. Usually, however, such situations do not arise, as most states of aggregation are associated with the presence of *defects* or impurities of some kind which, even at the start of a phase transition, constitute very large nuclei in which the role of the interface surface is overcome by stronger volume effects.

Let us, now, consider the case in which the first order derivatives of chemical potential are continuous along the critical line, whereas the second order derivatives are not. Also the latter are associated with important physical features of the system under consideration. From the definition of chemical potential and the main formulae of Thermodynamics it is possible to derive, for these second order derivatives, the important relationships:

$$(\partial^2 \mu / \partial P^2)_T = - v\, K_T \ , \ \ (\partial^2 \mu / \partial T^2)_P = - C_P / T \ , \ \ (\partial^2 \mu / \partial P \partial T) = v\, \alpha_P$$

where C_P is the heat capacity at constant pressure, K_T the isothermal compressibility and α_P the isobaric thermal volume expansion coefficient. The latter two quantities are defined as:

$$K_T = - (1/V) \, (\partial V / \partial P)_T \ , \ \ \alpha_P = (1/V) \, (\partial V / \partial T)_P$$

These phase transitions, called *second-order phase transitions*, are therefore characterized by the absence of a difference in the density of the two phases and of a latent heat emitted or absorbed in the transition. They can, however, be detected owing to the existence of discontinuities in heat capacity, in isothermal compressibility and in the isobaric volume expansion coefficient. These circumstances imply that a second-order phase transition occurs abruptly and simultaneously within the whole system under consideration. At the critical point there is no coexistence of both phases, the new and the old one, as they do not differ in molar volume and there is merely a continuous transition from one phase to the other, requiring neither a specific external energy supply nor the emission of a specific quantity of energy. In practice, this kind of transition consists only in an internal rearrangement of the system structure, occurring at the same time at all points within it. In other words, the transition occurs because the conditions necessary for the stable existence of the structure corresponding to the initial phase *cease to be valid* and a new stable structure replaces it.

These remarkable features attracted the interest of many physicists and system scientists, as well as mathematicians and philosophers, all feeling that second-order phase transitions were the prototypes of phenomena of profound structural change. In other terms, these transitions have been almost unanimously considered as typical examples of emergence and we could claim that most models of emergence have been built while keeping in mind the features of second-order phase transitions. On the other hand, the examples known so far of second-order phase transitions are so interesting and, sometimes, mysterious that such widespread interest is fully justified. These include the transition from the paramagnetic to the ferromagnetic state, the occurrence of superconductivity and of superfluidity, the order-disorder transitions in some kinds of crystals. The study of each of these phenomena marked the birth of a new branch of Physics. Before closing this subject, the reader must be warned that, beyond first- and second-order phase transitions, there are other kinds of phase transitions, sometimes improperly labeled as second-order phase transitions. Following the convention adopted by Ehrenfest, we define a phase transition as being of *k-th order* if all derivatives of chemical potential, up to the order $k - 1$, are free of discontinuities along the critical line, whereas the derivatives of k-th order are discontinuous. Thus, when $k > 2$, and there are no discontinuities in specific heat, we cannot use this variable to detect the critical point of this kind of phase transition.

This brief presentation shows how the contribution of classical macroscopic Thermodynamics to the study of phase transitions is mostly of a descriptive nature. Clearly, this does not mean that such a contribution is devoid of importance: without a knowledge of critical curves and of discontinuities associated with critical points we would lack any tool for detecting the occurrence of phase transitions and their macroscopic features. However, a deeper understanding of the phenomena associated with phase transitions is necessary and this requires the introduction of a *theory of phase transitions.* For this , we subdivide the theories so far proposed into two main categories: the *phenomenological* theories and the *microscopic* ones (following the categorization proposed by Patashinskij and Pokrovskij, 1979; other forms of subdivision are, however, possible; see, for instance, Liu, 1999). The former, while assuming the existence of a microstructure, deal mainly with averaging effects used to account for the universal features of phenomenological data. The latter, on the other hand, rely upon the building of more or less idealized models of microscopic phenomena trying to obtain from them, through the methods of statistical mechanics (classical or quantum), precise forecasting of macroscopic behaviors and features. This categorization cannot, obviously, be a rigid one as both share many hypotheses and technical approaches. In this section, we briefly illustrate the main points of some representative theories of phase transitions avoiding, however, any reference to theories using the formalism of Quantum Mechanics or Quantum Field Theory, which will be dealt with in the following sections 5.3 and 5.4.

However, before beginning our description, it must be stressed that every theory of phase transitions has to take into account a number of well-established empirical facts, most of which were observed for the first time within the context of the transition from the paramagnetic to the ferromagnetic state. Without referring to these facts it would be impossible to understand the goals and the motivation underlying the various theories of phase transitions. However, the reader is forewarned that currently it is unknown whether similar facts characterize other kinds of structural changes, occurring in biological, social, cognitive or economic domains. In the absence of such evidence (admittedly very difficult to obtain), the possibility of applying such methods proposed by the theories of phase transitions to the aforementioned structural changes remains only a hypothesis (alhough a very attractive one).

Here is a short list of the fundamental facts to be considered:

-       for a given pressure, every phase transition occurs at a given *critical temperature* T_c (usually denoted as the *Curie point* in the transition from the paramagnetic to the ferromagnetic state);

-       every second-order phase transition consists of passing from a *less ordered* to a *more ordered* phase (or vice versa); here the attribute "less ordered" is equivalent to "more symmetric", while "more ordered" is equivalent to "less symmetric". A look at the paramagnetic-ferromagnetic transition will clarify this. Consider, for instance, a paramagnetic material in the absence of an external magnetic field: its state does not change under rotation in 3-dimensional space. This phase can be therefore viewed as possessing a symmetry with respect to these geometrical transformations. Let us now consider the same substance in the ferromagnetic phase and assume that an external magnetic field was applied and then turned off. As the material is ferromagnetic, it remains magnetized and this "residual" magnetization has a well-defined orientation, defined by a particular axis lying in 3-dimensional space. Clearly, this time the symmetry with respect to 3-dimensional rotation is lost; namely the state of the material sample changes with generic rotations and such changes are detected through the work needed for performing these rotations. However, elements of symmetry still survive; namely all rotations having the residual magnetization symmetry axis do not require work. Thus, in the ferromagnetic phase the system is endowed with symmetry (rotations around a particular axis) which is only a particular subcase of a more general symmetry (generic 3-dimensional rotations) and in this sense we can speak of a transition from a "more symmetric" (paramagnetic) to a "less symmetric" (ferromagnetic) state;

every phase transition can be macroscopically described by an *order parameter*, that is a quantity measuring how much, after crossing the critical point, the new phase differs from the old one. In second-order phase transitions the order parameter measures the degree of symmetry loss, after the critical point, in the new ordered phase. This, of course, requires that in the more symmetric phase the order parameter is vanishingly small. There are few general rules for identifying the physical quantity best suited to act as order parameter and the best method is that based on Haken's Slaving Principle, already described in the previous section. Moreover, its mathematical nature can, in principle, be anything: a scalar, a vector, a tensor and so on. In the paramagnetic-ferromagnetic transition it is given by the total residual magnetization (which is a 3-dimensional vector);

To induce a variation in the order parameter, we must perform work on the system. In the case of infinitesimal variations dm in a scalar order parameter m we can write the infinitesimal work to be done dW as:

$$dW = H \, dm$$

the proportionality coefficient H between dW and dm defines the so-called "conjugate field" (the attribute "conjugate" refers, of course, to the particular choice adopted for m). It turns out that, in the ferromagnetic phase, the conjugate field coincides with the external magnetic field. This allows the introduction of a further quantity, measuring the "sensitivity" of the order parameter with respect to variations in the conjugate field (in most cases, as in ferromagnetism, of the external field). As the order parameter can undergo strong fluctuations near the critical point, the quantity taken into consideration will no longer be the order parameter itself, but its average with respect to fluctuations. Without entering into details on the complicated problem of the effective computation of this average, denoted as $<m>$, we limit ourselves to mentioning that the "sensitivity" introduced above is measured by a quantity called *generalized susceptibility* and defined by:

$$\chi = \partial <m> / \partial H$$

In the case of magnetic materials this coincides with the *magnetic susceptibility* χ_m, given by $\chi_m = M/H$, where M is the modulus of the total magnetization vector and H the modulus of the external magnetic field. Experiments showed that the generalized susceptibility is dependent upon temperature and its value diverges as the temperature approaches the critical temperature. In the case of the paramagnetic-ferromagnetic transition, the paramagnetic phase is stable when the temperature is greater than the critical temperature, while the ferromagnetic phase occurs below the critical temperature. Thus, we have two kinds of divergence of magnetic susceptibility, one for $T > T_c$ and another for $T < T_c$. These are described by two different laws, collectively denoted as the *Curie-Weiss law*, derived from experiments; they can be written as:

$$\chi_m = c_1 / (T - T_c) \quad \text{for } T > T_c$$

$$\chi_m = c_2 / (T_c - T) \quad \text{for } T < T_c$$

where c_1 and c_2 are suitable constants;
when the temperature approaches the critical temperature, the amplitude of fluctuations in the order parameter tends to increase; a manifestation of this fact is the anomalously strong scattering of electromagnetic waves close to the critical point, a phenomenon known as *critical opalescence*; for a greater understanding of this phenomenon, consider a scalar order parameter m, associated with a density which is dependent upon spatial location r and denoted by $q(r)$; if the symbol $<A>$ denotes, as before, the value of A averaged over the fluctuations, the order parameter can be written as:

$m = <\int q(r)\, dr>$

Let us now introduce the correlation function for $q(r)$, defined, for a system whose properties are invariant with respect to spatial translations, by the formula:

$\Gamma(r) = <q(r)\, q(0)> - <q(0)>^2$

This measures the degree to which the fluctuations in the order parameter at different points separated by a distance r are mutually correlated. Approximate theoretical computations show that $\Gamma(r)$ can be written in a form which is generally known as the *Ornstein-Zernike form*:

$\Gamma(r) = r^{-p} \exp(-r/\xi)$

where ξ denotes a quantity called the *correlation length*. Experiments have shown that ξ diverges as the temperature approaches the critical temperature. Regarding the exponent p appearing in the previous formula, both theoretical computations and experimental measurements showed that its value is very close to $d - 2$, where d denotes the dimensionality of the system in which the phase transition occurs. This allows an understanding of why it is impossible to have phase transitions in a 1-dimensional system and it is nearly impossible in a 2-dimensional system, a statement also known as the *Mermin-Wagner theorem* (Mermin and Wagner, 1966). Namely, when $d = 1$, p is very close to 1. If ξ diverges at the critical point, the correlation function $\Gamma(r)$ will take, at the critical point, more or less the form: $\Gamma(r) = r$. This implies that the correlation will increase with increasing distance and that, for very great distances, that is when r tends to infinity, the correlation will diverge. In other words, no long-range ordering is possible (every ordered structure would be destroyed by long range fluctuations) and thus no phase transition can occur. A similar reasoning can be applied to the case in which $d = 2$, where the only difference is that, as p is very close to zero, the value of the correlation function remains finite for r tending to infinity. This could allow ordering over a finite range, although on a macroscopic scale. We could thus say that we are lucky to live in a 3-dimensional world because, with a smaller number of dimensions, we would not have the possibility of observing phase transitions nor emergence phenomena;

when the order parameter is slightly perturbed (for instance by an external action), and once the cause of perturbation was turned off, the perturbation itself tends to vanish following a relaxation process towards the previous equilibrium value of the order parameter. This relaxation process is characterized by a *characteristic time* t_c and it has been experimentally observed that t_c increases as the temperature approaches the critical temperature. This phenomenon is sometimes called a *critical slowing down* and approximate computations show that this behavior is related to the fact that t_c is directly proportional to the generalized susceptibility χ ;

At the critical temperature of a first-order or a second-order phase transition the curve giving heat capacity (or specific heat) as a function of T shows a marked discontinuity. Two different kinds of discontinuity have been observed: *finite discontinuities*, in which the value of specific heat makes a finite jump at the critical temperature, and *infinite discontinuities*, in which there is evidence of a divergence in specific heat at the critical temperature. The latter phase transitions are sometimes called *lambda transitions*, a denomination deriving from the form of the experimental curve which vaguely resembles the capital Greek letter lambda. It must also be recalled that specific heat can be related to the properties of fluctuations occurring within the system. This, of course, holds not only for specific heat but, in general, for all quantities describing the macroscopic response of a system to external perturbations, as stated by a general theorem known as the *fluctuation-dissipation theorem*. A specific example is given by the generalized susceptibility, which is related to the correlation function $\Gamma(r)$ by the formula:

$$\chi = (1/kT) \int \Gamma(r) \, d^3r$$

where k denotes the usual Boltzmann constant. An analogous relationship holds for heat capacity at constant volume C_V , which is given by:

$$C_V = (<U^2> - <U>^2)/ kT^2$$

where U denotes the total internal energy. It should be remembered that this formula could be applied even in contexts where the energy conservation principle or the First Principle of Thermodynamics no longer applies. In these cases the energy U should be substituted by a suitable *Lyapunov function*. An example of the computation of the specific heat of a neural network based upon this procedure can be found in Smolensky (1986).

As these experimental facts are related to different kinds of divergences in a number of quantities of interest at the critical point, it is customary to introduce a number of parameters, called *critical exponents* , which specify, for each quantity of interest, the kind of divergence in the latter when the temperature approaches its critical value. In a sense, a knowledge of the values of critical exponents for a given phase transition provides a full macroscopic characterization of the transition itself. Should it occur that the values of the different critical exponents were the same *for all phase transitions*, then we would be in the presence of a form of *universality*, which would legitimize the attempt to build a general theory of phase transitions, holding for all kinds of phase transitions, independently of their context. Logically, we would be in the presence of a universality even if the latter circumstance held only for some critical exponents, or even for only one of them. In this regard we recall that some theories of phase transitions (such as the Landau theory cited below) postulate a form of universality, while experimental measurements of critical exponents showed their variations on going from one phase transition to another. How can such a circumstance be evaluated? The fact that the ranges of variation in critical exponent values obtained so far are quite narrow led to the suggestion that universality is an approximate concept, useful to build simple, and not-too-detailed, models of phase transitions. However, at the present stage of scientific research it is impossible to prove that this is correct.

Before listing the critical exponents introduced so far, in order to describe the approach of the temperature towards the critical temperature, a new variable, more useful than T, will be introduced, denoted by τ and given by:

$$\tau = (T - T_c)/T_c$$

Now the critical exponents can be defined as the exponents of power laws giving the dependence of the singular parts (that is, of those responsible for the divergence) of a number of quantities upon the modulus of τ, close to the critical point $\tau = 0$. More precisely, the first four critical exponents α, β, γ and ν are related, respectively, to heat capacity, order parameter, generalized susceptibility and correlation length through the following power laws:

$$C \approx |\tau|^{-\alpha} \; ; \; m \approx |\tau|^{\beta} \; ; \; \chi \approx |\tau|^{-\gamma} \; ; \; \xi \approx |\tau|^{-\nu}$$

Obviously the values of these critical exponents can be obtained from experimental curves giving each one of these quantities as a function of $|\tau|$. For instance, the value of β coincides with the slope of the regression line giving *log m* as a function of *log* $|\tau|$.

A further critical exponent, denoted by δ, is given by the power law connecting the order parameter with the conjugate field (that is the external field) at $\tau = 0$. This law takes the form:

$$m \approx H^{(1/\delta)}$$

Finally, the last critical exponent, denoted by η, is defined as the difference between the exponent p, appearing in the Ornstein-Zernike form of thecorrelation function $\Gamma(r)$, and the value of $d - 2$:

$$\eta = p - (d - 2)$$

We recall that these six critical exponents are not independent of one another. A set of relationships between them, derived from different theoretical considerations, allows us to obtain the values of some critical exponents from the values of others. Without entering into detail, we limit ourselves to mentioning that the validity of some relationships has been experimentally verified.

Having defined the critical exponents, we can now begin our presentation of the various theories of phase transitions with the most celebrated of the phenomenological ones: that proposed by L. D. Landau in 1937 (references on this theory can be found in every textbook on statistical mechanics or phase transitions: e.g., Landau and Lifshitz, 1959; Stanley, 1971; Huang, 1987; Goldenfeld, 1992; Domb, 1996; a book entirely devoted to this theory is Tolédano and Tolédano, 1987). The interest in Landau theory stems not from its validity (later we will show that the theory is incorrect) but rather from the kind of approach used to deal with phase transitions. Most theories, not only of phase transitions, but more in general, of emergent phenomena, are nothing but suitable variants of Landau theory. No-one who is interested in emergence can ignore the main points of this theory.

Landau theory

The theory is based upon three main hypotheses:
a. every phase transition is associated with a symmetry change (from a less symmetric to a more symmetric situation, in the sense explained above);
b. to describe the behavior of a system close to a phase transition the only relevant quantities are the order parameter m (or eventually its spatial

density $q(r)$) and the temperature. The only role played by microscopic interactions between a system's constituent parts is in fixing the values of some parameters entering into the equations which describe the system's thermodynamical behavior;

c. the physical behaviour of a system close to a phase transition is determined essentially by the average value of the order parameter. In turn, this average is computed by identifying $q(r)$ with the constant value q_0 which minimizes the total energy of the system and the average value of order parameter then becomes $m_0 = q_0 V$. This hypothesis is also referred to as the *mean field hypothesis* and we can say that Landau theory is an example of a *mean field theory*.

In accordance with these hypotheses, Landau assumes that the free energy density F of a system close to critical point is a function only of q (or, equivalently, of m). Owing to the small value of q close to the critical point, F can be developed in a power series in q taking the form:

$$F = F_0 - H q + a q^2 + (b/2) q^4 + \dots$$

The term $-H q$ describes the interaction with a possible external field H. The right hand side of the above formula not related to the external field contains an expansion in a power series including only even powers of q. This is due to the fact that we assume the free energy density to be independent of the sign of q, that is, endowed with a symmetry with respect to the change $q \rightarrow - q$. Such a requirement is fulfilled in most cases of physical interest, even though there are notable exceptions, which will not be discussed here. The coefficients a and b are assumed to depend, in general, upon the temperature and suitable features of microscopic interactions. For this, Landau proposed that a be dependent upon temperature according to the law:

$$a = a_0 \tau$$

where a_0 is a positive constant (which depends, of course upon microscopic features), whereas b is another positive constant, but independent of temperature. The expression for F introduced above is sometimes called the *Landau free energy*, to distinguish it from traditional thermodynamical free energy, given by $F = U - T S$. There are, however, several proofs which demonstrate that Landau free energy behaves like thermodynamical free energy.

Let us now find the values of q which minimize the Landau free energy. Elementary mathematics shows that they must fulfill the two conditions:

$$\partial F/\partial q = 0 \;\; ; \;\; \partial^2 F/\partial q^2 > 0$$

Putting $H = 0$, the first condition gives the equation for q:

$$a\,q + b\,q^3 = q\,(a_0\,\tau + b\,q^2) = 0$$

Clearly, when $T > T_c$ and $\tau > 0$, this equation has one and only one solution, given by:

$$q_0 = 0$$

A simple computation shows that in this case the second condition is also always automatically fulfilled close to the critical point, given that a_0 and b are both positive. This grants for the fact that this solution corresponds to the minimum of F. On the contrary, when $T < T_c$ and $\tau < 0$, we have three solutions, one corresponding, as before, to a zero value of q, and the other two given by:

$$q_0 = \pm(-a/b)^{\frac{1}{2}}$$

Although in this case the zero solution does not fulfill the condition of a positive second derivative (this solution corresponds not to a minimum of F, but to a maximum), this condition is, however, satisfied by both nonzero solutions (remember that in this case a is negative). The latter describe the new ordered state occurring after the phase transition, corresponding to $\tau = 0$, that is to the critical value $a = 0$. This state, as expected, is associated with a nonzero value of the order parameter (recalling that the order parameter is given by $m_0 = q_0 V$), while the latter is zero before the critical point, that is for $\tau > 0$. In conformity with the Landau hypotheses, after the phase transition we have a loss of symmetry (and an increase in order). Namely, although the Landau free energy is invariant with respect to the change $q \rightarrow -q$, this change now profoundly alters the physics of the system, as it gives rise to a shift from one of the two nonzero solutions to the other. They correspond, however, to two very different physical situations: one with "total magnetization" pointing up, and the other with "total magnetization" pointing down. To denote briefly this new situation, in which the free energy is endowed with a given symmetry, while the solutions of equations are not, the term *Symmetry Breaking* has been coined. The Landau theory describes just such a particular kind of Symmetry Breaking, which appears strongly related to the mechanism of bifurcation already described in the previous Chapter. Namely, Landau identifies a phase transition with a phenomenon of bifurcation associated with the change in a critical parameter, a, within the algebraic equation used to find the minima of F. Such a change is made possible by the fact that the critical parameter is assumed to be dependent upon temperature. These circumstances induced most people to accept the validity of a chain of equivalences such as:

(Emergence) *is equivalent to* (Phase transition) *which is equivalent to* (Bifurcation).

However, both in the previous Chapter as well in this one, we show how this chain (upon which very important models, such as those of Dissipative Structures, are based) is not completely correct.

Let us now begin to derive some quantitative consequences of Landau theory. First of all, it is clear that, after the phase transition, owing to the form of solution q_0, the order parameter will be given by:

$$m_0 = q_0 V = \pm(-a/b)^{1/2} V = \pm(-a_0\, \tau/b)^{1/2} V = \pm V (-a_0/b)^{1/2} \tau^{1/2}$$

A comparison between this formula and the power law defining the second critical exponent β, as written above, shows immediately that in Landau theory $\beta = \frac{1}{2}$, independently of the details of the phase transition considered. This is a first example of universality.

In order to derive the first critical exponent α, we must compute the heat capacity. Here, we limit ourselves to the heat capacity at constant volume C_V. In classical Thermodynamics the latter can be derived from entropy through the standard formula:

$$C_V = -T\,(\partial S/\partial T)_V$$

the entropy being derived from the free energy F through the relationship:

$$S = -(\partial F/\partial T)_V$$

By applying these formulae to Landau free energy, easy computations show that, while the entropy does not undergo discontinuities at the critical point, the heat capacity is characterized by a finite discontinuity at the critical temperature. More precisely, it can be shown that the jump in specific heat, adopting the units and conventions used so far, is given by:

$$\Delta C_V = (a_0)^2/b \, T_c$$

These findings illustrate how Landau theory describes only second-order phase transitions. Moreover, as there is no infinite discontinuity of specific heat at the critical point, we are forced to conclude that the first critical exponent is vanishingly small, that is $\alpha = 0$ for every phase transition.

To derive the other critical exponents, we need to take into account the external field H. When the latter is not vanishingly small, the equation determining the minima of Landau free energy becomes:

$$a \, q + b \, q^3 - H = 0.$$

Simple mathematical considerations show that the latter cannot have solutions equal to zero, so that the order parameter is always different from zero. In other words, we do not have a true phase transition. When the value of H is small enough, it is possible to prove, by resorting to the theory of third-degree algebraic equations, that this equation has only one nonzero solution for $\tau > 0$, whereas it has three for $\tau < 0$. However only two of them correspond to minima in F and, moreover, these two minima correspond to different values of F, so that one of them is a *global minimum* (corresponding to a stable phase), while the other is a *local minimum* (corresponding to a metastable phase). Besides, when H is large enough, there is only one minimum in F even for $\tau < 0$. Thus, in this case, at the critical point there is only a continuous transition from one phase to the other.

From the above equation it is possible to obtain an approximate expression for the generalized susceptibility χ as a function of τ. Namely, by taking the derivative with respect to H, we obtain, through elementary computations, that χ will be directly proportional to:

$$\chi \approx \partial q /\partial H = 1/(a + 3 \, b \, q^2)$$

Now, if we assume the external field H to be very small close to the critical point, even q can be neglected, as it practically tends to zero as the temperature approaches the critical temperature. By including this limit the expression for χ becomes:

$$\chi \approx 1/a = (1/a_0) \, (1/\tau)$$

which coincides with the Curie-Weiss law. It entails that the critical exponent γ takes a value of 1.

Without pursuing further the computation of critical exponents, we limit ourselves to remarking that in Landau theory there is a complete universality, as the values of all critical exponents are independent of the particular context (macroscopic or microscopic) characterizing the phase transition considered (provided it is a second order one). The complete list of critical exponents and of their associated values in Landau theory is the following:

$$\alpha = 0 \ , \ \beta = \tfrac{1}{2} \ , \ \gamma = 1 \ , \ \nu = \tfrac{1}{2} \ , \ \delta = 3 \ , \ \eta = 0$$

Some of these values differ greatly from those obtained from experimental data. This circumstance, however, has not been considered as the main weakness of Landau theory. Many authors believe that the main shortcoming of this theory lies in the fact that it does not take into account the role of fluctuations in the order parameter which should be very large close to the critical point, so as to make unreliable a rough average such as that used in the above discussion. Moreover, as also shown by fluctuation-dissipation theorem, it would be more correct to derive the critical exponents from arguments based upon dynamical features of fluctuations rather than from phenomenological approximations.

Many different strategies could be chosen to remedy the shortcomings of Landau theory, but the most simple consists in taking the average values of q as a zero-order approximation, with the fluctuations around these values being considered as small first-order approximations. In this way one can take into account the role of fluctuations, while keeping intact the advantages of Landau theory. This generalization of Landau theory is sometimes called the *Ginzburg-Landau theory*.

Ginzburg-Landau theory

Within this theory the order parameter density $q(r)$ is written as:

$$q(r) = q_0 + \varphi(r)$$

where q_0 is the average value introduced in Landau theory, and $\varphi(r)$ denotes a suitable *fluctuation field* (modern developments in the theory of phase transitions following this approach are reviewed in Olemskoi and Klepikov, 2000). Here, for the time being, we take into account only the dependence upon r, whereas later we also consider the dependence upon time t. The above expression implies a suitable generalization of Landau free energy, which must now also include terms containing the spatial derivatives of the field φ. Moreover, owing to the dependence upon r, the free energy must be explicitly written in the form of a spatial integral. If we now assume that the free energy be still invariant with respect to a change in sign of φ, it can be developed into a series of even powers both in φ and in $\nabla\varphi$ (the "gradient vector", whose components are the first spatial derivatives of φ). In the case of zero external field, considering only the contribution to free energy from terms containing φ, we obtain an expression of the form:

$$F\{\varphi\} = F_0 + (1/2) \int dV \, [c \, (\nabla\varphi)^2 + a \, \varphi(r)^2 + (b/2) \, \varphi(r)^4 + \ldots]$$

From this formula it appears that for each choice of the *form of the function* $\varphi(r)$ there corresponds a particular *value* of $F\{\varphi\}$. The latter, therefore, can no longer be considered as an ordinary function, as it associates forms of whole functions to output values. Mathematicians denote such objects as "functionals" and, in particular, the free energy functional written above is called the *Ginzburg-Landau functional*. The function $\varphi(r)$ minimizing the value of the latter can be found as the solution of a particular differential equation, obtained through the standard methods of Variational Calculus. The explicit form of this equation is:

$$- c \, \Delta\varphi + a \, \varphi + b \, \varphi^3 = 0$$

Here, as is customary, Δ denotes the usual Laplace operator. In the case in which φ can be considered as a small correction to q_0 we can derive from the first-order approximation of the solution to this equation, the explicit form of the correlation function $\Gamma(r)$ of the fluctuations $\varphi(r)$. The latter turns out to be exactly coincident with the Ornstein-Zernike form already introduced. This makes it clear how the latter, as well as the concept itself of correlation length, is valid only as long as the fluctuations in the order parameter can be considered small with respect to its average value. In this regard, we can introduce a criterion of validity for Landau theory, based upon the ratio between the variance of order parameter fluctuations and the average value of this order parameter, as computed within Landau theory. If we denote the variance of the field of fluctuations $\varphi(r)$ by $<\Delta q^2>$ and the average value of q by q_0, then this criterion can be expressed as:

$$<\Delta q^2>/(q_0)^2 \ll 1$$

This means that Landau theory can be still considered as reliable as long as the ratio of the variance of fluctuations to the average value of the order parameter can be considered as a very small quantity. In other words, fluctuations must be small enough to be neglected compared with the average value of the order parameter.

Trivial statistical considerations show that $<\Delta q^2>$ coincides with the correlation function $\Gamma(r)$. If we write the latter in the Ornstein-Zernike form and, close to the critical point, assume that the correlation length ξ is large enough to make the exponential term $exp\ (-r/\xi)$ practically coincident with unity, and adopt for the critical exponent η the value $\eta = 0$ computed in Landau theory (which agrees with experimental data), then the correlation function will take the form $\Gamma(r) = r^{2-d}$. A good estimate of the value of $\Gamma(r)$ can be obtained by putting $r = \xi$, so that we can finally obtain for $<\Delta q^2>$ the expression:

$$<\Delta q^2> = \Gamma(\xi) = \xi^{2-d}$$

For the value of $(q_0)^2$ we can use the value obtained in Landau theory when $\tau < 0$, that is:

$$(q_0)^2 = -a/b$$

If we now take into account that Landau theory, for ξ close to the critical point, predicts a power law dependence on τ of the form:

$$\xi = (2\ a_0\ \tau)^{-1/2}$$

and that $a = a_0\ \tau$, by substituting into the expression of the validity criterion, we obtain the latter as:

$$b\ (a_0\ \tau)^{(d-4)/2} \ll 1$$

This is one of the forms of the so-called *Ginzburg criterion*. It easy to see, that it is automatically satisfied for $\tau \to 0$ only when $d > 4$. Therefore, only for systems with a dimensionality greater than four can we assert that Landau theory is reliable. As the majority of physical systems has a dimensionality $d \le 3$, the Ginzburg criterion leads to the conclusion that in all these cases there exists a *critical region*, centered on critical temperature, such that within it Landau theory and mean field methods fail, due to long range fluctuations dominating the physical picture of phase transition. Outside this critical region, of course, Landau theory is still valid. Moreover, in some cases (such as, for instance, superconductors) the critical region is so small as to be so far inaccessible to experimental investigations.

Let us now spend some words on cases where order parameter fluctuations depend upon time as well as spatial coordinates. Within the context of Landau theory we can picture the time evolution of a fluctuation as a process of *relaxation* towards an equilibrium value minimizing the Ginzburg-Landau functional. The simplest way to describe such a process is by assuming that, for small fluctuations, the time derivative of the fluctuation field φ is directly proportional to the gradient of the Ginzburg-Landau functional with opposite sign, that is:

$$\partial \varphi / \partial t = - \varepsilon \, \delta F\{\varphi\} / \delta \varphi$$

where ε is a suitable proportionality factor and the symbols $\delta / \delta \varphi$ denote the "functional" derivative with respect to the function φ. By using the expression for the latter found above whilst searching for the minimization of the Ginzburg-Landau functional, from this relationship we obtain a general equation, driving the spatio-temporal evolution of the amplitude of order parameter fluctuations, taking the form:

$$\partial \varphi / \partial t = k_1 \, \Delta \varphi + k_2 \, \varphi + k_3 \, \varphi^3$$

This equation is generally known as the *Time-dependent Ginzburg-Landau equation* or, more simply, the *Ginzburg-Landau equation* (hereafter abbreviated as GLE). In other contexts it is sometimes called the *amplitude equation*. It is of the utmost importance in the theory of phase transitions, as well as in the whole theory of emergence and structural changes. It describes, in a universal form (even if only up to the third order), the time evolution of the order parameter after whatever kind of structural change, provided the Ginzburg criterion is satisfied. The only reference to the context of the specific structural change considered is contained in the values of the parameters k_1, k_2 and k_3. It is therefore not surprising that this equation and its generalizations have been widely used in many different contexts, mainly to describe the main features of pattern formation after the crossing of a critical point, not only in phase transitions but also in bifurcation phenomena. One important example is the fundamental work by Kuramoto and Tsuzuki (Kuramoto and Tsuzuki, 1975) who derived a GLE describing the dynamical evolution of the amplitude of the new emerging pattern after a bifurcation point in reaction-diffusion models. A somewhat different derivation of a generalized GLE was introduced by Haken within the context of Synergetics (Haken, 1983). Their analysis showed, in general, the parameters of a GLE to be complex numbers. This more general form of the GLE, sometimes called *complex-GLE*, was the starting point for a number of theoretical and numerical analyses of pattern formation features (e.g., Akhromeeva *et al.*, 1984; Chate and Manneville, 1996; a good review is given in Ipsen *et al.*, 2000) mainly within the context of periodic patterns (Sirovich and Newton, 1986). The work of Kuramoto and Tsuzuki was then extended to the case of models described by integro-differential equations (Toko and Yamafuji, 1990). In recent years even *vector-GLEs* have been introduced, to describe the competition between different patterns arising beyond a critical point (Hoyuelos *et al.*, 2003). Moreover, some methods have been proposed to derive the a GLE directly from available experimental data (Le Gal *et al.*, 2003). More recently, there has been a growing interest in the *cubic-quintic-GLE*, that is a GLE containing terms up to the fifth order. This interest stems from the fact that this equation allows for solitonic solutions (Crasovan *et al.* , 2001; Akhmediev and Soto-Crespo, 2003). Finally, a GLE can be used even to describe the phenomena occurring close to a critical point but in a region where the Ginzburg criterion is not satisfied. For this purpose it is sufficient to add a noise term to the GLE. A prototypical work based on this method is that by Nitzan and Ortoleva (Niztan and Ortoleva, 1980), already cited in the last Chapter amongst those illustrating the shortcomings of Dissipative Structures as models of emergent phenomena.

There is no point here in entering into further details about Ginzburg-Landau theory which, we underline, practically constitutes for most researchers the "prototypical" macroscopic model of emergence. However, before dealing with other topics, we draw the attention of the reader towards a particular consequence of GLE. For this purpose, we rewrite it, according to the findings of Landau theory, in the form:

$$\partial \varphi / \partial t = c \, \Delta \varphi - f \varphi + b \, \varphi^3$$

where c denotes a suitable diffusion coefficient, while b has the same meaning as in Landau theory and $f = -a$, so that $f = 0$ for $\tau = 0$ and $f > 0$ for $\tau < 0$. We further assume, for the sake of simplicity, that we are dealing with a 1-dimensional system. The homogeneous stationary equilibrium solution of this equation is readily obtained by putting equal to zero all derivatives and, for $\tau \leq 0$, is given by:

$$(\varphi_0)^2 = f/b$$

We can now perform a linearized stability analysis of this solution by resorting to small deviations, ψ, from it. Standard mathematics shows that the linearized equation, obeyed by ψ close to the equilibrium solution, takes the form:

$\partial \psi / \partial t = c \, \Delta \psi - 2 f \, \psi$

Let us now write the solution to the latter equation as:

$\psi = A \, exp(\lambda \, t) \, sin \, (k \, x)$

By substituting the latter into the linearized equation, we can immediately see that the eigenvalue λ will depend upon the "wave number" k through the relationship:

$\lambda(k) = - c \, k^2 - 2 f$

This implies that, at the critical point, that is for $f = 0$, $\lambda = 0$ for $k = 0$, while $\lambda < 0$ for $k \neq 0$. This can be interpreted as implying that, at the critical point, all perturbations tend to decay (as λ is negative) except those characterized by $k = 0$, that is, having an infinite spatial wavelength (we recall that k is inversely proportional to the latter). In other words, if we associate λ with the *energy* of perturbations and the *interaction range* with the spatial wavelength, at the critical point we have the occurrence of long-range (more exactly, infinite range) excitations carrying no energy. These can be viewed as *information carriers*, signalling to system components that the system itself lies in a situation of Symmetry Breaking, where a multiplicity of different equilibrium states occur. The zero energy is associated with the fact that, at the critical point, no energy is needed to go from one equilibrium solution to another. As in quantum theory, every excitation can be interpreted as a particle, whose mass m is connected to the interaction range ρ by the simple formula:

$m = \hbar / \rho \, c$

where $\hbar$ is the usual Planck constant divided by $2 \, \pi$ and c is the velocity of light in vacuum, and it can be immediately seen that these excitations correspond to particles with *zero mass*. In quantum theory they are usually called *Goldstone bosons*. Without entering into further detail on this topic, which will be dealt with later in the section on Symmetry Breaking, we mention here that the excitations found in our classical analysis of GLE, while not corresponding to their quantum counterparts, are called *classical Goldstone bosons*. We can therefore summarize our reasoning by stating that at the critical point of a phase transition the Ginzburg-Landau theory predicts the occurrence of classical Goldstone bosons. A number of authors have studied the conditions necessary for the occurrence of the latter even in equations different from the GLE, such as reaction-diffusion systems (Walgraef *et al.*, 1981; Thiesen and Thomas, 1987) or neural field equations (Pessa, 1988).

In the following we briefly mention some further approaches which, while starting from the propositions of Landau theory, generalize them in such a way as to take into account even strong fluctuations close to a transition point. These approaches have been very effective in providing new methods for carrying out concrete computations of critical exponents and clarifying the conditions for the occurrence of long-range ordering.

Scaling hypothesis and Renormalization Group

As we have seen above, the correlation length diverges at a critical point. This led to the formulation of the so-called *scaling hypothesis*, stating that, close to a critical point, the correlation length ξ constitutes the only characteristic length of a system and that all other physical quantities of interest scale, under a change in the unit of length, as if it were coincident with the correlation length. This hypothesis can be formulated in many different ways, the easiest to understand stating that a quantity having a physical dimension (length)^D be directly proportional to $(\xi)^D$ close to a critical point. For instance, if we write the correlation function of fluctuations $\Gamma(r)$ in the Ornstein-Zernike form and take into account that the exponential factor is adimensional, we easily obtain that $\Gamma(r)$ has the dimensions (length)$^{-p}$, where $p = (d - 2) + \eta$. According to the scaling hypothesis, therefore, $\Gamma(r)$ will be directly proportional to $(\xi)^{-p}$, a circumstance used above in the discussion about the Ginzburg criterion.

The scaling hypothesis has been very useful in deriving relationships between critical exponents. The latter have had some success in accounting for a number of experimental data. However, as the scaling hypothesis is based only upon phenomenological considerations, it lacks a sound theoretical foundation. This unsatisfactory situation induced a number of physicists, at the end of the 1960s, to adopt a different line of reasoning, starting directly from the microscopic structure of a system close to a critical point. For this purpose, one has to take into account that the divergence in correlation length produces a long-range correlation between different microscopic components, so that their individual degrees of freedom (for instance, their individual positions or velocities) are no longer important in determining the macroscopic features of the whole system close to a transition point. One could then conceive of introducing a method for the progressive elimination of irrelevant degrees of freedom which, starting from the microscopic level, would arrive at the macroscopic one, so as to derive the behavior of relevant physical quantities close to the critical point in an unambiguous way, without resorting to any phenomenological hypotheses. The best way to implement such a method would be to use *coarse-graining*, that is, a suitable averaging of the microscopic degrees of freedom over a spatial region of a suitable size. Unfortunately, such a method cannot work alone, as the coarse-graining operation changes the physical description of the system. Namely, if we have a set of microscopic components interacting according to given laws, the "averaged systems" obtained from the latter interact amongst themselves according to laws which are different from those of the original components. A simple example will clarify how this can occur. Consider a very simple system, containing only four different components, placed along a line in four different positions, given by the coordinates 1, 2, 3 and 4, respectively. Thus the component located at coordinate 2 will have, as nearest neighbors, the components at 1 and 3, but not that at 4. Let us denote the value of the degree of freedom (which can be imagined as a sort of "spin") of the component at i by the symbol s_i. We now assume that the total interaction energy of the system is given by an expression of the form:

$$E = (1/2) \ \Sigma_{ik} \ s_i \ s_k$$

the sum being taken only over neighboring components. Such a law is very common in a number of important systems, such as ferromagnets or neural networks. In our simple example it can even be written in an explicit way as:

$$E = s_1 \ s_2 + s_2 \ s_3 + s_3 \ s_4$$

Let us now introduce an averaging over our system. More precisely, we substitute the components at points 1 and 2 with an averaged component S_a and the components located in the points 3 and 4 with an averaged component S_b. The values of S_a and S_b are given by:

$$S_a = (s_1 + s_2)/2 \ ; \ S_b = (s_3 + s_4)/2$$

We can now ask whether, for the new averaged system containing only the two averaged components, the energy can still be expressed as a sum of products over the degrees of freedom of neighboring components. Unfortunately, the answer is negative. Namely, if we were to assume that the new energy E' could be still written as:

$$E' = (1/2) \ \Sigma_{ab} \ S_a \ S_b$$

we would obtain, by substituting in the latter the explicit expressions of S_a and S_b, the form:

$$E' = (s_1 s_3 + s_2 s_3 + s_1 s_4 + s_2 s_4)/4$$

which is manifestly different from that for E written above.

These considerations suggest that the procedure of coarse-graining, if we want it give rise to a new system having laws similar to those of the original system, must be associated with suitable *rescaling* or *renormalization* procedures acting both on characteristic lengths and on coupling constants. The concatenation of all successive operations of coarse-graining, length rescaling and coupling constant renormalization has been denoted by the single name of *renormalization*. As the composition of two successive renormalizations still constitutes a renormalization and this composition fulfils the associative property, the set of renormalization operations has the algebraic structure of a *semigroup*. For historical reasons, however, the set is universally known as a *Renormalization Group* (RG).

Through the use of the RG we can now investigate the process which, starting from a microscopic description of a system (usually based on a suitable Hamiltonian or on a suitable free energy, such as the Landau free energy), leads us, through successive rescaling of the length unit, to its macroscopic description. This process can be viewed as being equivalent to a path within the space of system parameters (that is, of parameters entering into the definition of its Hamiltonian or its free energy), as any renormalization induces a shift from one set of parameter values to a different set of values of the same parameters. It is natural to ask whether this path admits *equilibrium points*, or *fixed points*, that is, parameter values which are left invariant by renormalization transformations. In the case where the rescaling of the length unit takes the form:

$$L' = b_0 L$$

where b_0 is a suitable scaling factor, the search for fixed points of RG transformations is made easier by the fact that it is possible to find an approximate expression giving the infinitesimal change of parameters resulting from an infinitesimal rescaling, that is, a rescaling in which b_0 is infinitely close to 1. This allows one to write a set of *RG equations*, which are simply a set of differential equations, one for each parameter, giving its rate of change with respect to the independent variable b_0 or, better, to the auxiliary variable $\theta = \log b_0$. Thus, the mathematical machinery of differential equations and of bifurcation theory can be used to locate the fixed points, to ascertain whether they correspond or not to critical points, and to compute the critical exponents. It also becomes possible to answer in a definite way the question of universality, related to the hypothesis that macroscopic features of phase transitions are independent of the detailed form of interactions between microscopic components. The method of RG equations showed that different *universality classes* exist, each including different systems, associated with different microscopic features, but with identical behaviors near the critical point. This approach has had great success both in accounting for experimental data and in giving a solid theoretical basis to every investigation of phase transition phenomena. This is confirmed by the large number of important publications on this subject (excellent books include Amit, 1984; Sewell, 1986; Goldenfeld, 1992; Benfatto and Gallavotti, 1995; Cardy, 1996; briefer and useful reviews include: Jona-Lasinio, 2001; Zinn-Justin, 2001; still valid are the critical considerations contained in Griffiths, 1981).

Avoiding any further mathematical detail about the RG method, we limit ourselves to recalling that, historically, it originated from the *Renormalization method* applied, in the 1950s, in Quantum Field Theory to eliminate the occurrence of divergent results in calculating quantities of physical interest. Only in recent times, however, have RG techniques been reformulated in a unified fashion, so as to transform every phase transition problem into a field-theoretical problem and vice versa (excellent reviews on this topic are: Gallavotti, 1985; Shirkov and Kovalev, 2001). These connections between different theoretical frameworks will be briefly examined at the end of this Chapter.

Microscopic models

Before leaving phase transitions we cannot avoid briefly mentioning the principal microscopic models, that is explicit models of interactions between microscopic components of a system for which there is direct proof of the existence, or the absence, of a phase transition. In general, the methods used for investigating the macroscopic features of these models are based on Statistical Mechanics. This subject is too complex to be dealt with in such a limited amount of space, so that, while referring to the literature already cited above within this Chapter, only some general aspects of these methods will be cited here. Consider, for the sake of simplicity, a system defined by a fixed and given number of components (particles) N, each one associated with a given position and a given momentum, and contained within a finite and fixed volume V. The $6N$ microscopic variables associated with coordinates and momenta of the N components define the *phase space* of this system. Every microscopic state of the latter corresponds to a point in phase space. Let us now further suppose that our system be driven by Hamiltonian dynamics, conserving the total energy and associated with a Hamiltonian $H(q, p)$. Here the shortened notation q, p denotes all $6N$ microscopic variables. It is evident that, when N is very large, there will be many different microscopic states of the system fulfilling our conditions, that is lying on the hypersurface of phase space between E and $E + dE$. Conceptually, each of these states could be imagined as corresponding to a different copy of the same system, so that we can speak of the set of these states as defining an *ensemble* (more precisely, a *statistical ensemble*) associated with the Hamiltonian under consideration.

It is obvious that the main problem consists of finding the *density* of this ensemble in the phase space. However, it can be shown that, if our system is in thermal equilibrium with its surrounding environment consisting of a (practically infinite) heat reservoir at temperature T, then the corresponding ensemble constitutes a *canonical ensemble* and its phase space density is given by:

$$\rho(q, p) = exp\,[-\beta H(q, p)], \quad \beta = 1/k\,T$$

In this case k denotes the usual Boltzmann constant. The total phase volume occupied by a canonical ensemble within phase space is called *partition function* and is defined by:

$$Z = \int \{ exp\,[-\beta H(q, p)] \}(1/N!\ h^{3N})\ d^{3N}q\ d^{3N}p$$

where h denotes a suitable constant on which we will not enter into further detail.

Once the partition function is computed, all the macroscopic Thermodynamics of the system can then be easily derived from it. Namely, it can be shown that the free energy F is connected to Z by the relationship:

$$F = -(1/\beta)\,\log Z$$

Thus, after finding F, all other thermodynamical quantities can be easily computed from the latter through standard relationships.

This discussion illustrates how the bulk of an investigation about a microscopic model lies in the computation of its partition function. This circumstance raises two formidable problems:

1) in most cases the practical computation of Z is very difficult, if not impossible; this forces one to resort to approximate methods (whose computational complexity, however, is generally high) which could become unreliable when the system is close to critical points;

2) it is possible to show that, for finite values of volume V, the function Z is devoid of any singularity. Therefore, it is impossible to deduce from Z the well known singularities of thermodynamical quantities in correspondence to critical points. It has, however, been proven (see Lee and Yang, 1952) that Z could acquire singularities when the volume tends to infinity (for an initial approach to this subject see Ruelle, 1969; Sinai, 1982). This would seem to imply that a microscopic theory of phase transitions would be possible only for infinite volumes and to a number of researchers this appears as a sort of paradox (all phase transitions occurring in the physical world are associated with a finite volume).

Some celebrated microscopic models are listed below.

a) The Ising model

This model is a sort of oversimplified description of a ferromagnetic system. It consists of a regular (generally cubic) lattice of points in d dimensions, each of which contains a "dipole", lying in one of two possible states: $+1$ ("spin up") and -1 ("spin down"). If s_i is the state of the dipole located at the lattice point whose label is i, the total energy of an Ising system in the presence of an external field H is given by:

$$E = - \sum_{ik} \varepsilon_{ik} \, s_i \, s_k - H \sum_i s_i$$

where the coupling parameters ε_{ik} generally have the form:

$$\varepsilon_{ik} = \varepsilon \text{ if points } i \text{ and } k \text{ are nearest neighbors}$$
$$\varepsilon_{ik} = 0 \text{ otherwise (or if } i = k)$$

If $\varepsilon > 0$ we speak of a *ferromagnetic* model. In the contrary case the model is called *antiferromagnetic*. In the former case the ordered phase consists in a situation in which all dipoles lie in the same state. It is possible to show with relatively simple arguments that, when $d = 1$, no ordered phase can exist for $T > 0$. Therefore, in a unidimensional Ising model no phase transition is possible. The model is exactly soluble even when $d = 2$ and in this case a very complex proof that it allows for a phase transition leading to an ordered phase under a given critical temperature was given by L. Onsager many years ago. The method used by Onsager represents a milestone in the history of modern Statistical Mechanics.

b) The classical Heisenberg model

Similar to the Ising model described above, the only difference being that at any lattice site the dipole state is represented by a 3-dimensional vector which can point in any direction. A particular case of the Heisenberg model is given by the *classical XY-model*, in which the state of each dipole is represented by a 2-dimensional vector, constrained to lie within the x-y plane (hence the name). It has been shown that the two-dimensional version of the Heisenberg model, unlike the two-dimensional Ising model, does not allow for the existence of phase transitions to an ordered state. Even though there are many arguments supporting the conjecture that the 3-dimensional Heisenberg model allows for a phase transition to an ordered state at a given critical temperature, so far a rigorous proof of this assertion is still lacking.

The intrinsic limitations of the theory of phase transitions

The achievements in modern theory of critical phenomena, which we have tried to summarize in this section, show just how great a distance there is between the general principles, adopted in studying emergent phenomena, and the concrete possibilities of computing measurable quantities. Computation is possible by resorting to a mixture of approximate methods, heuristic arguments, often unreliable abstractions and experimental findings, which lead only to partial, context-dependent, imprecise results. It is therefore very dangerous to generalize the latter to situations very different from those for which the computations themselves and the associated models were designed. However, despite this, all theoretical and experimental effort so far has given rise to a general agreement about some general features of phase transitions which appear to be independent of particular contexts and theoretical frameworks. These features can be summarized as follows:

- the behavioral features of a system undergoing a phase transition are partly independent of the details of its microscopic structure;
- a phase transition can occur if and only if we are in the presence of a sort of cooperation between fluctuations, preventing them from spontaneous decay;
- a phase transition can be observed if and only if there are suitable conditions providing stability for the new phase;
- while it is possible to use the notions of "cause" and "effect" to study the behavior of a system close to a critical point, these same notions become progressively less useful as we approach it;
- perhaps most phase transitions correspond to symmetry breaking phenomena; however, the usual theory of phase transitions cannot describe what occurs at precisely a critical point of symmetry breaking;
- no phase transition is perfect, as it is influenced by defects, impurities, and so on; however, perhaps the presence of the latter is actually responsible for the occurrence of phase transition itself;
- the practical detection of a phase transition is possible only in the presence of the right indicators (heat capacity, for instance); however, in most systems there is still no rule for finding the best indicators to use,

so that some phase transitions go unnoticed; this occurs mainly in biological, cognitive and social systems;

- current models of phase transitions are heavily influenced by the actual structure of classical Physics, based on the Energy Conservation Principle; however, phenomena similar to phase transitions occur even in systems which do not fulfill this principle (such as Dissipative Structures); a generalization of the various current theories is therefore needed.

Following this list, it becomes clear that classical Physics (and classical Statistical Mechanics) are not compatible with many of these features. We are therefore forced to turn our attention to a very different theoretical framework: Quantum Theory and, in particular, Quantum Field Theory. The next section shows that within this framework it is possible to account for all phenomena associated with phase transitions, and hence all emergent phenomena, without the hindrance of the intrinsic limitations of classical Physics.

5.3 Quantum Field Theory

The expression "Quantum theory" is rather vague and, in concrete terms, corresponds to three different possible meanings: *semiclassical theories*, *Quantum Mechanics* (QM) and *Quantum Field Theory* (QFT). Semiclassical theories are very popular being known to most people, even to those devoid of a physico-mathematical education, and are often taught in high schools. They are merely the usual classical theories, supplemented with suitable quantization constraints. Amongst the most famous are the Planck formula:

$$E = N h \nu$$

where $N = 0, 1, 2, \ldots.$ is an integer, E the energy, ν the frequency and h the usual Planck's constant, and the *Bohr-Sommerfeld* quantization condition:

$$m \nu r = N \hbar$$

The attribute "quantum" itself arose within the context of semiclassical theories. The latter, however, have a limited range of validity and can mainly be considered only as practical recipes for approximate computations without a sound theoretical foundation.

Turning now to QM, (excellent textbooks among the multitude published on this subject include: Davydov, 1976: Cohen-Tannoudji *et al.*, 1977; Landau and Lifshitz, 1981; Sakurai, 1995; Merzbacher, 1998, Messiah, 2000) it can be characterized insofar as it deals with systems consisting of finite and fixed numbers of particles, contained within finite and fixed volumes of space (even though infinite volumes are allowed). Physical quantities characterizing these particles, however, cannot all be measured simultaneously with arbitrary precision. A first consequence of this uncertainty is that a complete characterization of a particle dynamical state with unlimited precision is impossible. Whilst avoiding any technical exposition of the mathematical formalism of QM, we limit ourselves to mentioning that this uncertainty is taken into account by describing physical quantities through suitable *operators*, instead of resorting to real or complex-valued functions, as in classical physics. If the system state φ is an *eigenstate* of operator $\hat{A}$ associated with a given physical quantity, we have the obvious relationship:

$$\hat{A}\,\varphi = a\,\varphi$$

where a is an *eigenvalue* of $\hat{A}$. Here a gives the value of the physical quantity considered which would be obtained by its measurement when the system is lying in a state coincident with eigenstate φ. Unfortunately, in most cases the system lies in states which are not eigenstates of the quantity being monitored. Its generic *state vector* ψ, however, can always be expanded in terms of a complete set of the eigenstates of a given operator describing a suitable physical quantity. More precisely, assuming this operator has a discrete spectrum, that is a countable (even though infinite) number of different eigenvalues and φ_i to be the eigenfunction of the operator describing the eigenstate associated with the i-th eigenvalue, we can express the generic state vector as:

$$\psi = \sum_i c_i\,\varphi_i$$

This formula defines a particular *representation* of the system's state vector. The absolute value squared of coefficient c_i gives the probability that a measurement of the physical quantity under consideration, when the system state is described by ψ, will give as outcome just the i-th eigenvalue. Moreover, it can be said that the state ψ is a (linear) *superposition* of states φ_i. In a sense, a representation can be viewed as the expression of a vector, in a suitable infinite-dimensional space, in terms of a suitable *basis*, given by the infinite set of vectors φ_i. Such a circumstance is possible owing to the fact that the eigenfunctions of every operator used in QM constitute an *orthonormal set*.

From these definitions it follows that the generic state vector of a quantum system can allow for a number of different representations. Namely, if the operators corresponding to two different physical quantities do not commute, that is if their commutator is different from zero, then they are associated with different sets of orthonormal eigenfunctions and hence can be used for obtaining two different representations of the state vector. This is the case, for instance, for the operators associated with the position and momentum of a particle, whose commutator is nonvanishing (a circumstance related to the fact that the uncertainties associated with these two quantities are related by the Heisenberg Uncertainty Principle). We thus have two different representations of the state vector, one based on eigenfunctions of the position operator (the so-called *q-representation*), and the other based on eigenfunctions of the momentum operator (the so-called *p-representation*).

In rough, non-technical terms, the choice of a representation consists of selecting a subset of dynamical variables describing the state of the system, such that all variables belonging to the subset can be measured simultaneously with arbitrary precision. In a sense, owing to the uncertainty principle and its generalizations, every representation can offer only a *partial* description of the system's dynamics. However, an important theorem, proven by Von Neumann (*cfr.* Von Neumann, 1955), states that in QM all possible representations are reciprocally *equivalent*. More precisely, we can go from one representation to another through the application of a *unitary transformation*, that is of a transformation not altering the length of the state vector. This means that different representations give rise to the same values of probabilities of occurrence of the results of all possible measurements related to the physical system under consideration, independently of the particular representation chosen (for a more recent discussion on this topic see Heylighen, 1990; Halvorson, 2001; Rédei and Stölzner, 2001; Halvorson, 2004).

A second consequence of uncertainty is that particles can no longer be considered as localized objects (as in classical physics), but rather as diffuse entities, virtually spread over entire space. Thus, given a particle, we can only speak of its most probable location at a given time remembering, however, that in any position in space (within the allowable space volume) there will be a nonzero probability (even though in most cases very small) of finding the particle. Besides, given any two particles, independently of the distance between their most probable locations, and a generic spatial location, there will be always a nonzero probability of finding them simultaneously at this location. Physicists often speak of *waves* associated with the particles, and of nonzero *overlap* of the waves associated with two different particles in every spatial location. What is important, is that such overlap gives rise to a correlation (even though very small) between the behaviors of the two particles. Such a correlation is, from any point of view, entirely equivalent to an *interaction* (of a quantum nature) between the two particles, independently of the fact that there exist other interactions between them mediated by suitable force fields.

The aforementioned consequences of uncertainty entail the occurrence of a typically quantum phenomenon, called the *Bose-Einstein condensation* (BEC). The latter occurs under suitable conditions (e.g., low temperatures) and consists in the fact that a macroscopic fraction of all particles belonging to a given system fall simultaneously into the same quantum state (coincident with system's ground state). The "condensation", however, does not refer to geometrical space but rather to momentum space; in other words, all particles belonging to the "condensate" have the same momentum, whose value is zero (or ground state momentum). This implies that their behaviors all become correlated, or synchronized, thus giving rise to a macroscopic state, which appears globally as *coherent*. BEC can be considered as the prototype of the formation of macroscopic entities emerging, as collective effects, from the laws governing the behavior of microscopic particles. It is to be stressed that BEC can occur only in a quantum system. Namely, within a classical system, the statistical distribution of particle energies would follow the Maxwell-Boltzmann distribution:

$$N_i = g_i \, exp[(\mu - \varepsilon_i)/T]$$

where N_i denotes the number of particles having energy ε_i, μ is the chemical potential and g_i a suitable statistical weight. By assuming that the ground state is associated with zero energy, one directly obtains from the above formula that the fraction of particles having zero energy is given, in classical physics, by:

$$N_0 / N = g_0 \, exp(\mu / T)/\{ \textstyle\sum_i g_i \, exp[(\mu - \varepsilon_i)/T] \} = g_0 / [\textstyle\sum_i g_i \, exp(- \varepsilon_i / T)]$$

Now, mathematical reasoning shows that the denominator of this equation takes the form:

$$\textstyle\sum_i g_i \, exp(- \varepsilon_i / T) = B \, V \, T^{3/2}$$

where B is a suitable constant and V is the total volume occupied by the system. As a consequence, the fraction of particles having zero energy will be given by:

$$N_0 / N = g_0 / B \, V \, T^{3/2}$$

We now assume, as is usual in Thermodynamics, that a quantity can be considered as *macroscopic* if its value remains finite when the volume tends to infinity. It is then easy to see that, for every nonzero value of temperature T, when we let $V \to \infty$ in the previous expression , the value of N_0 / N tends to zero. In other words, in classical physics, the fraction of zero energy particles is vanishingly small for every nonzero temperature and BEC cannot occur. A similar reasoning, which will not be repeated here, shows inversely, that the law of Bose-Einstein statistics, obeyed by quantum particles with integer spin, gives rise, in the limit $V \to \infty$, to a nonvanishing value of N_0 / N.

These advantages of QM are, however, counterbalanced by a number of shortcomings. The main ones can be listed as follows:

2   the Von Neumann theorem excludes any application of QM to the description of *structural changes*. Namely, as all representations are physically equivalent, it is impossible to have a model based on QM in which a system can exist in two different, not equivalent, forms corresponding to two different structural arrangements. Therefore, within QM, it is , in principle, impossible to formulate a theory of *phase transitions* and, *a fortiori*, of emergent phenomena;

3   the occurrence of macroscopically coherent states, as in BEC, is, in most cases, hindered by the interactions of the system under study with the external environment. As a matter of fact, such interactions act in such a way as to destroy the quantum coherence (see Giulini *et al.*, 1996) and, if the *decoherence time* is short enough, macroscopic coherence, due only to QM, becomes unobservable. In a sense, the states which are a superposition of basic states can no longer exist, as the interaction with the environment selects one particular basic state among the various possible states and the system falls into it with a probability equal to one, its dynamics thus losing its quantum character. This limits the usefulness of QM to specific cases (such as the world of atoms or molecules, very low temperatures, and so on).

This effect of decoherence is very strong. By resorting to standard formulae (for a detailed discussion, see Tegmark, 2000) it is possible to estimate that the decoherence time τ within a system whose elementary components undergo mutual collisions is given by:

$$\tau = 1/(N \, n <\sigma v>)$$

Here the brackets $<...>$ denote an average over the thermal velocity distribution at a given temperature T, σ is the scattering cross section for mutual collisions, v is the component velocity, N is the total number of components and n their volume density. We assume, for the sake of simplicity, that system components are charged particles, each having a charge equal to the electron charge e and a mass equal to the proton mass m. In this case the scattering cross section is given approximately by:

$$\sigma = (g\,e^2/\,m\,v^2)^2$$

The symbol g denotes the usual Coulomb constant. If we now assume that the average velocity is given by $v = (k\,T\,/\,m)^{1/2}$ (here k denotes Boltzmann's constant) and that the total number of components N coincides with Avogadro's number, then these formulae lead, after some algebraic and numerical computations, to the following final formula for the decoherence time:

$$\tau \approx 5\mathrm{x}10^{-17}\,T^{3/2}\,/\,n$$

Under normal conditions T is of the order of 300 K and n is about $3\mathrm{x}10^{25}$ particles m^{-3} and therefore $\tau \approx 8\mathrm{x}10^{-39}$ s. This value is to be compared with the average time taken by a semiclassical electron to make a single orbit around a nucleus, which is of the order of 10^{-16} s. We can thus see that the behavior of such a system is practically classical, as the decoherence time is much smaller than the characteristic time associated with the dynamical processes occurring within such a system. Of course, when dealing with systems whose density is very low, quantum coherence effects become important. If, for instance, while keeping constant the temperature, we assume that n is of the order of unity (as in interstellar gas), we obtain $\tau \approx 2.5\mathrm{x}10^{-13}$ s, much greater than the dynamical time quoted above. It should be noted, however, that a small decoherence time is not necessarily a criterion signalling the need for a classical treatment of the system under consideration. In a number of cases it merely shows the inadequacy of QM and the need for resorting to a better quantum framework, such as that provided by QFT (see Alfinito *et al.*, 2001).

These drawbacks prevent any application of QM to a description of intrinsic emergence, and suggest that perhaps QFT could be a better framework. The latter, contrarily to what happens in QM, assumes that the main physical entities are *fields* (of force) and not particles. Such a standpoint has a long tradition in physics, from Faraday and Maxwell onwards, and gave rise to the most powerful architectures ever built in theoretical physics, such as Einstein's General Relativity and unified gauge theories. Within this framework, the world is populated by fields (of force), and the concept of particle is considered as nothing but an auxiliary concept to denote regions of space where a field has a particularly high intensity. The laws ruling physical phenomena coincide with *field equations*, driving the space-time evolution of fields.

Following such an approach, QFT attempts to deal with fields as defined by uncertain quantities. As the fields, in principle, are not restricted to definite volumes, QFT typically deals with infinite volumes. In this way it becomes easier to introduce a sharp distinction between *macroscopic* phenomena (those surviving when the volume tends to infinity) and *microscopic* phenomena (which appear as fluctuations when the scale of observation is large enough). Of course, the approach followed by QFT is very difficult to implement in a practical manner, much more difficult than in the case of QM. Amongst the sources of these difficulties we can cite the fact that, while quantum uncertainty introduces a typically *nonlocal* form of interaction (partly responsible for quantum coherence phenomena), field dynamics must obey a causality principle based only on *local* interactions. For this and other reasons, QFT can still be considered an incomplete theory, of which only particular cases have so far been worked out, valid in specific domains, and at the expense of introducing some very complex mathematical machinery. Moreover, some general conceptual problems raised by this approach still remain unsolved (see, for instance, Cao, 1999; Celeghini and Rasetti, 2000; Clifton and Halvorson, 2001; Halvorson, 2002; Kuhlmann *et al.*, 2002).

Despite these difficulties (which have prevented a widespread diffusion of the ideas and methods of QFT within domains different from the physics of many-body systems), QFT, first proposed in 1926 by P. A. M. Dirac, has obtained over the last fifty years remarkable success in describing and forecasting phenomena occurring within the domains of particle physics and condensed matter physics (a collection of reprints of the main papers which marked the historical development of QFT is contained in Schwinger, 1958). Quantum electrodynamics, the explanation of the laser effect, the unified theory of weak and electromagnetic forces are amongst the most remarkable achievements of QFT, marking the whole history of physical research in the twentieth century (important QFT textbooks include Itzykson and Zuber, 1986; Umezawa, 1993; Peskin and Schroeder, 1995; Weinberg, 1995; 1996; Huang, 1988; Kiselev *et al.*, 2000; Stone, 2000; Lahiri and Pal, 2001; Maggiore, 2005). Here we limit ourselves to mentioning one important feature of QFT, which is central to our discussion about emergence: within QFT, unlike in QM, there is the possibility of having different, *non-equivalent*, representations of the same physical system (*cfr.* Haag, 1961; Hepp, 1972). As each representation is associated with a particular class of macroscopic states of the system (via quantum statistical mechanics) and this class, in turn, can be identified with a particular thermodynamical *phase* of the system (for a proof of the correctness of such an identification, see Sewell, 1986), we are forced to conclude that *only* QFT allows for the existence of different phases of the system itself. A further consequence is that only within QFT can we deal with *phase transitions*, i.e., with global structural changes of the system under study. Before closing our brief discussion of QFT, we stress that the existence of non-equivalent representations is strictly connected to the fact that, if we interpret quantum fields as equivalent to sets of suitable particles lying in suitable dynamical states, then QFT describes, even in its simplest implementations, situations in which the total number of particles is *no longer conserved*. In other words, within QFT (and *only* within QFT) are the processes of *creation* and *destruction* of particles allowed. This gives QFT a descriptive power enormously greater than that of QM, where the number of particles is constant.

On this point, it is instructive to understand the basic reason why QFT implies the existence of non-equivalent representations. Here we use a simple argument, introduced by Umezawa *et al.* (Umezawa *et al.,*1982, pp. 18-19) which is based on a particular picture of the state of a quantum system consisting of a number of interacting elementary components: the so-called *occupation number picture*. This assumes the existence, once a given representation (in the sense previously discussed) has been adopted, of an infinite (and countable) number of different possible states for the system under consideration, each one being an eigenstate of a given operator, associated with the chosen representation. Each of these states describes a possible state of an elementary component (that is, of a particle) of the system considered, labelled with the eigenvalues of the operators associated with the physical quantities (for instance position, spin or momentum) chosen for the adopted representation. An instantaneous state of the system can then be described through an ordered list of values (that is a vector) specifying, for each state, the number of elementary components lying in that state. Usually the instantaneous state is symbolically denoted (following the conventions introduced by Dirac) as:

$$|n_1, n_2, \ldots, n_i, \ldots\rangle$$

In principle, this vector has an infinite and countable number of elements. The symbol n_i denotes the number of elementary components (particles) lying, at the chosen instant of time, in the i-th state. This number is generally referred to as the *occupation number* of the considered state.

Recalling the fact that the possible values of occupation numbers depend upon the statistics obeyed by the (quantum) elementary components, we have two possibilities: either these components follow *Bose-Einstein statistics* (and in this case the occupation numbers can assume any non-negative integer value) or they follow *Fermi-Dirac statistics* (in which case the occupation numbers can assume only two possible values: 1 or 0). Let us now assume, for the sake of simplicity, that the elementary components obey Fermi-Dirac statistics (as is the case, for instance, for electrons or protons). In this case the instantaneous state vector will consist of a list (of infinite length) of binary numbers (that is 1 or 0).

Let us now ask ourselves how many different state vectors of this kind could in principle exist. The answer is easy: their number is equal to the number of different possible infinite sequences of 1s and 0s. But, from Cantor's theory of infinite sets, we know that this number is not countable, having the power of continuum. On the other hand, we must take into account that each representation used in QM has only the power of a countable set. Namely, each one of them is associated with a countable set of eigenfunctions. We are therefore forced to conclude that the whole set of these vectors will never be fully represented through countable sequences, and this entails the possibility of the existence of different representations of the set of state vectors which are not unitarily equivalent to one another. Namely, as each representation can capture only a limited subset (theoretically of zero measure) of the whole space of state vectors, it is possible to have two representations which, referring to different subsets, cannot be related to one another through suitable unitary transformations.

We can now summarize the previous discussion by listing two main arguments for choosing models based on QFT for modelling intrinsic emergence:

- every interaction between the microscopic components of a given system can be viewed as being mediated by suitable fields (of force). Besides, at the microscopic level, the presence of unavoidable uncertainties, due both to the very nature of the measurement operations and to interactions with an unpredictable and noisy environment, forces us to adopt a quantum, rather than a classical, description. As a consequence, QFT is the only correct framework for describing systems consisting of many microscopic components (such as the brain, neural networks, multiagent societies, ferromagnets, and their like);
- only within QFT do we have the existence of physically non-equivalent representations of the same system, and hence the possibility of describing structural changes such as, for instance, phase transitions. Each phase can then be viewed as a macroscopically coherent entity which, providing suitable conditions be met, could even be stable with respect to external perturbations.

The *a priori* arguments offered above for choosing QFT as the natural framework for modelling intrinsic emergence will be now substantiated by an *a posteriori* proof: the exhibition of a model of intrinsic emergence based on QFT.

5.4 Spontaneous Symmetry Breaking

Let us start by stressing that in most models we can introduce a function which plays a role analogous to that of *energy* in physical systems, so that stable and metastable equilibrium states are directly associated with the local minima of such a function. The occurrence of intrinsic emergence can then be identified with a transition, triggered by a change in the value of a given parameter, in which (at least) one local energy minimum is split into a number (finite or infinite, but always greater than one) of different local energy minima, all of which are *equivalent*, i.e., characterized by the same value of minimum energy (we refer to these states as *ground states*). Intrinsic emergence derives from the fact that, if the system was, before the transition, in the state corresponding to the old energy minimum, the transition will certainly provoke the settling of the system into one of the new energy minima although we cannot forecast *which* of them will be chosen on the basis of the model we have, because all minima are equivalent to one another. This kind of transition is usually called *spontaneous symmetry breaking* (SSB). We have already encountered this expression when dealing with the Landau theory of phase transitions. In a sense, SSB is merely the phenomenon placed by Landau at the basis of every phase transition. The attribute "spontaneous" was introduced to stress the fact that SSB arises only as a consequence of the change in a parameter value, and not as a consequence of a change in the form itself of the expression defining system energy deriving, for instance, from an external input or from the insertion of a new interaction term.

Regarding SSB, some remarks are in order. The first is that, in most cases, the multiplicity of ground states exists only if we go to the infinite volume limit. Namely in finite systems such an effect is often hidden by the existence of suitable boundary conditions. The most well-known case is that of the transition from paramagnetic to ferromagnetic phase, in which the unique ground state of paramagnetic phase is split, in the ferromagnetic phase, into two equivalent ground states: one with all spins parallel to the inducing external field, and the other with all spins antiparallel. However, the equivalence of these two ground states holds only in the infinite volume limit, and disappears when the volume of magnets becomes finite (as is the case in the real world), so that the ground state with all spins parallel becomes the favoured one.

Secondly, both in classical and QFT-based descriptions of SSB, the system will in any case be forced to choose one particular ground state. States corresponding to linear combinations of different ground states are not allowed, even in QFT, because it can be proven that any operator connecting two different ground states vanishes at the infinite volume limit (*cfr.* Huang, 1987).

A third remark is that, if we describe SSB within the context of QFT, the occurrence of SSB implies the appearance of collective excitations, which can be viewed as zero-mass particles carrying long-range interactions. They are generally called *Goldstone bosons* (*cfr.* Goldstone, Salam and Weinberg, 1962; for a more general approach see Umezawa, 1993). Such a circumstance endows systems, in which SSB takes place, with a sort of *generalized rigidity*, in the sense that, acting upon one side of the system with an external perturbation, such a perturbation can be transmitted to a very distant location, essentially unaltered. The reason for the appearance of Goldstone bosons is that they act as *order-preserving* messengers, preventing the system from changing the particular ground state chosen at the moment of the SSB transition. On the other hand, it should be remembered that this kind of change would be equivalent to a further phase transition. Namely, it can be shown that, within QFT, every ground state is associated with a particular representation (in the sense specified in the previous Section) and, in the case of the existence of a multiplicity of ground states, two representations associated with two different ground states are unitarily non-equivalent.

Moreover, we stress that Goldstone bosons are a direct manifestation of intrinsic emergence, as none of the forces acting between a system's elementary constituents is able to produce generalized rigidity. We thus have that only within a QFT-based description of SSB can we observe the occurrence of macroscopic coherent entities which are stable with respect to external perturbations. We further recall that Goldstone bosons themselves can undergo BEC, thus giving rise to new macroscopic objects and to new emergent symmetries (*cfr.* Wadati *et al.*, 1978a, 1978b; Matsumoto *et al.*, 1979; Umezawa, 1993).

It is instructive to consider a simple example, in order to understand both the limitations inherent in the framework of classical physics and the fact that Goldstone bosons appear only if we adopt a quantum framework. Consider a simple field, described by a single scalar quantity φ which is a function of space-time coordinates but, unlike the previous examples, associated with values represented by *complex numbers*. This assumption makes φ very similar to the well-known *wavefunctions* used in standard QM. In order to describe the dynamics of this field we resort, rather than to explicit dynamical equations, to a Lagrangian function L. As already illustrated in Section 2 of Chapter 4, the latter coincides with the difference between the kinetic energy and the potential energy of the system under consideration. Within our example we choose for L an explicit form of the kind:

$$L = \sum_\mu (\partial \varphi^*/\partial x_\mu)(\partial \varphi/\partial x_\mu) - a\,\varphi^*\,\varphi - b\,(\varphi^*\,\varphi)^2$$

Some explanations of the symbols introduced are in order. First of all, the symbol φ^* denotes the complex conjugate of φ. In the second place, the space-time coordinates are represented in a unified fashion (following notational conventions often used in Special Relativity Theory). More precisely and for the sake of simplicity, we allow for only two kinds of coordinates: a spatial coordinate x (so that our system exists in 1-dimensional space) and a time coordinate t. In the symbol x_μ the index μ takes one of only two possible values: 0 and 1. When $\mu = 0$, we assume that $x_0 = i\,c\,t$, where i is the imaginary unit and c the velocity of light in vacuum, while, when $\mu = 1$, we assume that $x_1 = x$. As regards the parameter b, we assume its value to be real and positive, whereas a is allowed to vary, as in Landau theory (for instance, as a function of some other parameter such as temperature). In particular, the value of a, although still being real, can be positive, null or negative. It is evident how it plays the role of a *critical parameter*.

From the previous Lagrangian it is possible to derive suitable dynamical field equations. However, as the field φ is a system differing profoundly from simple systems such as a point particle or a pendulum, being associated with an infinite (and continuous) number of degrees of freedom (as we must specify the value of the field in each space point), we are forced to derive the field equations by resorting to a suitable generalization of Lagrange's equations, as already introduced in Section 2 of Chapter 4. These generalized Lagrange equations can be readily found in any textbook on the mechanics of continua and have the form:

$$(\partial L/\partial \psi) - \sum_\mu \partial/\partial x_\mu [\partial L/\partial(\partial \psi/\partial x_\mu)] = 0$$

Here the symbol ψ denotes a generic independent component of the field describing the continuum under consideration. In our case we have two such independent components: φ^* and φ. We will thus have two different field equations, one taking $\psi = \varphi^*$ and the other taking $\psi = \varphi$. Trivial computations, starting from the Lagrangian introduced above and generalized Lagrange equations, show that these two equations are equivalent to one another, as one is the complex conjugate of the other. This allows us to retain only one of them, with which all the following analyses will be carried out. Its explicit form is:

$$- \Sigma_\mu \, \partial^2 \varphi / \partial x_\mu^{\,2} - a \, \varphi - 2 \, b \, (\varphi^* \, \varphi) \, \varphi = 0$$

As our model is rather abstract, we will not deal with the question of the physical interpretation of this equation. We recall, however, that the terms without derivatives strongly resemble those appearing in the right-hand side of the Ginzburg-Landau equation, already encountered above in Section 2 of this Chapter. Such a resemblance is not fortuitous, as the kind of non-linearity represented by these terms constitutes the best-known and most studied recipe for producing bifurcation and symmetry-breaking phenomena.

It is possible to prove, through direct substitution, that this equation is invariant in form, with respect to the following transformations of the field variables:

$$\varphi' = \varphi \, exp \, (i \, \theta) \, , \quad (\varphi^*)' = \varphi^* \, exp \, (- i \, \theta)$$

Here θ denotes the so-called *parameter* of the transformation, specifying the amount of change to be performed on the indicated variables. Usually these transformations are called *global gauge transformations* and we can say that the dynamical equation (or the Lagrangian) of our model is endowed with a *symmetry* with respect to these transformations.

Now, let us search for the ground states of our system. The latter coincide with the minima in potential energy and are therefore given by the minimum values of homogeneous stationary equilibrium solutions of the dynamical equation. In order to obtain them, we must put all derivatives equal to zero and solve the equation:

$$- a \, \varphi - 2 \, b \, (\varphi^* \, \varphi) \, \varphi = 0$$

It can be immediately seenthat, depending upon the sign of parameter a, two different situations can occur: 1) $a > 0$; 2) $a < 0$. The value $a = 0$ corresponds to the *critical value* separating 1) from 2). In situation 1) the previous equation allows for one and only one solution, given by $\varphi_0 = 0$. This solution corresponds to the absolute minimum of potential energy and it can be clearly seen that its value is invariant with respect to global gauge transformations. There is, therefore, no symmetry breaking, as both the dynamical equation and the ground state are endowed with the *same symmetry*. Moreover, in this case the ground state is not degenerate, but unique.

On the contrary, situation 2) is very different. In this case we have two possible solutions of the above equation: one still given by $\varphi_0 = 0$ and the other by the (infinite) set of values satisfying the relationship:

$$\varphi^*_0 \, \varphi_0 = |a| / 2 \, b$$

It is possible to prove that only the latter are absolute minima in potential energy, as the solution $\varphi_0 = 0$ corresponds to a local maximum in this energy. We have thus obtained that, when $a < 0$, there is an infinite set (a continuum) of different ground states, all associated with the same energy. Within the complex plane they lie on a circumference having the centre on the origin and a radius given by $\{|a| / 2 \, b\}^{1/2}$. They are not invariant with respect to global gauge transformations, as each of them transforms one ground state into another, different, one. Thus, in this case we have a *symmetry breaking* as, although the dynamical equation is symmetric (invariant in form) with respect to global gauge transformations, its ground state solutions are not. Moreover, the ground state is characterized by an infinite number of degenerate states.

So far, our reasoning has been entirely classical and no quantum feature has been taken into consideration. This would seem to suggest that a classical framework would be enough to gain all relevant information on the dynamical behavior of our system when moved away from a given ground state by an external perturbation. In order to test such a hypothesis, let us start by writing the dynamical equation describing a small perturbation ξ around the unique ground state $\varphi_0 = 0$ in situation 1), that is when $a > 0$. After suitable computations and a linearization, the equation takes the form:

$$\partial^2 \xi / \partial t^2 = c^2 \, \partial^2 \xi / \partial x^2 + c^2 \, a \, \xi$$

Recalling that, owing to the complex nature of the field φ, even ξ must be considered as a complex-valued quantity, this allows us to search for the solution of this equation through an *ansatz* of the form:

$$\xi = exp \, (\lambda \, t) \, exp \, (i \, k \, x)$$

By substituting the latter into the above equation it is possible to obtain the following relationship between λ and k :

$$\lambda^2 = - c^2 \, k^2 + c^2 \, a$$

When a assumes the critical value $a = 0$, it can be clearly seen that, for $k = 0$, then $\lambda = 0$. Remembering what has already been said in Section 2 of this

Chapter on the Ginzburg-Landau equation, we can state that here we have the appearance of classical Goldstone bosons. This is, however, a modest result. Namely, when $a > 0$, the above relationship shows that for $k = 0$, we have $\lambda^2 > 0$ and that, when $k^2 > a$, then $\lambda^2 < 0$, that is, λ is purely imaginary. Such conditions do not tell us whether the ground state is linearly stable or not. All we can say is that there are constant amplitude oscillations around the ground state for great enough values of k and an unknown behavior (in principle the sign of λ, when $k^2 < a$, is indeterminate) associated with long-wavelength perturbations (remember that k is the inverse of a wavelength).

Entirely similar results can be obtained in the case of perturbations of one of the ground states arising in situation 2), that is when $a < 0$. Namely the dynamical equation describing small perturbations has the form:

$$\partial^2 \xi / \partial t^2 = c^2 \, \partial^2 \xi / \partial x^2 + 2 \, c^2 \, |a| \, \xi$$

which is qualitatively similar to that of the equation holding for small perturbations of the unique ground state in situation 1). In both situations, therefore, the traditional methods used when adopting the classical framework are poorly suited to investigate the dynamical behavior of a ground state following a small perturbation.

We now follow a different strategy to investigate what happens in situation 2). Let us express the field φ, close to a given ground state, as a sum of the ground state itself plus a suitable perturbation. As we are free to choose any ground state we like, provided it fulfills the relationship $\varphi^*_0 \, \varphi_0 = |a| / 2 \, b$, let us adopt as our reference ground state the one corresponding to the case in which φ_0 has no imaginary part and is therefore given by $\varphi_0 = \{|a| / 2 \, b\}^{1/2}$. Moreover, as the perturbation can still be considered as a complex quantity, it can be written by separating in an explicit way the real part from the imaginary one. Thus, φ can be written as:

$$\varphi = [\{|a| / 2 \, b\}^{1/2} + \varphi_1 + i \, \varphi_2]$$

Here φ_1 and φ_2 denote, respectively, the coefficient of the real and imaginary parts of the perturbation. Let us now substitute this development directly into the dynamical field equation describing φ. After straightforward computations we obtain, by separating the real from the imaginary terms, that the two auxiliary fields φ_1 and φ_2 can be described by the following two dynamical equations:

$$- \sum_\mu \partial^2 \varphi_1 / \partial x_\mu^2 - 2 \, |a| \, \varphi_1 = 0 \, ; \quad \sum_\mu \partial^2 \varphi_2 / \partial x_\mu^2 = 0$$

If we take into account the notational conventions adopted at the beginning of this example, it is easy to see that the second equation coincides with the well-known *D'Alembert equation* describing, amongst other things, the propagation of electromagnetic or sound waves as well as a vibrating string. In order to interpret the meaning of the latter, we now adopt a quantum framework. Within this framework, the wavefunction ψ of a particle of zero spin and mass M, obeying Bose-Einstein statistics and the principles of Special Relativity, is a solution of the *Klein-Gordon equation*:

$$\sum_{\mu} \partial^2 \psi/\partial x_{\mu}^2 - (M^2 c^2/\hbar^2) \psi = 0$$

Then a comparison between the Klein-Gordon and the D'Alembert equation shows that the latter describes the behavior of bosonic particles with zero mass (and zero spin in our case).

We must now remember that within a quantum framework every interaction can be seen as the exchange of suitable *quanta*, consisting of bosonic particles. The celebrated Heisenberg Uncertainty Principle allows us to derive a simple relationship between the mass M of each quantum and the range R of the interaction produced by these exchanges. Such a relationship has the simple form:

$$R = \hbar/ M c$$

This entails zero mass particles being associated with infinite-range interactions. We can thus conclude that the equation for φ_2 is merely describing zero mass bosonic particles carrying infinite-range interactions: the *Goldstone bosons*!

We have thus seen that the adoption of a quantum interpretation brings to light the presence of Goldstone bosons which could never be detected, as we have seen, by resorting only to a classical approach. Of course, they arise as a consequence of the symmetry breaking, as their presence is associated only with situation 2). Namely, in situation 1), by repeating the previous computations, they would be characterized by a finite, but imaginary mass. We underline that, at first sight, the Goldstone bosons could appear as devoid of any physical meaning, being described by the imaginary part of a field. As a matter of fact, this was the opinion of a number of physicists for many years. However, when looking at the form of the Lagrangian, it can be immediately seen that the field energy (and hence any physical effect) does not depend upon the fields themselves, but rather on products such as $\varphi^* \varphi$, which always take real values. Moreover, they are affected by the value of the coefficient of the imaginary part, which thus directly contributes to produce physical effects: Goldstone bosons are no less real than the particles constituting a given quantum system. They provide the main support for the phenomena of macroscopic coherence arising as a consequence of a symmetry breaking (for a review see, for instance, Burgess, 2000). More precisely, within QFT it can be shown that all these phenomena can be viewed as being due to a Bose-Einstein condensation of Goldstone bosons. Moreover, the interaction between Goldstone bosons explains the existence of macroscopic quantum objects (Umezawa, 1993; Leutwyler, 1997).

To be complete, however, we must recall that simple descriptions of SSB phenomena exist based solely upon classical physics (see, for instance, Greenberger, 1978; Sivardière, 1983; Drugowich de Felício and Hipólito, 1985). We are therefore forced to ask ourselves whether classical models of SSB, not based upon the framework of QFT, can give rise to Goldstone bosons and to generalized rigidity. In order to answer such a question, let us consider a classical system of individual particles described by the laws of classical mechanics, whose global potential energy has a form such that it undergoes a SSB at a suitable value of a given critical parameter. Let now focus on what happens when, after the occurrence of a SSB transition, the system relaxes into one of its many, equivalent, ground states. Such a relaxation process is described, from a microscopic point of view, by a part of theoretical physics usually called *Kinetics*. The latter provides a description of relaxation process when individual particles obey classical mechanics and, as a consequence, their statistical behavior is ruled by the well known *Maxwell-Boltzmann* probability distribution. These results agree fairly well with experimental data on the behavior of particular types of ideal gases, and provide a derivation of the macroscopic form of the celebrated *Second Principle of Thermodynamics*, holding for these gases (now known as the *H-theorem*).

Notwithstanding the partial successes achieved by kinetics based upon classical statistical mechanics, for a long time many physicists were deeply dissatisfied as these successes were based on equations (such as the well-known *Boltzmann transport equation*) relying heavily on simplifying approximations introduced merely on the basis of phenomenological or *bona fide* arguments. In other words, classical kinetics was not grounded on general principles. Such a situation changed after the Second World War, when N. Bogoliubov was able to show that all classical kinetics, including the form itself of the Maxwell-Boltzmann distribution, could be derived only on the basis of two general principles:

pb.1) the validity of the laws of classical mechanics;

pb.2) a principle ruling the space-time behavior of an aggregate of classical particles, known as the *correlation weakening principle*.

Without entering into technical details (*cfr.* Rumer and Ryvkin, 1980; Akhiezer and Péletminski, 1980; De Boer and Uhlenbeck, 1962) the correlation weakening principle states that, during a relaxation process, all long-range correlations between the individual motions of single particles tend to vanish when the volume tends to infinity (occurring, of course, when time tends to infinity). This can be interpreted as being equivalent to the disappearance of Goldstone bosons (or of classical long-range collective excitations) in the relaxation process of a system of classical particles. As a consequence, classical systems *cannot* be endowed with generalized rigidity. This, in turn, implies that, when the volume tends to infinity, the choice of a particular ground state, amongst the many possible after a SSB transition, can be destroyed even by very small perturbations. Namely, the order-preserving messengers in the classical case are absent. As a conclusion, SSB, although possible in a classical case, is *unstable* with respect to (thermal) perturbations.

Models of SSB based upon QFT provide a coherent description even of collective effects, identified with the long-range correlations supported by the Goldstone bosons arising after a system, crossing the critical point of a SSB, chooses a particular ground state amongst the many available. However, despite our previous argument about the absence of Goldstone bosons in a classical description of SSB, the QFT-based picture of collective effects seems to be in open contradiction with the fact that long-range correlations are observed even in many classical systems.

Such systems include the Fermi-Pasta-Ulam chain of anharmonic oscillators (for a review *cfr.* Carati and Galgani, 2001), and partial differential equations having solitonic solutions, already discussed in the previous Chapter. In order to account for the apparent contrast between the behavior of these systems and the predictions of QFT, we must first remark that all classical systems allowing for long-range correlations are described by non-ideal models. Namely, they exist within a finite volume and/or in the presence of suitable boundary conditions. This forces us to study what would occur when an ideal model, based on QFT, has to be modified in order to take into account finite volume effects (due to the appearance of domain boundaries, defects, impurities, etc.). Here, we recall that Koma and Tasaki (1994) built a theory of a QFT-based description of a SSB 'obscured' by finite volume effects. They were able to show that the latter gave rise to a transformation of the multiplicity of equivalent ground states into a single ground state, associated with a number of metastable 'low-lying' states orthogonal to it. The latter were characterized by the fact that their minimum energy differed from that of the ground state by an amount of the order of thermal energy. In a sense, these states thus represented a sort of remnant, or trace, of the original SSB transition. But, what type of collective effects could be associated with these remnants? It must first be said that their associated Goldstone bosons can no longer support long-range correlations, as the condition of infinite volume no longer holds. This is equivalent to saying that we work with finite-mass Goldstone bosons and that all correlations have a limited range (even though very large with respect to the dimensions of the system). In such a situation, we live in a world in which QFT is substituted by QM. Moreover, the very existence of such a world is supported by *topological constraints*, which must necessarily propagate within the system in some form (e.g., solitary waves). The propagation of these constraints and of the correlations must therefore be described by QM but now, interaction with the environment can trigger thephenomenon of decoherence and thus we finally obtain that, except for a number of particular cases, all propagations of constraints and of correlations within dimensions of the order of a system's spatial extent have a *classical* nature. In other words, we obtain long-range correlations which can be described in an entirely classical way. However, their origin is rooted within QFT. They represent nothing but a classical remnant of the collective effects described by QFT in the case of a finite volume and/or special boundary conditions. We can thus conclude that a complete understanding of these phenomena can be reached, again, only by resorting to QFT.

Here, it must be stressed that the long-range correlations associated with a SSB arise as a consequence of a disentanglement between the condensate mode and the rest of system (Shi, 2003). This finding, related to the fact that within QFT we have a stronger form of entanglement than within QM (Clifton and Halvorson, 2001), explains how the structures arising from a SSB in QFT have a very different origin from those arising from a SSB in classical physics. Namely, while in the latter case we need an exact, and delicate, balance between short-range activation (due to non-linearity) and long-range inhibition (due, for instance, to diffusion), a balance which can be broken even by a small perturbation, in the former case we have systems which, already from the starting, lie in an entangled state, with strong correlations which cannot be altered as much by the introduction of perturbations. However, in suitable cases, the action of these perturbations or the change in value of a control parameter can induce a form of disentanglement which puts the system in a nearly-classical situation, the features of which, however, are not of a classical nature and, despite their classical appearance, keep a trace of their quantum origin (for a comprehensive treatment of this topic, see Sewell, 2002).

Moreover, as SSB and phase transitions occur within a finite time; this allows one to study the dynamics of the events characterizing them. Such dynamics has a quasi-classical character and may be of a chaotic deterministic kind (Pessa and Vitiello, 2004a; 2004b). This is precisely the circumstance which allows for the occurrence of ordered structures as a result of a SSB, owing to the fact that chaotic deterministic processes are significantly easier to control than other kinds of linear or non-linear systems (see, for instance, Crook *et al.*, 2003). This control, exerted through suitable interactions with classical objects, drives the system towards ordered behaviors close to the new ground state. In this regard, it is to be underlined that the analysis of phase transition dynamics (Rivers, 2000; Lombardo *et al.*, 2002; Rivers *et al.*, 2002) has shown how these classical objects, occurring in the form of *defects*, are an unavoidable by-product of the same dynamics (see also Zurek, 1996; Rajantie, 2002). The predictions from these studies have not only been confirmed by experimental findings, but also provide a means to check whether a SSB has occurred or not merely by verifying whether or not we have the occurrence of defects and, when possible, by measuring their distribution.

From the above discussion on the QFT-based description of SSB we could deduce that the latter seems to be the best tool, so far available, to model intrinsic emergence. Such a claim, however, appears excessive when we take into account, on one hand, a number of unsolved technical problems in modeling concrete cases of SSB, and on the other, a number of conceptual difficulties raised by the identification of SSB with intrinsic emergence (on this subject, see Brown and Cao, 1991; Kosso, 2000; Balashov, 2002; Brading and Castellani, 2003; Liu, 2003; Ruetsche, 2003). On this point, in the following Sections of this Chapter we describe a number of alternative routes used to model emergent phenomena without making explicit reference to QFT or to SSB.

5.5 Non-ideal models of emergence

The previous Section discussed a number of models identifying intrinsic emergence with phase transition phenomena associated with SSB. All these models can be classified as ideal and homogeneity-based, in conformity with the criteria introduced in the first Section of the previous Chapter. Having been introduced mainly to describe phenomena occurring in elementary particle or condensed matter physics, they also share three other important features:

a) They assume that in the system under study we can always identify a quantity playing the role of *energy*, so as to take advantage of the conceptual and mathematical framework of classical mechanics (whose formal structure has been adopted by QM and QFT without substantial alterations);

b) They assume that the concept of *phase* can be defined in a clear and unambiguous way; this assumption implies in turn that, far from the critical point, fluctuations are negligible;

c) They neglect the role played by the external environment either because, as in QFT, we are working with an infinite volume limit, or because the environment is reduced to a simple thermal bath.

These features prevent the practical application of these models to the study of a number of situations, occurring in biological, economical, social, cognitive and, of course, physical domains in which we deal with phenomena which, at first sight, appear to be endowed with the characteristics of intrinsic emergence but in which features a), b) and c) are absent. A simple way out would be to state that the latter phenomena are beyond the domain of science or modelling possibilities. Although such an extreme position was often adopted in the past by a number of scholars, mainly in the humanities, philosophy, psychology and social sciences (but even within Systemics itself), it gives rise to such unproductive stance that the best approach seems to be to deal directly with the problem of introducing new modelling principles to account for the phenomena mentioned above. In this regard, two different (but not contrasting) attitudes have been adopted:

d) Exploiting some *analogies* between these phenomena and SSB phase transitions, in order to better understand the behavior of suitable non-ideal models of the former;

e) Attempting to translate these non-ideal models within a framework characterized by features a), b) and c).

This Section deals only with the former choice, while the latter will be briefly discussed in the last Section of this Chapter. Of course, when approaching the world of non-ideal models of emergence, there is such a large body of modelling activity that it is virtually impossible to review even the most significant achievements. We are therefore forced to limit our discussion to only three representative domains: *Neural Networks, Cellular Automata* and *Artificial Life*. This choice was made to exemplify the problems arising in non-ideal modelling.

5.5.1 Neural Networks

The expression *Neural Network* or, better, *Artificial Neural Network* (ANN) (for a review see textbooks such as Bishop, 1995; Rojas, 1996: Haykin, 1998) denotes generically a system containing two components:

1. *units* (sometimes called *neurons*, or *nodes*), each of which is an input-output device, with N input lines and one output line, characterized, at each time instant t, by an *output state* $u(t)$ (also called *activation*), by an *inner state* $p(t)$ (the so-called *activation potential*), and by an *input state* $x(t) \equiv [x_1(t) , ... , x_N(t)]$, where the

symbol $x_i(t)$ denotes the activation state of the i-th input line; suitable laws allow the computation of $u(t)$ from a knowledge of $p(t)$, as well as the computation of $p(t)$ as a function of the input state present at time t or, eventually, at previous points in time;

2. *connection lines* between the units; these feed the activation state from the output line of one unit to the input line of another; each connection is associated with a *connection weight* (in most cases a real number), which modulates the activation signal crossing it.

Despite the generality of such a definition, and the unlimited number of possible choices it allows, in practice almost the totality of ANN designers have made use of a very narrow range of possibilities regarding the laws for computing $u(t)$ and $p(t)$; we denote these choices through the attribute *neural-like*, because they derive from the attempt to model the gross behavioral features of biological neurons. Between the neural-like laws for computing $p(t)$ surely the most popular is given by:

$$p(t) = \Sigma_i w_i x_i(t) - s$$

where w_i denotes the connection weight associated with the i-th input line, and s is a parameter called *threshold*. Amongst the laws for computing $u(t)$ one of the most widespread is:

$$du/dt = -u + F[p(t)]$$

where F is a function which often takes the form (the so-called *sigmoidal function*):

$$F(y) = 1/[1 + exp(-y)]$$

This law characterizes *McCulloch-Pitts neurons*. Its time-discretized version (almost universally used in computer simulations) is written as:

$$u(t+1) = F[p(t)]$$

There are also cases, however, in which the law for computing $u(t)$ is expressed through a second-order differential equation such as, for instance, *Freeman's neuron model* (see Whittle, 1998; Freeman, 2000). Properly speaking, in this model, 'neuron' rather than denoting a single biological neuron, refers to a more abstract entity whose behavior describes the average behavior of a neuronal *population*. Such a kind of neuron is characterized by a law for computing $u(t)$ of the following kind:

$$u = f(p) \quad , \qquad L\,p = x$$

where u denotes the neuron firing rate, p the current within the cell body, x the external input pulse rate, and L a differential operator taking the form:

$$L = Dtt + a\,Dt + b \quad , \qquad Dt = d/dt$$

in which the two parameters a and b are chosen in such a way that the polynomial $s^2 + a\,s + b$ has two negative real roots.

Over the past few years there has been an intensive search for models of more biologically realistic ANN units, that is, endowed with features more similar to those characterizing the operation of biological neurons, without, at the same time, becoming exceedingly complicated from the mathematical point of view, as is the case for detailed models of biological neuron dynamics (see Koch, 1998). These more biologically-oriented ANN units are known as *spiking neurons*. It can be shown that it is possible to design networks of spiking neurons endowed with the same abilities characterizing traditional ANNs with less biologically realistic units (for reviews of this subject, see Maass, 1997; Maass and Bishop, 1999). Amongst the simplest models of ANN units belonging to this category we consider that proposed by Izhikevich (Izhikevich, 2003; 2004) consisting of the simple pair of first order differential equations:

$$dv/dt = 0.04\,v^2 + 5\,v + 140 - u + I \quad , \quad du/dt = a\,(b\,v - u)$$

supplemented with the auxiliary condition:

if $v \geq 30$ then the value of v is set to c and the value of u is set to $u + d$

Here a, b, c, d are suitable parameters, while v represents neuron membrane potential (measured in mV) and u denotes an auxiliary variable describing the membrane recovery process. The variable I denotes the external input. It can be shown that this model neuron is endowed with all the features characterizing the operation of a real biological neuron.

The foregoing examples show that an ANN can be viewed as a special kind of dynamical system, in which the *state vector*, whose components are the instantaneous output states of the single network units, starting from a given initial state, changes with time as a consequence both of the form of the activation laws adopted and of the distribution of connection weight values. There is, however, another sense in which an ANN can be considered as a dynamical system: when its parameters (mainly the connection weights) change with time as a consequence of a *learning process*. Namely, the most interesting aspect of ANNs is their ability to *learn only from examples* to approximate any form of input-output relationship, to categorize the input patterns, to act as associative memories (see Bartlett and Anthony, 1999).

A survey of the field of neural network learning, however, is a very complicated task, owing to the huge number of learning methods so far proposed. Generally, these methods consist of algorithms which dictate the modifications in connection weights (or other parameters) to be made as a function of suitable external influences. The algorithms can be grouped into the five fundamental categories described below.

- *Supervised learning*, based on the presentation to the network of examples, each consisting of a pattern pair: the *input pattern*, and the corresponding *wanted*, or *correct*, *output pattern*. The modification of connection weights is determined by the *output error*, that is by a suitable measure of the difference existing between the desired output pattern and the ouput pattern effectively produced by the network. This kind of learning is largely, and most often successfully, used to obtain neural networks realizing a given input-output transfer function.

- *Unsupervised learning*, in which each example consists only in an input pattern, and the concept itself of correct output pattern is devoid of any meaning. The modification of connection weights at a given instant in time is determined by the local pattern presented in input at the same time, and by the result of the competition between suitable *categorization units*. This kind of learning is very useful to *clusterize* (that is, group into categories) a complex data set.

- *Reinforcement learning*, in which the network is presented with *time sequences of examples*, each consisting of an *input pattern*, the corresponding *output pattern* (*action* or *decision*) and sometimes also the *reinforcement signal* received from the environment as a consequence of the output pattern produced. The modification of connection weights aims to maximize the total amount of positive reinforcement (the *reward*) received as a consequence of the entire behavioral sequence. Such a type of learning is very useful to design systems able to adapt to a rapidly changing, unpredictable, and noisy environment.

- *Rote learning*, in which the connection weights are modified in a brutal way, according to a fixed law, under the direct influence of input patterns, or of suitable input-output associations. This kind of learning is useful for designing *associative memories*, in which external patterns (or input-output associations) are stored, and subsequently recalled following the presentation of suitable cues.

- *Evolutionary learning*, in which the connection weights, and most often the whole network architecture change as a function of the *fitness value* associated with suitable behavioral features of the network itself. This value is related to a given measure of the efficiency of the network in dealing with problems posed by external environment, a measure which, obviously, is chosen by the designer on the basis of needs or goals.

Each kind of learning is associated with a number of preferred architectural implementations, even though there is no necessary connection between a learning process and its network realization. Besides, by resorting to suitable prototypical architectures, some kinds of learning have been studied from a theoretical point of view, in order to characterize them with respect to the convergence of adopted learning laws. Most studies have dealt with supervised learning, mostly using methods of optimization theory (Frasconi *et al.*, 1997; Bianchini *et al.*, 1998; Bartlett and Anthony, 1999). A significant number of other studies, traditionally within the context of *spin-glass physics*, have been devoted to rote learning (Amit, 1989; Kamp and Hasler, 1990; Domany *et al.*, 1996). A smaller number of papers and books have dealt with unsupervised learning (see, for instance, Kohonen, 1995; Hinton and Sejnowski, 1999). Both reinforcement learning and evolutionary learning, although extensively studied in many respects, actually appear to be very difficult topics when one is interested in assessing their convergence features (on reinforcement learning see textbooks such as Jang *et al.*, 1997; Sutton and Barto, 1998; Sun and Giles, 2000).

One of the main results of these studies is the mathematical proof that ANNs can act as universal approximators of any kind of input-output relationship. This means they are able to reproduce a given relationship with arbitrary precision, even if the relationship itself is defined not in an explicit way but only through a finite set of input-output pairs. The first proof of such a property had already been given in 1989 (see Hornik *et al.*, 1989; White, 1992; for a more recent paper on this topic, see Castro *et al.*, 2000). The approximation capabilities of neural networks, however, go far beyond input-output relationships. These systems can, in fact, mimic with any degree of approximation, even the behaviour of *classical continua*, whose dynamics can be described using partial differential equations. The easiest way to reproduce the behaviour of continuous media is to resort to so-called Cellular Neural Networks (CNNs), first introduced by Chua and Yang (see Chua and Yang, 1988). These can be described as systems of nonlinear units arranged in one or more layers on a regular grid (for a short review see. Chua and Roska, 1993). CNNs differ from other neural networks since the interconnections between units are local and translationally invariant. The latter property means that both the type and the strength of the connection from the i-th to the j-th unit depend only upon the relative position of j with respect to *i*. At every time instant, two values are associated with each unit of a CNN: its (internal) *state*, denoted by $v^m_i(t)$, and its *output*, denoted by $u^m_i(t)$. Here the index *m* denotes the layer to which the unit belongs and the index *i* denotes the spatial location of the same unit within the layer. The general form of the dynamical laws governing the time evolution of these functions is:

$$dv^m_i(t)/dt = -g[v^m_i(t)] + \sum_q \sum_k a^{qm}_k[u^q_{i+k}(t), u^m_i(t); P^{qm}_{ak}]$$

$$u^m_i(t) = f[v^m_i(t)]$$

In these formulae the function *g* describes the *inner dynamics* of a single unit, whereas *f* denotes the *output function* (often of sigmoidal type). Besides, the summation over index *q* covers all layer indices, and the summation over index *k* covers all values such that *i + k* lies in the neighbourhood of *i*. Finally, the symbol P^{qm}_{ak} denotes a set of suitable constant parameter values entering into the explicit expression of the *connection function* a^{qm}_k. Such parameters, whose values are independent of *i*, are referred to as *connection weights* or *templates*.

In the case of a single-layered CNN, in the absence of inner dynamics, and when the output function coincides with the identity function, the previous laws can be reduced to the simpler form:

$$dv_i(t)/dt = \sum_k a_k[v_{i+k}(t), v_i(t); P_{ak}]$$

Let us now show, through a simple example, how a continuum described by partial differential equations can be approximated by a suitable CNN (see Kozek *et al.*, 1995; Roska *et al.*, 1995). To this end, let us choose a 1-dimensional medium described by the celebrated Korteweg-De Vries equation, already introduced in the previous Chapter when dealing with solitary waves and, for reasons of convenience, written as:

$$\partial\varphi/\partial t + \varphi(\partial\varphi/\partial x) + \delta^2(\partial^3\varphi/\partial x^3) = 0$$

where δ^2 is a suitable parameter. If we introduce a discretization of spatial coordinates, based on a fixed space step Δx, the field function $\varphi(x, t)$ is replaced by a set of time-dependent functions $v_i(t)$ (*i* = 1, 2, ... , *N*), and the previous equation is replaced, in turn, by the set of ordinary differential equations:

$$dv_i/dt = -v_i[(v_{i+1} - v_{i-1})/\Delta x] - \delta^2[(v_{i+2} - 2v_{i+1} + 2v_{i-1} - v_{i-2})/(\Delta x)^3]$$

It is then easy to see that this set of equations describes the dynamics of a CNN, whose connection function is given by:

$$a_k = w_k \, v_{i+k} + r_k \, v_i \, v_{i+k}$$

In such a CNN the neighbourhood of the i-th unit goes from the $(i + 2)$-th unit to the $(i - 2)$-th unit, so that index k can take only the values $+2, +1, 0, -1, -2$. Direct inspection shows that the values of connection weights w_k and r_k are given by:

$$w_2 = -\delta^2 / (\Delta x)^3 \;, \quad w_1 = 2\,\delta^2 / (\Delta x)^3 \;, \quad w_0 = 0 \;, \quad w_{-1} = -2\,\delta^2 / (\Delta x)^3 \;,$$
$$w_{-2} = \delta^2 / (\Delta x)^3 \;,$$

$$r_2 = 0 \;, \quad r_1 = -1 / \Delta x \;, \quad r_0 = 0 \;, \quad r_{-1} = 1 / \Delta x \;, \quad r_{-2} = 0$$

A more complex and still unanswered question is whether or not neural networks are universal approximators of quantum fields. Here, we recall that physicists have, for a long time, stressed a close analogy between the lattice version of the so-called Euclidean formulation of QFT and the statistical mechanics of a system of spin (that is, neuron-like) particles, coupled via next-neighbour interactions (see Mézard *et al.*, 1987; Montvay and Münster, 1997). The existence of such an analogy, of course, provides a strong motivation for studying neural networks as possible implementations of QFT. Roughly speaking, neuronal unit activations should correspond to fermionic source fields and connection weights should represent bosonic gauge fields mediating the interactions between fermions. Suitable restrictions on the possible neuronal interconnections (limited to neighbouring units) should allow for the fulfilment of requirements such as microscopic causality. Besides, by taking both a thermodynamical and a continuum limit on neural network descriptions, we should recover the usual QFT ones. However, despite the attractiveness of such a picture, the detailed specification of a precise equivalence relationship between neural networks and QFT is still lacking. On the contrary, there are a number of no-go theorems (see, for instance,. Meyer, 1996), which show how the search for such an equivalence is a very difficult task.

The fact that ANNs are dynamical systems should constitute the basis for substantiating the claim that they are models of intrinsic emergence, even though of a non-ideal kind. As a matter of fact, if we could show that the dynamical evolution of an ANN is associated with a sort of *phase transition*, similar to those discussed above in physical systems, then we could take advantage of the theoretical apparatus already developed by physicists and, within this context, the attribute 'emergent' would acquire a well-defined meaning. Unfortunately, proof of the existence of a phase transition associated with the dynamical evolution of an ANN is very difficult (in some cases impossible) to obtain using current techniques. However, when the attribute 'dynamical' refers only to time evolution of a network state vector towards an attractor, a possible way out of this problem, provided the dynamics be of a stochastic nature, could be found through the following steps:

- identification of a suitable measure of the degree of stochasticity of ANN time evolution, here denoted by T and called *temperature*. The average value of every macroscopic quantity A, denoted by $<A>$, is computed on all available values of A corresponding to a situation in which T is kept constant;
- identification of a suitable Lyapunov function, denoted by H, governing the dynamics close to the attractor, the evolution towards the minimum value of H being modelled as a relaxation process associated with a gradual lowering of T;
- for each step of the relaxation process, associated with a momentarily constant value of T, computation of the value of the following quantity, denoted as *specific heat* (in agreement with the considerations made above in the second section of this Chapter) and defined by:

$$C(T) = (<H^2> - <H>^2)/T^2$$

- on examining the plot of $C(T)$ vs. T, if the plot contains jumps or discontinuities, these would provide evidence for the presence of a phase transition.

Such a technique, if it could be implemented, would at least allow for a reconstruction of the thermodynamics of ANN learning processes starting from the trends experimentally observed in computer simulations. Unfortunately, it works only in a limited number of cases, which include the celebrated *Hopfield model* (*cfr.* Hopfield, 1982) and *Smolensky's Harmony Theory* (Smolensky, 1986). Even in these cases, however, most plots of $C(T)$ vs. T fail to show any evidence of a phase transition. Matters are complicated because in these kinds of ANN, despite the law of dynamical state evolution having been designed *a priori* to describe a relaxation process towards a local minimum of a given global function playing a role analogous to that of energy, the landscape of this global function has a very different form compared to that in the usual cases of Symmetry Breaking, both in classical bifurcation theory or in QFT. Namely this landscape, instead of consisting of a multiplicity (possibly infinite) of ground states, all equivalent to one another, associated with well-defined basins of attraction, has a very complex structure, including a very high number of different ground states, most of which are only local minima (and therefore *metastable*), whose basins of attraction form a complicated puzzle. In such a situation the dynamics is practically unpredictable, as it can be trapped within a local minimum which, however, it can leave, owing to the stochastic character of the evolution rules. The problem is that all these phenomena depend in a strong way upon the choice of the initial state of the dynamics. Given that the number of different possible choices, however, is so high as to prevent any practical investigation to find a relationship between the initial state and the features of its subsequent dynamical evolution, we are in a situation where the usual methods of QFT or of theory of phase transitions cannot work. To give an idea of the difficulties involved, let us consider a network having N units, each associated with two possible states (for instance 'on' or 'off'). The total number of different possible states, each of which can be chosen as an initial state, of such a system is 2^N. For a simple network of only 100 units, this number is given by 2^{100}, which is close to 10^{30}. Assuming we have a computer so fast that it can gain all information relative to the dynamical evolution starting from a given initial state in 10^{-9} seconds, in order to obtain a complete information about all possible dynamical evolutions of this network, the above computer would then need 10^{21} seconds. Unfortunately, the estimated life of our Universe, starting from Big Bang, is less than 10^{18} seconds!

Usually such a situation is described by saying that ANNs of this kind are examples of *disordered systems*. In the past, many attempts have been made to build a general theory of these systems (*cfr.* Mézard *et al.*, 1987; Amit, 1989; Dotsenko, 1994; Newman, 1997; Mikhailov e Calenbuhr, 2002, Chap.8). The aim of these efforts is to show the existence of a bridge between disordered systems and QFT (see, for instance, Parisi, 1988; Cardy, 1996). We spare a word on this fascinating topic in the last Section of this Chapter.

The situation, of course, becomes even more complicated when the attribute 'dynamical' refers to a time evolution of the connection weights (or, more generally, of the parameters) of an ANN as a consequence of a learning process. In this case the previously described procedure cannot be applied (see, however, the methods proposed in Pessa, 2000). Recently, a number of authors (see Saad, 1998; Marinaro and Scarpetta, 2000) have tried to describe the learning process within particular kinds of ANNs by resorting to the methods of Statistical Mechanics. This field of research is known also as *on-line learning*. Results obtained so far are very encouraging, as they concern the individuation of order parameters and of the macroscopic evolution equations associated with these processes. Here is a brief example, to provide a more precise understanding of the advantages of this approach. It deals with the case of the so-called *Ising Perceptron*, an ANN with an output layer containing only one unit (its architecture is depicted in Figure 5.1), whose output is given by:

$$S(x) = sgn(J\,x)$$

x denotes the input vector, and J is the vector of the connection weights associated with the feedforward connections between the input layer and the one single unit present in the output layer. Consider the learning process through which, by changing the values of the weights J, the Perceptron will be able to produce as output, in presence of the input vector x, the value given by:

$$T(x) = sgn(W\,x)$$

where W denotes a previously given vector, which, for the sake of convenience, we choose in such a way as to fulfil the constraint $|W| = 1$.

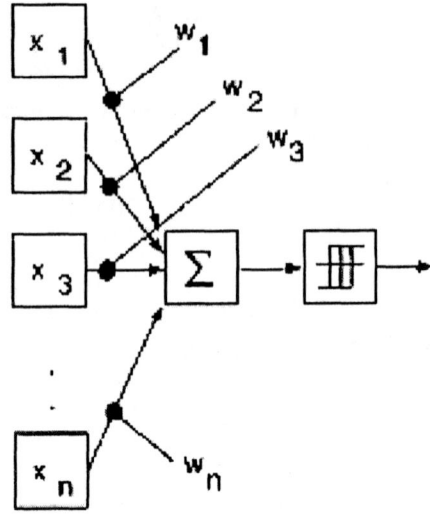

Figure 5-1. The architecture of a simple Perceptron. On the left are the input units, on the right the output unit, as indicated by the symbols. The Perceptron takes the weighted sum of the input values and outputs 1 if the sum is positive, or −1 if negative.

We can conceive the learning process as an attempt to *imitate*, by our Perceptron, the behavior of a *teacher*, who knows in advance the correct answer $T(x)$ for each input pattern x. Let us also assume that the learning process be driven by a weight change rule (which is a suitable generalization of the usual Perceptron learning rule currently employed in practical applications of this kind of ANN), so that for every presentation of a given input vector x , the change of the connection weight vector is given by:

$$\Delta J = (1/2) \left[T(x) - sgn(J\,x) \right] x$$

It is useful, now, to introduce a unitary vector V such that:

$$J = Q\,V \;,\;\; |V| = 1$$

(where Q is a scalar) and a scalar quantity ω defined by:
$$\omega = W\,V$$

It is now possible to show that if:

- the input vectors x are chosen in a random way within the set $\{-1, 1\}^N$
  (N is the number of input units) with a uniform probability distribution
  $p(x) = 2^{-N}$;
- we take the limit $N \to \infty$;

then the learning dynamics generated by the learning rule described above is governed, at a macroscopic level, by two order parameters, that is Q and ω, whose macroscopic evolution is dictated by the following system of ordinary differential equations:

$$dQ/dt = (1 - \omega)/\sqrt{2\pi} \ , \quad d\omega/dt = (1 - \omega^2)/(Q \sqrt{2\pi})$$

Of course, an analysis of the solutions of these equations could, in principle, reveal the existence of sudden jumps, to be interpreted as cues for the presence of a phase transition.

Unfortunately, despite their attractiveness, these results cannot be applied to concrete ANNs because they violate both assumptions *a)* and *b)*. As a matter of fact, not only is the number of input units always finite (and generally small), but the input vectors are never derived in a random way from a probability distribution. On the contrary, they are chosen according to well-defined rules, in turn related to the very nature of the data domain from which these vectors were extracted. In other words, within the realm of ANNs, as in the biological one, the predominant role is played by *individuality*, rather than by *homogeneity* or *randomness*.

5.5.2 Cellular Automata

The expression "Cellular Automata" denotes a class of formal systems introduced in the 1950s by J. Von Neumann (see Von Neumann, 1966 for a posthumous publication of his results). His main goal was to build an abstract model of self-reproducing automata. As a matter of fact, he was able to design a cellular automaton simulating a universal Turing machine and able to self-replicate. The original construction of Von Neumann was very complicated, but later contributors introduced considerable simplifications of his method (see, for instance, Codd, 1968). In the 1960s this triggered a number of studies on these new systems (a representative collection is contained in Burks, 1970).

From a general point of view, a cellular automaton is merely a formal system consisting of:

4   a set (typically infinite) of *cells*, arranged within a *lattice* in a suitable *n*-dimensional space (very popular examples are cellular automata consisting of infinite square lattices of cells in a 1- or 2-dimensional

Euclidean space); each cell is, in turn, associated with a finite number of different possible *states*;

5   a *neighbourhood relationship*, stating which cells are to be considered as *neighbors* of a given cell; this relationship usually holds in a *homogeneous* way, as all cells are associated with the same relationship;

6   a suitable *local transition rule*, operating on a discretized time scale; this gives the state of a cell at time $t + 1$ as a function of the states of its neighbouring cells at time t; the state updating process is *parallel*, as all cells of a cellular automaton update their states simultaneously.

In the case of 2-dimensional cellular automata with square cell lattices some kinds of neighbourhood relationships are very popular. We show here the *Moore* neighbourhood, including, besides the cell considered, also its 8 neighbouring cells sharing at least one vertex with it (see Figure 5.2a). Equally popular is the *Von Neumann* neighbourhood where the cell considered shares at least one edge with its 4 neighbours (see Figure 5.2b).

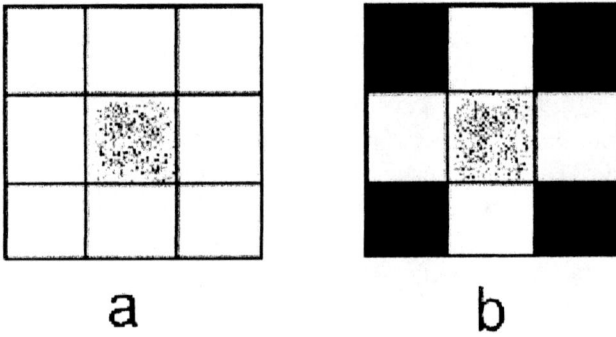

a   b

Figure 5-2. The Moore (a) and Von Neumann (b) neighbourhoods. The grey reference cell is at the centre.

The number, *N*, of different possible local transition rules increases rapidly with the number, *k*, of different allowed states per cell and with the number, *r*, of cells included within the neighbourhood of a given cell. Namely it can be easily proven that:

$$N = k^q \text{ , where } q = k^r$$

It can immediately be seen that, even with small values of *k* and *r*, *N* can be very large. For instance, when dealing with 2-dimensional cellular automata with a Moore neighbourhood (in which $r = 9$) and with $k = 10$, $N = 10^{1.000.000.000}$. For comparison, the estimated total number of atoms within the Universe is *only* 10^{80}! We stress, however, that the choice of $k = 10$ is not particularly unusual as, for instance, the cellular automaton originally introduced by Von Neumann to simulate a universal Turing machine had $k = 29$.

Thus, cellular automata are systems which, from many aspects, are currently beyond the reach of mathematics. Namely:

— as they can simulate a universal Turing machine, their computational power is at least equal to that of such a machine; therefore there must be aspects of their evolutionary behavior which are undecidable, just as in the case of Turing machines; as a matter of fact, such undecidability results were found very early in the history of cellular automata theory (see, for instance, Burks, 1970); the question of whether their computational power is *greater* than that of Turing machines remains unanswered; probably the current form of Computation Theory is not suited for dealing with this kind of problem;

— even exploration of the space of possibilities offered by the different transition rules is practically precluded, so that we have no idea about the existence of *classes of dynamical behaviours* of these systems.

In the early 1970's, cellular automata became very popular, owing to the introduction by J. H. Conway of a particular kind of 2-dimensional cellular automaton named *Game of Life* or, more shortly, *Life* (its popularity was initially due to the work of Martin Gardner; see Gardner, 1970). It was characterized by a square lattice of cells, with only two states allowable for every cell (here denoted as "1" and "0") and a Moore neighbourhood. Its transition rule was very simple:

— if a cell at time t lies in state "1" and if less than two of its neighbours are in state "1", then at time $t + 1$ the cell changes to state "0";

— if a cell at time t lies in state "1" and if more than three of its neighbours are in state "1", then at time $t + 1$ the cell changes to state "0";

— if a cell at time t lies in state "0" and if three of its neighbours are in state "1", then at time $t + 1$ the cell changes to state "1";

— in all other cases the state of the cell remains unchanged.

If one interprets the configurations of cells lying in state "1" as 2-dimensional spatial patterns, then this rule produces very interesting (and partly unpredictable) time evolutions of these patterns, strongly resembling evolutionary behaviors observed in the biological realm (whence the name "Life"). It is possible to observe invariant patterns, as shown in Figure 5.3a), oscillating patterns, as the one shown in Figure 5.3b), or patterns displacing themselves with constant speed, as the one shown in Figure 5.3c). Many other interesting (and complex) patterns have been discovered but for sake of brevity will not be mentioned here.

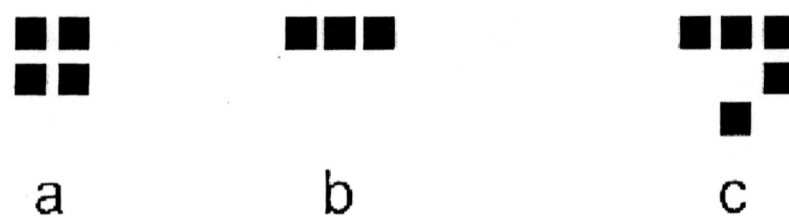

Figure 5-3. Examples of simple, interesting patterns in "Life"(see text). The cells in state "1"
are in black. The pattern in (c) is called a *glider*.

An initial mathematical analysis of cellular automata, based also upon the methods of statistical mechanics, was carried out in the 1980s by Wolfram (see Wolfram, 1983; 1984). This analysis, performed mostly on 1-dimensional cellular automata endowed with *totalistic* transition rules (that is, rules in which the state transition depends only upon the *sum* of state values of neighbouring cells), led to a subdivision of cellular automata rules into four classes:

Class I. Characterized by evolution towards a spatially homogeneous equilibrium state.

Class II. Characterized by evolution toward stable or periodic attractors, always of finite spatial extent.

Class III. Characterized by the possibility of chaotic evolutions, with unlimited spatial growth of initial patterns.

Class IV. Characterized by the occurrence of localized patterns of great complexity, which can alternatively grow and contract.

Cellular automata belonging to class IV are considered the most interesting. They appear as the best candidates for proving the validity of a conjecture made by Langton (see Langton, 1990), according to which systems able to show significant complex behaviors are to be found only on the edge between order and chaos (see the attempt to test this in Mitchell *et al.*, 1993).

Over the last few years, studies on cellular automata have undergone rapid development (for an overview, see Ilachinski, 2001). This development was triggered by the discovery that these systems were excellent candidates for simulating, on discrete temporal and spatial scales, the behavior of many physical systems governed by partial differential equations (a typical instance is given by the so-called *lattice gas automata*; see Hardy *et al.*, 1976; Frisch *et al.*, 1986). Such a line of research, in concomitance with the growth in interest for *Quantum Computing*, led to the introduction of

Quantum Cellular Automata (Grössing and Zeilinger, 1988; Tougaw and Lent, 1996), physically realized through quantum dot devices (for a review, see Porod, 1997). A typical quantum cellular automaton still has a finite number of allowable states per cell (for instance two, as in the case of *Life*), but their evolution is governed by a quantum-mechanical, rather than deterministic, law. On the contrary, the interactions between different cells follow deterministic rules, a circumstance allowing for the design of logical gates based upon suitable arrangements of quantum cells (for a simple description, see Snider *et al.*, 1999). Thus, a quantum cellular automaton can be used, in principle, to implement a *quantum computer*, the only technical difficulties, so far, being the need for the absence of impurities and very low temperatures, in order to avoid decoherence effects.

With regard to emergence, given that cellular automata are the discretized counterpart of systems described by partial differential equations allowing for the occurrence of bifurcation phenomena, it is not surprising to find in them evidence of structural changes (among the earliest examples see, for instance, Derrida and Stauffer, 1986; Wolfram, 1986). Perhaps the most striking example is offered by a class of cellular automata in which, starting from any initial pattern whatsoever, even when chosen at random, their evolution always gives rise to self-replication processes (the first cellular automaton of this kind was introduced by Chou and Reggia, 1997). There is, however, no argument supporting the claim that these systems exhibit intrinsic emergence. On the other hand, one must take into account that, if quantum cellular automata are a sort of discrete implementation of QFT, then they should share with the latter all the characteristics associated with spontaneous symmetry breaking (for speculative discussions on this topic see Svozil, 1986; t'Hooft, 2003). For this reason, it is interesting to remark that classical solitons, which can be considered as deriving from quantum tunnelling between two different ground states at the critical point of a phase transition, can be represented using quantum cellular automata (Vahala *et al.*, 2003).

5.5.3 Artificial Life

With "Artificial Life" (AL) we label a large body of research activities aiming at some design of artificial systems endowed with behavioral features similar to those of existing biological systems. The history of AL as an autonomous discipline began in 1987, owing to the initiative of C. Langton (the first contributions by Langton and other researchers are contained in the Proceedings of the first AL conferences; see Langton, 1989; Langton *et al.*, 1992; Langton, 1994). Despite the debates raised by the introduction of this discipline, it soon acquired widespread popularity, owing also to the availability of software allowing users to implement a "synthetic biology" on home computers.

All activities so far done within AL (for a synthesis, see Langton, 1997; Johnson, 2002) can be classified according to three main goals:

1. the direct modelling of biological processes, such as those relating to the origin of life and evolution, taking into account specific biological features, such as, for instance, those associated with the synthesis of DNA/RNA;

2. the identification of principles underlying intelligent behavior observed in biological beings (Camazine *et al.*, 2001; Mikhailov and Calenbuhr, 2002); this requires suitable modelling of emergence and self-organization processes (as often recalled within this book, the recent resurgence of interest in the topic of emergence is due to the activity of AL researchers); in this domain most studies deal with so-called *distributed intelligence* (sometimes called *swarm intelligence*; see Bonabeau *et al.*, 1999; Kennedy *et al.*, 2001); these expressions refer to a set of agents, each endowed with very limited cognitive abilities, interacting amongst themselves and with the environment through simple rules, in which we have the emergence of "intelligent" collective behaviors without the existence of centralized coordination rules; such a situation characterizes many biological phenomena, related to the formation of swarms, flocks, herds and similar collective entities, as well as to the collective behaviors of ants and termites;

3. the exploitation of suitable biological features to solve mathematical or engineering problems; typical examples are *genetic algorithms* (see, for instance, Goldberg, 1989; Whitley, 1993), used to solve optimisation problems, or *immune networks* (Stewart and Varela, 1991; Perelson and Weisbuch, 1997; Dasgupta, 1999), used in intelligent data mining.

Often these goals are pursued by resorting to a *synthetic* approach, in which the behavior of a system is generated by specifying only individual features of its components, and their interaction rules. The most widely used tools consist of:

- computer simulation of studied models;
- observation of autonomous robot behaviors;
- observation of behavior of *hybrid* systems, partly of biological nature (for instance bio-molecules, living cells or simple animals; for the latter case, see the interesting attempt by Tsuda *et al.*, 2004) and partly artificial.

The number of AL models is so great that here it is virtually impossible to review even the most representative ones. We thus limit ourselves to describe only one model, that of *virtual ants* (sometimes abbreviated as *vants*). On this model, first introduced by Langton, there is widespread research activity (for a synthesis, see Gale, 1998; Bonabeau *et al.*, 1999). It is based on an abstract representation of real ant behavior, whose relevant features are:

- they have only a local perception of environment;
- they are incapable of any kind of planning;
- their motion is based on following local gradients of chemical substances;
- they, in turn, secrete chemical substances (pheromones).

The simplest ant model assumes an environment consisting of a 2-dimensional square lattice of cells (as in most cellular automata). Each cell can lie in one of two possible states, conventionally denoted as "1" and "0". The time is discretized and, at each time step, each ant is associated with a precise direction (typically a limited number of different directions is allowed, for instance only two: left or right) and a constant speed (one cell per time step). When the ant is on a cell in state "0", it turns right, otherwise it turns left (typically through a right angle). Immediately after the ant leaves a cell, the state of the latter changes. It can be shown mathematically (for a formal analysis of this model see, for instance, Moreira *et al.*, 2001) that the trajectory of an ant is always unlimited. Moreover, by resorting to these rules, it is possible, provided we introduce a suitable previous assignment of states to certain cells of the environment, to use the ant motion to compute Boolean functions. More precisely, the assignment of states mentioned above defines a *logical gate* which, roughly speaking, can be viewed as a sort of "corridor", located in the environment, into which the ant enters from one side and exits from the opposite one. On one wall of this corridor are located the cells implementing the *inputs* of the gate, while on the opposite wall we have the cells implementing its *outputs*. The initial assignment of states to the cells of the gate is made in such a way as to force the ant to follow a precise path within the gate itself. When the ant enters the gate, the states of all cells implementing the outputs are set to state "0" (corresponding to truth value "False"), while the cells implementing the inputs are set to states defined by truth values of actual inputs. After the ant has entered the gate, the internal arrangement of cell states constrains the ant to touch all input cells and, eventually, some output cells (which output cells are visited depends upon the kind of Boolean function implemented by the gate considered). Having followed the desired path, the ant exits from the gate. The states of the gate output cells after the ant's visit give the output of the Boolean function corresponding to the current input coded through the states of the gate input cells. From this description it follows that virtual ants can simulate any kind of logical machine. Not only, but arguments based upon Computation Theory lead to the conclusion that a virtual ant can even simulate any kind of Turing machine. Therefore, despite the simplicity of their behavioural rules, virtual ants appear as very complex systems.

A slight modification of the previous model gives rise to a set of virtual ants exhibiting a sort of collective intelligence. In particular, they can:
−   pick up food in the environment and store it in the nest;
−   gather together (cluster) scattered material.

These abilities are still based upon simple local rules. In the case of food collection the algorithm can be described under the form of a set of simple directives:

IF (the ant has no food) THEN
−   if it finds food, it must pick up it
−   if it does not find food, it must move in the direction of maximum food pheromone concentration
−   at each step it must secrete the food pheromone

IF (the ant has food) AND (is not in the nest) THEN
−   it must move in the direction of maximum nest pheromone concentration
−   at each step it must secrete the nest pheromone

The clustering algorithm, on the other hand, is of a probabilistic nature and is based upon the following rules:
−   the ant can collect an object if and only if it is located in the cell immediately ahead along its path;
−   the probability of collecting an object is given by:

$$P(Coll) = [ka/(ka + f)]^2$$

−   the probability of putting down a collected object is given by:

$$P(put) = [f/(kb + f)]^2$$

Here f denotes the fraction of neighbouring cells containing objects of the same kind, while ka and kb are suitable parameters.

The computer simulations showed how, starting from a disordered set of ants, each following a random path, these rules give rise to ordered queues of ants which, by always following the same paths, pick up in a fast and efficient way the food present in the environment to store it in an ordered way inside the nest, just as in the case of real ants. However these simulations, although fascinating, do not help us to understand why the rules just introduced produce the observed effects. And, without such an understanding, all notions of *emergent computation* so far produced within AL (see, amongst others, Forrest, 1990; Crutchfield e Mitchell, 1995; Gunji, 1995; Banzhaf *et al.*, 1996; Cariani, 1997; Lundh *et al.*, 1997; Sipper, 1997; Teuscher *et al.*, 2003; Tsuda *et al.*, 2004) are devoid of any meaning. Unfortunately, this understanding requires sophisticated mathematical methods, accessible only to researchers having a specific expertise. Namely, as is often the case in modelling emergence and self-organization, these methods are those of theoretical physics, in particular of statistical mechanics, precisely those enabling the study of phase transitions. Here, without entering into mathematical detail, we work out a simple example in order to illustrate the framework adopted when using analytical methods for studying systems such as ant swarms. The symbols and the procedure follow those appearing in a paper by Millonas (see Millonas, 1994).

Let us begin by denoting with $n_i(t)$ the number of ants occupying, at time t, the i-th cell of the environment (here we use a single label to distinguish the different cells, avoiding any reference to their two-dimensional coordinates) and with μ_i its volume (being in a 2-dimensional space, it would be better to call it an area, but we use the generic term 'volume' because we neglect in a deliberate way any reference to the dimensionality of the environment). The ant density at time t in the i-th cell will be given by:

$$\rho_i(t) = n_i(t) / \mu_i$$

We can now write a dynamical equation describing the change of ant density in the i-th cell owing to the input and output flow of ants. This equation has the form:

$$\mu_i \, d\rho_i / dt = v_0 \sum_j (\omega_{ij} \rho_j - \omega_{ji} \rho_i)$$

where v_0 is a parameter which can be identified with ant velocity, and the symbol ω_{ij} denotes the probability of an ant moving from the j-th cell to the i-th cell. The first term in parentheses on the right hand of this equation describes the input flow while the second term describes the output flow. Processes of escape from the environment or of injection of new ants into the environment have been neglected.

Of course the environment does not contain only ants. There are also chemical substances, which, for the sake of simplicity, are assumed to consist only of a single kind of pheromone, whose density at time t in the i-th cell will be denoted by $\sigma_i(t)$. This density will change with time, owing to two processes: spontaneous decay due to evaporation and secretion by ants in the considered cell. We assume that the equation governing the time evolution of pheromone density takes the form:

$$d\sigma_i /dt = - k \, \sigma_i + \eta \, \rho_i$$

where k and η are suitable parameters. If we could solve the system of the two coupled equations for ρ_i and σ_i we would obtain a complete description of ant dynamics. Unfortunately this is practically impossible, as, owing to the interactions among different cells (due to ant transition probabilities), we would be forced to consider as many pairs of equations as there are cells in the environment.

In order to proceed, we assume, at this point, that the time scale of pheromone evolution is very long with respect to the time scale of ant density evolution, so that, in the long run, the ant density will relax to its equilibrium value and the pheromone density evolution will depend, rather than upon ant density, upon this equilibrium value (this is merely an application of Haken's Slaving Principle described in Section 1 of this Chapter). On the basis of such an assumption, having introduced the auxiliary variables:

$$\tau = k \, t \quad , \quad s_i = (k/\eta) \, \sigma_i$$

the dynamical equation for pheromone density evolution takes the form:

$$ds_i /d\tau = - s_i + \rho_i *$$

where $\rho_i *$ is the aforementioned equilibrium value. In order to compute the latter we must specify the form of transition probabilities entering into the equation for ant density. More precisely, we assume these are given by an expression of the form:

$$\omega_{ij} = f(\sigma_i) \, \Omega_{ij} / \sum_r f(\sigma_r) \, \Omega_{rj}$$

where $f(\sigma_i)$ is a function of pheromone density to be specified (this is the one driving ant motion). The symbol Ω_{ij} takes a value of 1 if the i-th cell is a neighbour of the j-th cell and a value of zero otherwise. It is possible to show that this choice allows for the introduction of a statistical mechanics of the ant system along the lines used when dealing with systems of particles. Namely, it is possible to introduce an energy $U(\sigma)$ and a temperature T through the relationship:

$$f(\sigma) = exp[- \beta \, U(\sigma)] \quad , \quad \beta = 1/T$$

This, in turn, allows us to introduce a partition function and to use it to compute the equilibrium value $\rho_i *$. In order to perform the computations, however, we need to resort to a further simplification. More precisely, once we have introduced the mean density of ants in the environment, given by:

$$\gamma = N/V$$

where N is the total number of ants and V the total volume of environment, we introduce two further global variables, that is μ^+, given by the sum of the volumes of all cells in which s_i is greater than γ, and μ^-, given by the sum of the volumes of all cells in which s_i is less than γ. Such a definition allows the introduction of two other global variables, denoted as s^+ and s^-: the former is given by the sum of the values of s_i in all cells in which s_i is greater than γ, divided by μ^+, whereas the latter is the sum of the values of s_i in all cells in which s_i is less than γ, divided by μ^-. In short, s^+ is the average value of s_i in all cells in which s_i is greater than γ, and s^- is the average value of s_i in all cells in which s_i is less than γ. Now the simplification comes from assuming that the individual differences between the values of s_i can be neglected, so that s_i can assume only two possible values: s^+ or s^-. This hypothesis is usually called a *mean field hypothesis*.

All the previous hypotheses lead, after a number of computations which we omit for sake of brevity, to the following pair of dynamical equations for the time evolution of s^+ and s^-:

$$d s^+/d\tau = - s^+ + [\gamma (1 + v) R^{\beta} /(v + R^{\beta})]$$

$$d s^-/d\tau = - s^- + [\gamma (1 + v)/(v + R^{\beta})]$$

where:

$$R(s^+, s^-) = f(s^+)/f(s^-) \quad , \quad v = \mu^-/\mu^+$$

Even these equations, however, are fairly complicated. To further simplify things we introduce two auxiliary variables, which act as order parameters, given by:

$$m = s^+ - s^- \quad , \quad s = [(s^+ + v s^-)/(1 + v)] - \gamma$$

It can be shown that the equation for s has the form:

$$ds/d\tau = - s$$

so that s tends to zero with increasing time. If we make the adiabatic approximation (holding asymptotically) $s = 0$, then the equation for the order parameter m takes the form:

$$dm/d\tau = - m + \gamma(1 + v) \left[(R^{\beta} - 1)/(R^{\beta} + v)\right]$$

where:

$$R(m) = R\{\gamma + [v\, m/(1 + v)] , \gamma - [m/(1 + v)]\}$$

Let us now consider what occurs with particular choices for the form of $f(\sigma_i)$. If, for instance, we put:

$$f(\sigma_i) = A\,\sigma_i$$

where A is a suitable parameter, and putting $T = 1$, after a number of straightforward computations, we obtain the following equation for the order parameter m:

$$dm/d\tau = \{m \left[v^{2} (\gamma - 1)\right] - \gamma v\}/(\gamma + 2\,\gamma v + v^{2})$$

An equilibrium solution of this equation, obtained by putting $dm/d\tau = 0$, is given by:

$$m = \gamma / \left[v(\gamma - 1)\right]$$

This solution describes an inhomogeneous phase, as the homogeneous phase corresponds to the value $m = 0$. However, it will be stable only if γ is less than 1. In the contrary case it will be unstable. We thus obtain that there is a critical value $\gamma = 1$ where there is the possibility of a phase transition. In our model, of course, such a transition is not possible, as there is no other stable phase available to the ant system for $\gamma > 1$. If we had made a different choice for the form of the function f, we would have obtained a symmetry breaking phase transition. This, however, is not of interest to us, as the example was introduced only to illustrate the kind of arguments used to deal with a mathematical analysis of the ant system.

Methods partly differing from those presented above have been used to investigate other kinds of swarms (modelling, for instance, bird flocking behavior) whose components move actively in space and the individual motions are coordinated so as to obtain a collective coherent motion of the swarm as a whole. In these cases (for a review see Flierl *et al.*, 1999; Mikhailov and Calenbuhr, 2002), owing to the fact that space has the nature of a continuum, resort is made to the introduction of suitable *fields*, functions of spatial and temporal coordinates, describing densities and velocities of swarm components. Thus it is possible obtain formulations which generalize those usually adopted in hydrodynamics and field theory (see, for instance, Toner and Tu, 1998; Minati *et al.*, 2004). In this context the occurrence of a collective coherent behavior is also due to a phase transition in a model system obeying very simple rules. We can consider, for instance, the rule introduced by Vicsek *et al.* (see Vicsek *et al.*, 1995). It operates in a swarm whose components are endowed with the same absolute value of velocity, but whose individual directions are different. Adopting a discretized time scale, the rule states that, at any point in time *t*, the individual direction of each component is updated so that, the new direction is the average of the directions of neighbouring components plus the previous direction of the component considered. Computer simulations show that, with suitable choices of neighbourhood radius and absolute value of velocity, such a simple rule produces the occurrence of a common direction for all components, without the existence of a centralized coordination rule. However, in order to explain why such a rule works, one needs to resort to graph theory and to the mathematical theory of infinite products of matrices (Jadbabaie *et al.*, 2003).

From what has been said it can be seen that the phase transitions occurring in AL models can be of two different kinds: either they correspond to pattern formation associated with a bifurcation, just as in the case of dissipative structures (our example fits into this category), or they correspond to some sort of intrinsic emergence associated with the stochastic character of the considered models (when taking into account the role of temperature *T*). It would be interesting to know whether the latter kind of phase transition might be similar to symmetry-breaking phase transitions in QFT. In order to answer this question, however, we need to consider the role of noise, by resorting to the theory of stochastic processes. The next section of this Chapter is devoted to a brief introduction on this topic.

5.6 The role of noise

As in many domains of science, even the concept of *noise* is nothing but an abstraction, inspired by experimental observation of seemingly random phenomena, but never completely realized in nature. For reasons of mathematical convenience the most widely used kind of noise introduced in models is the so-called *white noise*. This expression denotes a stationary stochastic process whose spectral power density is constant over all frequencies. This implies that its autocorrelation function is directly proportional to a Dirac delta function or, in other words, that the output of this process at time *t* is completely uncorrelated with the output of the same process at time $t + \tau$, whatever the value of τ, provided it is not zero. The same property also implies that the variance of this process tends to infinity, a circumstance obviously conflicting with the constraints fulfilled by all real processes occurring in nature, be they random or not. People often make use of *Gaussian white noise*, that is white noise in which the output value at every instant is produced with a Gaussian probability distribution. This turns out to be very convenient, as a Gaussian distribution is completely defined by first and second order moments , that is by its mean and its variance, all other higher-order moments being identically zero. It is possible to obtain an intuitive idea of Gaussian white noise by resorting to a Markovian dichotomic process, which can be understood in more simple terms. The latter is a stochastic process $x(t)$ thus defined: it can take only two values, denoted by $+ a$ and $- a$ and, over any time interval of length *dt*, there is a probability $k\,dt$ of a jump between these two values. Let us now take the two limits $a \to \infty$ and $k \to \infty$, taking care to keep constant and finite the value of the quantity a^2 / k : we obtain precisely a Gaussian white noise process.

From previous considerations it is clear that any theory about the role of noise in emergent phenomena must rely heavily on the theory of stochastic processes. Here, however, being impossible to enter into mathematical details on this complex subject, we limit ourselves to mentioning some relevant findings on noise-induced transitions and on noise-induced phase transitions. We further restrict our considerations to the case in which the models are described through ordinary differential equations which, in the case of a single dependent variable *x*, will have the simple form:

$$dx/dt = f(x)$$

In this case we assume that noise can arise in two ways: either added directly as an external input which sums to the right hand member of the above equation (*additive noise*), or as a parameter entering in a multiplicative way into the form of function *f* which becomes noisy (*multiplicative noise*). In both cases, in the presence of noise, the previous equation changes its form and, in most cases, becomes:

$$dx/dt = g(x) + h(x)\,\xi(t)$$

where $\xi(t)$ denotes a noise process. The first term on the right hand side is the *deterministic part*, while the second term is the *noisy part*.

A simple example can illustrate this distinction. Assume that x denotes the population of a given living species and that the evolution of this population can be described by the well known Verhulst equation:

$$dx/dt = a\,x - b\,x^2$$

whose solution is the celebrated *logistic function*, often recalled within this book. If the value of parameter b is fluctuating with time, we can write it as:

$$b = b_0 + c\,\xi(t)$$

where b_0 denotes its average value and c is the fluctuation amplitude, given that $\xi(t)$ represents a noise process. By substituting the expression for b into the original Verhulst equation we then obtain:

$$dx/dt = (a\,x - b_0\,x^2) - c\,x^2\,\xi(t)$$

and the origin of the separation of deterministic and noisy parts is evident.

Differential equations containing a noisy part are called *stochastic differential equations*. Their theory, although far more difficult than that of normal differential equations, is now based on solid ground (valuable references are Soong, 1973; Arnold, 1974; more recent textbooks are Gard, 1988; Mao, 1998; Oksendal, 2003; Protter, 2003). In this regard two main points are worth noting:

- the solution of a stochastic differential equation does not consist of a function, but rather a stochastic process; all we can extract from the mathematical theory is how to find the main features of the transition probability distribution function associated with this process; we need, however, further correspondence rules (not contained in the mathematical formalism) in order to interpret the meaning of these features;

- in the same way as the solution of ordinary differential equations requires the introduction of an operation of *integration*, the solution of a stochastic differential equation requires the introduction of a *stochastic integral*; unfortunately, unlike normal integrals, there are various ways for defining a stochastic integral, the main ones being the *Ito integral* and the *Stratonovich integral*; the choice made for practical computations largely depends upon the adopted interpretation and, mainly, upon the context considered; generally speaking, we can say that most contexts encountered in modelling emergence require the Stratonovich interpretation and the associated stochastic integral; in the following examples we always adopt the latter.

Let $P(x, t)$ be the probability of transition from an initial value x_0, present at time $t = 0$, to a value x at time t. This describes the probability that the stochastic process $x(t)$ gives rise, at time t, just to the value x. Now it can be shown that, if the stochastic process obeys the stochastic differential equation:

$$dx/dt = g(x) + h(x) \, \xi(t)$$

where $\xi(t)$ is a Gaussian white noise process, then, by adopting the Stratonovich interpretation, the associated transition probability $P(x, t)$ obeys the *Fokker-Planck equation* (FPE):

$$\partial_t P = - \partial_x [g(x) \, P] + (\sigma^2/2) \, \partial_x \{h(x) \, \partial_x [h(x) \, P]\}$$

for which, to save symbols, we use the shortened notations:

$$\partial_t = \partial/\partial t \, , \quad \partial_x = \partial/\partial x$$

and σ^2 is the noise intensity. As can be seen, the FPE is a partial differential equation in the unknown function P. Its right-hand member is the sum of two different terms, the first called a *drift term*, and the second a *diffusion term*. The FPE was first introduced to describe Markov processes, such as Brownian motion. As a matter of fact, on introducing it, we have cast the solution process of our stochastic differential equation in the form of a Markov process.

The FPE is the main tool for studying the effects of noise on deterministic processes (for a review of these techniques see Van Kampen, 1981; Gardiner, 1983; Risken, 1984; Grasman, 1989; San Miguel and Toral, 2000). In principle it is rather difficult to solve, especially when dealing with multidimensional cases, that is with systems of stochastic differential equations. However, to extract the relevant information about the process solutions of stochastic differential equations we do not need to know the full solution of the FPE, it being sufficient to know the location of the maxima of the steady-state transition probability. Namely the latter are merely the stochastic counterpart of equilibrium points of ordinary differential equations. In the case of a single stochastic differential equation it is even possible to write the explicit form of a steady-state solution of the FPE.

Namely, if we set to zero the time derivative of P (as required if we are searching for the steady-state solution), then the FPE becomes (after a trivial transformation) a first-order linear ordinary differential equation in the dependent variable $P(x)$ (of course, the reference to the time variable has been suppressed, as we are dealing with a steady-state solution). By resorting to the theory of first-order linear differential equations (known to every undergraduate student of a scientific or technical faculty) it is possible to write in an explicit form this steady-state solution as:

$$P(x) = N \exp \left(\int_0^x \left\{ [\, g(y) - (\sigma^2/2)\, h(y)\, h'(y)]/[(\sigma^2/2)\, h^2(y)] \right\} dy \right)$$

where $h'(y)$ denotes in shorthand the derivative of $h(y)$ with respect to its argument and N is a suitable normalization constant. From this expression it follows that the values of x corresponding to the maxima of the steady-state solution are obtained by putting the denominator under the integral equal to zero, that is, they are solutions of the equation:

$$g(x) - (\sigma^2/2)\, h(x)\, h'(x) = 0$$

With these solutions two different cases can occur: either they coincide (except for a suitable shift) with the equilibrium points of the deterministic equation in the absence of noise, or we have the appearance of *new solutions* not corresponding to the equilibrium points of the deterministic equation. In the latter case we have a so-called *noise-induced transition* (for a more detailed discussion see Horsthemke and Lefever, 1984).

In order to understand how a noise-induced transition can occur, let us introduce a suitable modification of the Verhulst equation. More precisely, we assume that we are dealing with a population which, in absence of environmental influences, tends toward an equilibrium value α but, in presence of environmental influence, grows according to a logistic law. The differential equation obeyed by population x will therefore be:

$$dx/dt = \alpha - x + \beta x\,(1-x)$$

where β is a suitable parameter, describing the environmental influence. Elementary algebra shows that this equation allows for two equilibrium points:

$$x_1 = [(\beta-1) + D^{1/2}]/2\beta \ , \quad x_2 = [(\beta-1) - D^{1/2}]/2\beta$$

where:

$$D = (1 - \beta)^2 + 4\,\alpha\,\beta$$

A linear stability analysis shows also that, for every value of β, x_1 is always stable, whereas x_2 is always unstable.

Let us now assume that β fluctuates around an equilibrium value, so that we can write:

$$\beta = \beta_0 + \xi(t)$$

where $\xi(t)$ is a Gaussian white noise process. One can immediately see that in this case the functions $g(x)$ and $h(x)$ present in the previous equations are given by:

$$g(x) = \alpha - x + \beta_0\, x\,(1 - x) \;,\;\; h(x) = x\,(1 - x)$$

As a consequence the equation for the determination of the maxima of the steady-state solution assumes the explicit form:

$$\alpha - x + \beta_0\, x\,(1 - x) - (\sigma^2/2)\, x\,(1 - x)\,(1 - 2\,x) = 0$$

In order to simplify the calculations let us assume that $\alpha = \frac{1}{2}$ and $\beta_0 = 0$. The equation then allows for three roots:

$$x = \tfrac{1}{2} \;,\;\; x = (1/2)\{1 \pm [1 - (4/\sigma^2)]^{\frac{1}{2}}\}$$

For $\sigma^2 < 4$ only the first root is real. It corresponds to the unique stable equilibrium point of the deterministic equation. However, if $\sigma^2 > 4$, the other two roots become real and correspond to maxima of the steady-state solution of the FPE, while the first root corresponds to a minimum. In short, at the critical noise intensity $\sigma^2 = 4$, we have a symmetry-breaking bifurcation: the unique maximum of the steady-state solution splits into two different maxima. As this phenomenon occurs only if the noise intensity is greater than a critical value, the two maxima can be considered as the result of a noise-induced transition. We stress that in this case all the conditions characterizing intrinsic emergence are present: namely, owing to the stochastic character of the phenomenon, it is virtually impossible to foresee, for a given evolution of our population, whether it will relax toward one or the other of the two maxima of steady-state probability. Contrary to what happens in the case of bifurcation phenomena in differential equations, where an external influence can transform an ideal bifurcation into an imperfect bifurcation, here even an external influence cannot foretell which equilibrium state will be chosen, as the only effect of this influence is a shift in their values (provided, of course, the noise intensity is great enough).

These examples lead to the suspicion that noise-induced transitions bear some resemblance to spontaneous symmetry breaking in QFT. Recent researches have shown that this is more than a suspicion: perhaps the two theories are merely two sides of the same coin. As a matter of fact, when noise is introduced into spatially extended systems described, for instance, by partial differential equations, it has been observed how suitable noise intensities can give rise to *noise-induced phase transitions*, that is phase transitions toward more ordered states which would be impossible in the absence of noisy fluctuations (see Van den Broeck *et al.*, 1997; García-Ojalvo and Sancho, 1999; San Miguel and Toral, 2000; Carrillo *et al.*, 2003). It is interesting to remark that the formalism introduced to deal with the effects of noise in spatially extended systems is very similar to that already used in QFT. This raises a difficult question: is there some sort of equivalence relationship between QFT models and other, even traditional models of emergence?

5.7 The relationship between traditional and non-traditional models

One initial, trivial observation is that the FPE, introduced in the previous section, has a form strongly resembling that of the Schrödinger equation used in QM. The two can be made identical by introducing an imaginary time given by $\tau = i\,t$ into the Schrödinger equation which immediately becomes a sort of FPE. Of course, this is only a mathematical trick and, to proceed further, we need a physical interpretation. This consists of assuming that the probabilistic features of QM are simply a consequence of the fact that the ground state of the Universe is merely a noisy state, preventing the existence of truly deterministic phenomena. As a matter of fact, such a hypothesis was introduced in the 1970s by Nelson and De La Peña-Auerbach, who mathematically showed that the formalism of QM could be recovered when describing a system using classical mechanics, by embedding it into a stochastic, noisy, background (see Nelson, 1967). Then, in 1976, Doi showed how a stochastic many-body system could be described by resorting to the formalism of *second quantization*, one of the main technical tools of QFT (see Doi, 1976; Peliti, 1985). This line of research gave rise to so-called *Statistical Field Theory* (see Parisi, 1988; Itzykson and

Drouffe, 1989a; 1989b; Parisi, 1998; Chaichian and Demichev, 2001). Without entering into too much detail, we can say that this approach studies quantum fields by viewing them as deterministic entities influenced by noise in a context in which time is imaginary. The advantage lies in the fact that, in this way, we can start only from dynamical equations, without the need for a Lagrangian or Hamiltonian structure, which in many models is absent. But other results have shown how the analysis of noisy systems could even give rise to a profound revision of QFT. For instance, Fogedby (Fogedby, 1998; Fogedby and Brandenburg, 2002), when studying the noisy Burgers equation:

$$\partial u/\partial t = \nu \, \nabla^2 u + \lambda \, u \, \nabla u + \nabla \eta \ ,$$

where ν is a damping constant, or a viscosity, λ a coupling coefficient, and η a Gaussian white noise satisfying:

$$<\eta(x\ ,t)\ \eta(x'\ ,t')> = \Delta\ \delta(x-x')\ \delta(t-t')\ ,$$

found that the theory of this equation could be recast exactly in the form of an equivalent QFT, provided the usual Planck constant $\hbar$ is identified with the quantity Δ/ν. In other words, the Planck constant could lose its unique value and, in a noisy context, we could have different "Planck constants" as a function of noise intensity. Moreover, other authors (see Cardy, 1996) were able to recast a stochastic reaction-diffusion model described by a suitable master equation in terms of an equivalent QFT, in which the value of the 'Planck constant' was equal to 1. All these results point to the fact that a number of stochastic models could, from a formal point of view, be reformulated in such a way as to take on the appearance of a QFT-based model. Of course, in each case we would have to redefine in a suitable way the 'Planck constant' of the system. Such circumstances seem to suggest that, once granted the presence of the three fundamental ingredients of intrinsic emergence, that is non-linearity, spatial extension and fluctuations, all theories can be found equivalent to one another, at least with regard to their formal structure. This opens a new perspective on non-ideal models of emergence. If the above claim were true, then a non-ideal model, provided it is endowed with noisy fluctuations, should have a good probability of being already equivalent to a QFT model, without the need for quantizing it.

While it is difficult to verify whether or not these arguments could be applied to Collective Beings, we stress that the methods illustrated in this Chapter must be considered useful even within this context. Namely, in many situations the model of a Collective Being can be reduced to one of the cases we have illustrated above, or to a suitable combination or generalization of them. The main difficulties, lie not so much in generalizing the above considerations for describing emergence in Collective Beings, but in the fact that most simple models, at a lower level than Collective Beings, are still waiting for a theoretical analysis. Throughout this Chapter we have stressed the mathematical aspects of the different approaches precisely for this reason: the mathematical description provides an understanding of the true limitations of current models of emergence along with their practical significance and their effective extent. As we have shown, the models introduced so far are very poor and highly problematic. However, current efforts and the progress obtained, lead us to the assumption that many of these limitations will be overcome in the near future.

References

Akhiezer, A. and Péletminski, S., 1980. *Les méthodes de la physique statistique.* Mir, Moscow.

Akhmediev, N. and Soto-Crespo, J.M., 2003. Exploding solitons and Shil'nikov's theorem. *Physics Letters A*, **317**: 287-292.

Akhromeeva, T.S., Kurdyumov, S.P., Malinetskiĭ, G.G. and Samarskiĭ, A.A., 1984. Classification of two-component systems near a bifurcation point. *Soviet Physics Doklady*, **29**: 911-913.

Alfinito, E., Viglione, R.G. and Vitiello, G., 2001. The decoherence criterion. *Modern Physics Letters B*, **15**: 127-136.

Amit, D.J., 1984. *Field Theory, the Renormalization Group, and Critical Phenomena.* World Scientific, Singapore.

Amit, D.J., 1989. *Modeling Brain Function. The world of Attractor Neural Networks.* Cambridge University Press, Cambridge, UK.

Arnold, L., 1974. *Stochastic Differential Equations: Theory and applications.* Wiley, New York.

Balashov, Y., 2002. What is a law of nature? The broken-symmetry story. *Southern Journal of Philosophy*, **40**: 459-473.

Banzhaf, W., Dittrich, P. and Rauhe, H., 1996. Emergent computation by catalytic reactions. *Nanotechnology*, **7**: 1-8.

Bartlett, P.L. and Anthony, M.M., 1999. *Neural Network Learning: Theoretical Foundations.* Cambridge University Press, Cambridge, UK.

Benfatto, G. and Gallavotti, G., 1995. *Renormalization Group.* Princeton University Press, Princeton, NJ.

Bianchini, M., Fiasconi, P., Gori, M. and Maggini, M., 1998. Optimal learning in artificial neural networks: A theoretical view. In C.Leondes (Ed.), *Neural Network Systems Techniques and Applications* (pp. 1-51). Academic Press, New York.

Bishop, C.M., 1995. *Neural networks for pattern recognition.* Oxford University Press, Oxford, UK.

Bonabeau, E., Dorigo, M. and Theraulaz, G., 1999. *Swarm Intelligence: From natural to artificial systems.* Oxford University Press, Oxford, UK.

Brading, K. and Castellani, E. (Eds.), 2003. *Symmetries in physics: philosophical reflections.* Cambridge University Press, Cambridge, UK.

Brown, L.M. and Cao, T.Y., 1991. Spontaneous breakdown of symmetry: its rediscovery and integration into quantum field theory. *Historical Studies in the Physical and Biological Sciences*, **21**: 211-235.

Burgess, C.P., 2000. Goldstone and pseudo-Goldstone bosons in nuclear, particle and condensed-matter physics. *Physics Reports*, **330**: 193-261.

Burks, A.W. (Ed.), 1970. *Essays on Cellular Automata.* Illinois University Press, Urbana, IL.

Callen, H., 1960. *Thermodynamics.* Wiley, New York.

Camazine, S., Deneubourg, J.L., Franks, N.R., Sneyd, J., Theraulaz, G. and Bonabeau, E., 2001. *Self-Organization in biological systems.* Princeton University Press, Princeton, NJ.

Cao, T.Y. (Ed.), 1999. *Conceptual Foundations of Quantum Field Theory.* Cambridge University Press, Cambridge, UK.

Carati, A. and Galgani, L., 2001. Theory of dynamical systems and the relations between classical and quantum mechanics. *Foundations of Physics*, **31**: 69-87.

Cardy, J., 1996. *Scaling and Renormalization in Statistical Physics.* Cambridge University Press, Cambridge, UK.

Cariani, P., 1997. Emergence of new signal-primitives in neural networks. *Intellectica*, **2**: 95-143.

Carrillo, O., Ibañes, M., García-Ojalvo, J., Casademunt, J. and Sancho, J.M., 2003. Intrinsic noise-induced phase transitions: Beyond the noise interpretation. *Physical Review E,* **67**: 046110, 1-9.

Castro, J.L., Mantas, C.J. and Benítez, J.M., 2000. Neural networks with a continuous squashing function in the output are universal approximators. *Neural Networks*, **13**: 561-563.

Celeghini, E. and Rasetti, M., 2000. Bosons and environment. In R.C.Hilborn and G.M.Tino (Eds.). *Spin-Statistics connection and commutation relations* (pp. 59-66). American Institute of Physics, New York.

Chaichian, M. and Demichev, A., 2001. *Path Integrals in Physics. Volume 2: Quantum Field Theory, Statistical Physics and other modern applications.* IOP Press, Bristol, UK.

Chate, H. and Manneville, P., 1996. Phase diagram of the two-dimensional complex Ginzburg-Landau equation. *Physica A*, **224**: 348-368.

Chou, H.-H. and Reggia, J.A., 1997. Emergence of self-replicating structures in a cellular automata space. *Physica D*, **110**: 252-276.

Chua, L.O. and Roska, T., 1993. The CNN Paradigm. *IEEE Transactions on Circuits and Systems*, **40**: 147-156.

Chua, L.O. and Yang, L., 1988. Cellular Neural Networks: Theory and applications. *IEEE Transactions on Circuits and Systems*, **35**: 1257-1290.

Clifton, R.K. and Halvorson, H.P. , 2001. Entanglement and open systems in algebraic quantum field theory. *Studies in the History and Philosophy of Modern Physics*, **32**: 1-31.

Codd, E.F., 1968. *Cellular Automata.* Academic Press, New York.

Cohen-Tannoudji, C., Diu, B. and Laloë, F., 1977. *Quantum Mechanics* (2 voll.). Wiley, New York.

Crasovan, L-C., Malomed, B.A. and Mihalache, D., 2001. Spinning solitons in cubic-quintic nonlinear media. *PRAMANA – Journal of Physics*, **57**: 1041-1059.

Crook, N., olde Scheper, T. and Pathirana, V., 2003. Self-organised dynamic recognition states for chaotic neural networks. *Information Sciences*, **150**: 59-75.

Crutchfield, J.P. and Mitchell, M., 1995. The evolution of emergent computation. *Proceedings of the National Academy of Sciences USA*, **92**: 10742-10746.

Dasgupta, D. (Ed.), 1999. *Artificial Immune Systems and their applications*. Springer, Berlin-Heidelberg-New York.

Davydov, A.S., 1976. *Quantum Mechanics*. Pergamon Press, Oxford, UK.

De Boer, J. and Uhlenbeck, G.E. (Eds.), 1962. *Studies in statistical mechanics, vol.I*. North Holland, Amsterdam.

Derrida, B. and Stauffer, D., 1986. Phase transition in two dimensional Kaufmann cellular automata. *Europhysics Letters*, **2**: 739-745.

Doi, M., 1976. Second quantization representation for classical many-particle system. *Journal of Physics A*, **9**: 1465-1477.

Domany, E., Van Hemmen, J.L. and Schulten, K. (Eds.), 1996. *Models of Neural Networks III: Association, Generalization, and Representation (Physics of Neural Networks)*. Springer, Berlin-Heidelberg-New York.

Domb, C., 1996. *The critical point*. Taylor and Francis, London.

Dotsenko, V., 1994. *An introduction to the theory of Spin Glasses and Neural Networks*. World Scientific, Singapore.

Drugowich de Felício, J.R. and Hipólito, O., 1985. Spontaneous symmetry breaking in a simple mechanical model. *American Journal of Physics*, **53**: 690-693.

Fiasconi, P., Gori, M. and Tesi, A., 1997. Successes and failures of backpropagation: A theoretical investigation. In O.Omidvar and C.Wilson (Eds.). *Progress in Neural Networks* (pp. 205-242). Alex Publishing, Norwood, NJ.

Flierl, G., Grunbaum, D., Levin, S. and Olson, D., 1999. From individuals to aggregations: The interplay between behavior and physics. *Journal of Theoretical Biology*, **196**: 397-454.

Fogedby, H.C., 1998. Soliton approach to the noisy Burgers equation. Steepest descent method. *Physical Review E*, **57**: 4943-4968.

Fogedby, H.C. and Brandenburg, A., 2002. Solitons in the noisy Burgers equation. *Physical Review E*, **66**: 016604, 1-9.

Forrest, S., 1990. *Emergent computation*. MIT Press, Cambridge, MA.

Freeman, W.J., 2000. *Neurodynamics: An Exploration of Mesoscopic Brain Dynamics*. Springer, Berlin-Heidelberg-New York.

Frisch, U., Hasslacher, B. and Pomeau, Y., 1986. Lattice-gas automata for the Navier-Stokes equation. *Physical Review Letters*, **56**: 1505-1508.

Gale, D., 1998. *Tracking the automatic ant and other mathematical explorations*. Springer, New York.

Gallavotti, G., 1985. Renormalization theory and ultraviolet stability via renormalization group methods. *Reviews of Modern Physics*, **57**: 471-569.

García-Ojalvo, J. and Sancho, J.M., 1999. *Noise in spatially extended systems*. Springer, New York.

Gard, T.G., 1988. *Introduction to Stochastic Differential Equations*. Marcel Dekker, New York.

Gardiner, C.W., 1983. *Handbook of Stochastic Methods for physics, chemistry and the natural sciences*. Springer, Berlin-Heidelberg-New York.

Gardner, M., 1970. Mathematical Games. The fantastic combinations of John Conway's new solitaire game "life". *Scientific American*, **223**, n.4: 120-123.

Giulini, D., Joos, E., Kiefer, C., Kupsch, J., Stamatescu, I.-O. and Zeh, H.D., 1996. *Decoherence and the appearance of a classical world in Quantum Theory*. Springer, Berlin.

Goldberg, D., 1989. *Genetic algorithms in search, optimization, and machine learning*. Addison-Wesley, Reading, MA.

Goldenfeld, N., 1992. *Lectures on phase transitions and the renormalization group.* Addison-Wesley, Reading, MA.

Goldstone, J., Salam, A. and Weinberg, S., 1962. Broken symmetries. *Physical Review,* **127**: 965-970.

Grasman, J., 1989. Asymptotic analysis of nonlinear systems with small stochastic perturbations. *Mathematics and Computers in Simulation,* **31**: 41-54.

Greenberger, D.M., 1978. Esoteric elementary particle phenomena in undergraduate physics – spontaneous symmtery breaking and scale invariance. *American Journal of Physics,* **46**: 394-398.

Griffiths, R.B., 1981. What's wrong with real-space renormalisation group transformations. In J.Fritz, J.L.Lebowitz and D.Szasz (Eds.). *Random fields,* vol. I (pp. 463-479). North Holland, Amsterdam.

Grössing, G. and Zeilinger, A., 1988. Quantum cellular automata. *Complex Systems,* **2**: 197-208.

Gunji, Y.-P., 1995. Global logic resulting from disequilibrium process. *BioSystems,* **38**: 127-133.

Haag, R., 1961. Canonical commutation relations in Quantum Field Theory and Functional Integration. In W.E.Brittin, B.W.Downs and J.Downs (Eds.). *Lectures in Theoretical Physics, vol. 3* (pp. 353-381). Wiley, New York.

Haken, H., 1978. *Synergetics. An introduction.* Springer, Berlin.

Haken, H., 1983. *Advanced Synergetics.* Springer, Berlin.

Haken, H., 1988. *Information and Self-Organization. A macroscopic approach to complex systems.* Springer, Berlin.

Halvorson, H.P., 2001. On the nature of continuous physical quantities in classical and quantum mechanics. *Journal of Philosophical Logic,* **37**: 27-50.

Halvorson, H.P., 2002. No place for particles in relativistic quantum theories? *Philosophy of Science,* **69**: 1-28.

Halvorson, H.P., 2004. Complementarity of representations in quantum mechanics. *Studies in History and Philosophy of Modern Physics,* **35**: 45-56.

Hardy, J., de Pazzis, O. and Pomeau, Y., 1976. Molecular dynamics of a classical lattice gas: transport properties and time correlation functions. *Physical Review A,* **13**: 1949-1961.

Haykin, S., 1998. *Neural networks; A comprehensive foundation* (2nd edition). Prentice Hall, Upper Saddle River, NJ.

Heylighen, F., 1990. Classical and non-classical representations in Physics II: Quantum Mechanics. *Cybernetics and Systems,* **21**: 477-502.

Hepp, K., 1972. Quantum theory of measurement and macroscopic observables. *Helvetica Physica Acta,* **45**: 237-248.

Hinton, G.E. and Sejnowski, T.J. (Eds.), 1999. *Unsupervised Learning (Computational Neuroscience).* MIT Press, Cambridge, MA.

Hopfield, J.J., 1982. Neural networks and physical systems with emergent collective computational abilities. *Proceedings of the National Academy of Sciences of USA,* **79**: 2554-2558.

Hornik, K., Stinchcombe, M. and White, H., 1989. Multilayer feedforward networks are universal approximators. *Neural Networks,* **2**: 359-366.

Horsthemke, W. and Lefever, R., 1984. *Noise-induced transitions.* Springer, Berlin-Heidelberg-New York.

Hoyuelos, M., Hernández-García, E., Colet, P. and San Miguel, M., 2003. Dynamics of defects in the vector complex Ginzburg-Landau equation. *Physica D,* **174**: 176-197.

Huang, K., 1987. *Statistical mechanics,* 2nd edition. Wiley, New York.

Huang, K., 1998. *Quantum Field Theory: From operators to path integrals.* Wiley, New York.

Ilachinski, A., 2001. *Cellular Automata: A discrete Universe.* World Scientific, Singapore.

Ipsen, M., Kramer, L. and Sørensen, P.G., 2000. Amplitude equations for description of chemical reaction-diffusion systems. *Physics Reports,* **337**: 193-235.

Itzykson, C. and Drouffe, J.-M., 1989a. *Statistical Field Theory: Volume 1, from Brownian motion to Renormalization and Lattice Gauge Theory.* Cambridge University Press, Cambridge, UK.

Itzykson, C. and Drouffe, J.-M., 1989b. *Statistical Field Theory: Volume 2, Strong Coupling, Monte Carlo methods, Conformal Field Theory and Random Systems.* Cambridge University Press, Cambridge, UK.

Itzykson, C. and Zuber, J.B., 1986. *Quantum Field Theory.* McGraw-Hill, Singapore.

Izhikevich, E.M., 2003. Simple model of spiking neurons. *IEEE Transactions on Neural Networks,* **14**: 1569-1572.

Izhikevich, E.M., 2004. Which model to use for cortical spiking neurons? *IEEE Transactions on Neural Networks.* **15**: 1063-1070.

Jadbabaie, A., Lin, J. and Morse, A.S., 2003. Coordination of groups of mobile autonomous agents using nearest neighbor rules. *IEEE Transactions on Automatic Control,* **48**: 988-1001.

Jang, J.-S.R., Sun, C.-T. and Mizutani, E., 1997. *Neuro-fuzzy ans soft computing. A computational approach to learning and machine intelligence.* Prentice Hall, Upper Saddle River, NJ.

Johnson, S., 2002. *Emergence: The connected lives of Ants, Brains, Cities and Software.* Touchstone, New York.

Jona-Lasinio, G., 2001. Renormalization group and probability theory. *Physics Reports,* **352**: 439-458.

Kamp, Y. and Hasler, M., 1990. *Recursive neural networks for associative memory.* Wiley, Chichester, UK.

Kennedy, J., Eberhart, R.C. and Shi, Y., 2001. *Swarm intelligence.* Morgan Kaufmann, San Francisco, CA.

Kiselev, V.G., Shnir, Ya.M. and Tregubovich, A.Ya., 2000. *Introduction to Quantum Field Theory.* Gordon and Breach, Amsterdam.

Koch, C., 1998. *Biophysics of Computation: Information Processing in Single Neurons.* Oxford University Press, Oxford, UK.

Kohonen, T., 1995. *Self-organizing maps.* Springer, Berlin-Heidelberg-New York.

Koma, T. and Tasaki, H., 1994. Symmetry breaking and finite size effects in quantum many-body systems. *Journal of Statistical Physics,* **76**: 745-803.

Kosso, P., 2000. The epistemology of spontaneously broken symmetries. *Synthese,* **122**: 359-376.

Kozek, T., Chua, L.O., Roska, T., Wolf, D., Tezlaff, R., Puffer, F. and Lotz, K., 1995. Simulating nonlinear waves and partial differential equations via CNN – Part II: Typical examples. *IEEE Transactions on Circuits and Systems,* **42**: 816-820.

Kuhlmann, M., Lyre, H. and Wayne, A. (Eds.), 2002. *Ontological aspects of Quantum Field Theory.* World Scientific, Singapore.

Kuramoto, Y. and Tsuzuki, T., 1975. On the formation of dissipative structures in reaction-diffusion systems: Reductive perturbation approach. *Progress of Theoretical Physics,* **54**: 687-699.

Lahiri, A. and Pal, P.B., 2001. *A first book of Quantum Field Theory.* CRC Press, Boca Raton, FL.

Landau, L.D. and Lifshitz, E.M., 1959. *Statistical Physics*. Pergamon Press, London.

Landau, L.D. and Lifshitz, E.M., 1981. *Quantum Mechanics (Non-relativistic theory)*. Butterworth Heinemann, Oxford, UK.

Langton, C.G. (Ed.), 1989. *Artificial Life*. Addison-Wesley, Reading, MA.

Langton, C.G., 1990. Computation at the edge of chaos: Phase transitions and emergent computation. *Physica D*, **42**: 12-37.

Langton, C.G. (Ed.), 1994. *Artificial Life III*. Addison-Wesley, Reading, MA.

Langton, C.G. (Ed.), 1997. *Artificial Life: An overview (Complex adaptive systems)*. MIT Press, Cambridge, MA.

Langton, C.G., Taylor, C., Doyne Farmer, J. and Rasmussen, S. (Eds.), 1992. *Artificial Life II*. Addison-Wesley, Reading, MA.

Le Gal, P., Ravoux, J.F., Floriani, E. and Dudok de Wit, T. , 2003. Recovering coefficients of the complex Ginzburg-Landau equation from experimental spatio-temporal data: two examples from hydrodynamics. *Physica D*, **174**: 114-133.

Lee, T.D. and Yang, C.N., 1952. Statistical theory of equations of state and phase transitions. *Physical Review*, **87**: 410-419.

Leutwyler, H., 1997. Phonons as Goldstone bosons. *Helvetica Physica Acta*, **70**: 275-286.

Liu, C., 1999. Explaining the emergence of cooperative phenomena. *Philosophy of Science*, **66**: S92-S106.

Liu, C., 2003. Spontaneous symmetry breaking and chance in classical world. *Philosophy of Science*, **70**: 590-608.

Lombardo, F.C., Rivers, R.J., and Mazzitelli, F.D., 2002. Classical behavior after a phase transition: I. Classical order parameters. *International Journal of Theoretical Physics*, **41**: 2121-2144.

Lundh, D., Narayanan, A. and Olsson, B., 1997. *Biocomputing and emergent computation*. World Scientific, Singapore.

Maass, W., 1997. Networks of spiking neurons: The third generation of neural network models. *Neural Networks*, **10**: 1659-1671.

Maass, W. and Bishop, C.M., 1999. *Pulsed Neural Networks*. MIT Press, Cambridge, MA.

Maggiore, M., 2005. *A modern introduction to Quantum Field Theory*. Oxford University Press, Oxford, UK.

Mao, X., 1998. *Stochastic Differential Equations and applications*. Albion/Horwood Publishing, Chichester, UK.

Marinaro, M. and Scarpetta, S., 2000. On-line learning in RBF neural networks: a stochastic approach. *Neural Networks*, **13**: 719-729.

Matsumoto, H., Sodano, P. and Umezawa, H., 1979. Extended objects in quantum systems and soliton solutions. *Physical Review D*, **19**: 511-516.

Mermin, N.D. and Wagner, H., 1966. Absence of ferromagnetism or antiferromagnetism in one- or two-dimensional isotropic Heisenberg models. *Physical Review Letters*, **17**: 1133-1136.

Merzbacher, E., 1998. *Quantum Mechanics* (3rd edition). Wiley, New York.

Messiah, A., 2000. *Quantum Mechanics* (2 voll.). Dover, New York.

Meyer, D.A., 1996. From quantum cellular automata to quantum lattice gases. *Journal of Statistical Physics*, **85**: 551-574.

Mézard, M., Parisi, G. and Virasoro, M.A., 1987. *Spin Glass Theory and Beyond*. World Scientific, Singapore.

Mikhailov, A.S. , 1990. *Foundations of Synergetics I. Distributed active systems*. Springer, Berlin.

Mikhailov, A.S. and Calenbuhr, V., 2002. *From cells to societies. Models of complex coherent actions*. Springer, Berlin.

Mikhailov, A.S. and Loskutov, A,Yu., 1996. *Foundations of Synergetics II. Chaos and Noise*, 2nd revised edition. Springer, Berlin.

Millonas, M.M., 1994. Swarms, phase transitions, and collective intelligence. In C.G.Langton (Ed.). *Artificial Life III* (pp. 417-445). Addison-Weley, Reading, MA..

Minati, G., Penna, M.P. and Pessa, E., 2004. Collective phenomena in living systems and in social organizations. *Chaos and Complexity Letters*, 1: 173-184.

Mitchell, M., Hraber, P.T. and Crutchfield, J.P., 1993. Revisiting the edge of chaos: Evolving cellular automata to perform computations. *Complex Systems*, 7: 89-130.

Montvay, I. and Münster, G., 1997. *Quantum Fields on a Lattice*. Cambridge University Press, Cambridge, UK.

Moreira, A., Gajardo, A. and Goles, E., 2001. Dynamical behavior and complexity of Langton's ant. *Complexity*, 6: 46-51.

Nelson E., 1967. *Dynamical theories of Brownian motion*. Princeton University Press, Princeton, NJ.

Newman, C.M., 1997. *Topics in disordered systems*. Birkhäuser, Basel, CH.

Nitzan, A. and Ortoleva, P., 1980. Scaling and Ginzburg criteria for critical bifurcations in nonequilibrium reacting systems. *Physical Review A*, 21: 1735-1755.

Oksendal, B., 2003. *Stochastic Differential Equations: An introduction with applications* (5th ed.). Springer, Berlin-Heidelberg-New York.

Olemskoi, A.I. and Klepikov, V.F., 2000. The theory of spatiotemporal pattern in nonequilibrium systems. *Physics Reports*, 338: 571-677.

Parisi, G., 1988. *Statistical Field Theory*. Addison-Wesley, Redwood City, CA.

Parisi, G., 1998. *Statistical Field Theory*.(New edition). Perseus Books, New York.

Patashinskij, A.Z. and Pokrovskij, V.L., 1979. *Fluctuation theory of phase transitions*. Pergamon Press, Oxford, UK.

Peliti, L., 1985. Path integral approach to birth-death processes on a lattice. *Journal de Physique*, 46 : 1469-1483.

Perelson, A.S. and Weisbuch, G., 1997. Immunology for physicists. *Reviews of Modern Physics*, 69: 1219-1267.

Peskin, M.E. and Schroeder, D.V., 1995. *An introduction to Quantum Field Theory*. Addison-Wesley, Reading, MA.

Pessa, E., 1988. Symmetry breaking in neural nets. *Biological Cybernetics*, 59: 277-281.

Pessa, E., 2000. Connectionist Psychology and Synergetics of Cognition. In A.Carsetti (Ed.), *Functional Models of Cognition. Self-Organizing Dynamics and Semantic Structures in Cognitive Systems* (pp. 67-90). Kluwer, Dordrecht.

Pessa, E. and Vitiello, G., 2004a. Quantum noise, entanglement and chaos in the Quantum Field Theory of Mind/Brain states. *Mind and Matter*, 1: 59-79.

Pessa, E. and Vitiello, G., 2004b. Quantum noise induced entanglement and chaos in the dissipative quantum model of brain. *International Journal of Modern Physics B*. 18: 841-858.

Porod, W., 1997. Quantum-dot devices and quantum-dot cellular automata. *Journal of Franklin Institute*, 334B: 1147-1175.

Protter, P.E., 2003. *Stochastic integration and differential equations* (2nd ed.). Springer, Berlin-Heidelberg-New York.

Rajantie, A., 2002. Formation of topological defects in gauge field theories. *International Journal of Modern Physics A*, 17: 1-44.

Rédei, M. and Stölzner, M. (Eds.), 2001. *John Von Neumann and the Foundations of Quantum Physics.* Kluwer, New York.

Risken, H., 1984. *The Fokker-Planck equation.* Springer, Berlin-Heidelberg-New York.

Rivers, R.J., 2000. Fluctuations and phase transition dynamics. *International Journal of Theoretical Physics*, **39**: 1779-1802.

Rivers, R.J., Lombardo, F.C. and Mazzitelli, F.D. 2002. Classical behavior after a phase transition: II. The formation of classical defects. *International Journal of Theoretical Physics*, **41**: 2145-2160.

Rojas, R., 1996. *Neural networks. A systematic introduction.* Springer, Berlin-Heidelberg-New York.

Roska, T., Chua, L.O., Wolf, D., Kozek, T., Tezlaff, R. and Puffer, F., 1995. Simulating nonlinear waves and partial differential equations via CNN – Part I: Basic techniques. *IEEE Transactions on Circuits and Systems*, **42**: 807-815.

Ruelle, D., 1969. *Statistical mechanics.* Benjamin, New York.

Ruetsche, L., 2003. A matter of degree: Putting unitary inequivalence to work. *Philosophy of Science*, **70**: 1329-1342.

Rumer, Yu.B. and Rivkyn, M.Sh., 1980. *Thermodynamics, Statistical Physics, and Kinetics.* Mir, Moscow.

Saad, D. (Ed.), 1998. *On-line Learning in Neural Networks.* Cambridge University Press, Cambridge, UK.

Sakurai, J.J., 1995. *Modern Quantum Mechanics* (revised edition by S.F.Tuan). Benjamin/Cummings, Menlo Park, CA.

San Miguel, M. and Toral, R., 2000. Stochastic effects in physical systems. In E.Tirapegui, J.Martinez, R.Tiemann (Eds.). *Instabilities and Nonequilibrium Structures VI* (pp. 35-130). Kluwer, New York.

Schwinger, J. (Ed.), 1958. *Selected papers on Quantum Electrodynamics.* Dover, New York.

Sears, F.W., 1955. *Thermodynamics* , 2nd ed.. Addison-Wesley, Reading, MA.

Sewell, G.L., 1986. *Quantum Theory of Collective Phenomena.* Oxford University Press, Oxford, UK.

Sewell, G.L., 2002. *Quantum Mechanics and its emergent macrophysics.* Princeton University Press, Princeton, NJ.

Shi, Y., 2003. Quantum disentanglement in long-range orders and spontaneous symmetry breaking. *Physics Letters A*, **309**: 254-261.

Shirkov, D.V. and Kovalev, V.F., 2001. The Bogoliubov renormalization group and solution symmetry in mathematical physics. *Physics Reports*, **352**: 219-249.

Sinai, Ya.G., 1982. *Theory of Phase Transitions: Rigorous Results.* Pergamon Press, London.

Sipper, M., 1997. *Evolution of Parallel Cellular Machines.* Springer, Berlin.

Sirovich, L. and Newton, P.K., 1986. Periodic solutions of the Ginzburg-Landau equation. *Physica D*, **21**: 115-125.

Sivardière, J., 1983. A simple mechanical model exhibiting a spontaneous symmetry breaking. *American Journal of Physics*, **51**: 1016-1018.

Smolensky, P., 1986. Information processing in dynamical systems: Foundations of Harmony Theory. In D.E.Rumelhart, J.L. McClelland (Eds.). *Parallel Distributed Processing. Explorations in the microstructure of cognition*, vol. 1, pp. 194-281. MIT Press, Cambridge, MA.

Snider, G.L., Orlov, A.O., Amlani, I., Zuo, X., Bernstein, G.H., Lent, C.S., Merz, J.L. and Porod, W., 1999. Quantum-dot cellular automata: Review and recent experiments. *Journal of Applied Physics*, **85**: 4283-4285.

Soong, T.T., 1973. *Random Differential Equations in science and engineering.* Academic Press, New York.

Stanley, H.E., 1971. *Introduction to phase transitions and critical phenomena.* Clarendon Press, Oxford, UK.

Stewart, J. and Varela, F.J., 1991. Morphogenesis in shape-space. Elementary meta-dynamics in a model of the Immune Network. *Journal of Theoretical Biology*, **153**: 477-498.

Stone, M., 2000. *The Physics of Quantum Fields.* Springer, Berlin.

Sun, R. and Giles, C. (Eds.), 2000. *Sequence learning.* Springer, Berlin.

Sutton, R.S. and Barto, A.G., 1998. *Reinforcement learning. An introduction.* MIT Press, Cambridge, MA.

Svozil, K., 1986. Are quantum fields cellular automata? *Physics Letters A*, **119**: 153-156.

t'Hooft, G., 2003. Can quantum mechanics be reconciled with cellular automata? *International Journal of Theoretical Physics*, **42**: 349-354.

Tegmark, M., 2000. Why the brain is probably not a quantum computer. *Information Sciences*, **128**: 155-179.

Teuscher, C., Mange, D., Stauffer, A. and Tempesti, G., 2003. Bio-inspired computing tissues: towards machines that evolve, grow, and learn. *BioSystems*, **68**: 235-244.

Thiesen, S. and Thomas, H., 1987. Bifurcation, stability, and symmetry of nonlinear waves. *Zeitschrift für Physik B – Condensed Matter*, **65**: 397-408.

Toko, K. and Yamafuji, K., 1990. Bifurcation characteristics and spatial patterns in an integro-differential equation. *Physica D*, **44**: 459-470.

Tolédano, J.C. and Tolédano, P., 1987. *The Landau theory of phase transitions.* World Scientific, Singapore.

Toner, J. and Tu, Y., 1998. Flocks, herds, and schools: A quantitative theory of flocking. *Physical Review E*, **58**: 4828-4858.

Tougaw, P.D. and Lent, C.S., 1996. Dynamic behavior of quantum cellular automata. *Journal of Applied Physics*, **80**: 4722-4736.

Tsuda, S., Aono, M. and Gunji, Y.-P., 2004. Robust and emergent *Physarum* logical-computing. *BioSystems*, **73**: 45-55.

Umezawa, H., 1993. *Advanced Field Theory. Micro, Macro, and Thermal Physics.* American Institute of Physics, New York.

Umezawa, H., Matsumoto, H. and Tachiki, M., 1982. *Thermo Field Dynamics and Condensed States.* North Holland, Amsterdam.

Vahala, G., Yepez, J. and Vahala, L., 2003. Quantum lattice gas representation of some classical solitons. *Physics Letters A*, **310**: 187-196.

Van den Broeck, C., Parrondo, J.M.R., Toral, R. and Kawai, R., 1997. Nonequilibrium phase transitions induced by multiplicative noise. *Physical Review E*, **55**: 4084-4094.

Van Kampen, N.G., 1981. *Stochastic processes in physics and chemistry.* North Holland, Amsterdam.

Vicsek, T., Czirok, A., Ben Jacob, E., Cohen, I. and Schochet, O., 1995. Novel type of phase transitions in a system of self-driven particles. *Physical Review Letters*, **75**: 1226-1229.

Von Neumann, J., 1955. *Mathematical foundations of quantum mechanics.* Princeton University Press, Princeton, NJ.

Von Neumann, J., 1966. *Theory of Self-Reproducing Automata.* University of Illinois Press, Urbana, Il.

Wadati, M., Matsumoto, H. and Umezawa, H., 1978a. Extended objects in crystals. *Physical Review B*, **18**: 4077-4095.

Wadati, M., Matsumoto, H. and Umezawa, H., 1978b. Extended objects created by Goldstone bosons. *Physical Review D*, **18**: 520-531.

Walgraef, D., Dewel, G. and Borckmans, P., 1981. Chemical instabilities and broken symmetry: The hard mode case. In L.Arnold and R.Lefever (Eds.). *Stochastic nonlinear systems in physics, chemistry, and biology* (pp. 72-83). Springer, Berlin-Heidelberg-New York.

Weinberg, S., 1995. *The Quantum Theory of Fields*, vol.1. Cambridge University Press, Cambridge, UK.

Weinberg, S., 1996. *The Quantum Theory of Fields*, vol.2. Cambridge University Press, Cambridge, UK.

White, H., 1992. *Artificial Neural Networks – Approximation and Learning Theory*. Blackwell, Oxford, UK.

Whitley, D. (Ed.), 1993. *Foundations of Genetic Algorithms 2*. Morgan Kaufmann, San Mateo, CA.

Whittle, P., 1998. *Neural Nets and Chaotic Carriers*. Wiley, Chichester, UK.

Wolfram, S., 1983. Statistical mechanics of cellular automata. *Reviews of Modern Physics*, **55**: 601-644.

Wolfram, S., 1984. Universality and complexity in cellular automata. *Physica D*, **10**: 1-35.

Wolfram, S. (Ed.), 1986. *Theory and applications of cellular automata*. World Scientific, Singapore.

Zinn-Justin, J., 2001. Precise determination of critical exponents and equation of state by field theory methods. *Physics Reports*, **344**: 159-178.

Zurek, W.H., 1996. Cosmological experiments in condensed matter systems. *Physics Reports*, **276**: 177-221.

Chapter 6

THE ROLE OF ERGODICITY

As discussed in the previous chapters, the emergence of *systems* is related to the concept of *interaction*. In physics an *interaction* is assumed to occur when the behavior of one element affects the behavior of another, as stated by Von Bertalanffy in the original definition of system (Von Bertalanffy, 1968).

We just mention the difference between *interaction* and *composition*. The latter term refers to processes occurring when elements are characterized by *reacting*, *merging* or *diluting*, by taking *new positions* or *new roles in a structure*, the most simple examples being those of crystal or molecular structure formation.

Composition gives rise to the appearance of new stable or unstable entities having properties different from those of their components (such as a molecule of water H_2O, which arises from the composition of two atoms of hydrogen and one atom of oxygen bonded together).

In this case the interactions take place *during* the process of formation and new properties are established *as, stable or unstable, outcomes of the underlying interactions*. In the process of emergence these new properties are established not only *during* the formation of the entity, but also due to continuing interactions. Well known examples can be found within the domain of chemical reactions (Belousov, 1959; Zhabotinsky, 1964) or oscillating biological phenomena (Zaikin and Zhabotinsky, 1970), see

Appendix 1. The story of the research which converged to the *Belousov-Zhabotinsky* reaction may be found in Winfree (Winfree, 1984).

Disciplinary and inter-disciplinary research deals with processes of transition *between* phases such as non-system and system, disorder and order, non-living and living, between different phases of matter as studied in the physics of phase transitions.

The crucial point of such research is that it deals with processes taking place *during* the transition, to understand the emergence of coherence and how to model, control, keep, regulate, simulate and manage them. In short, we can say that such research is the main topic of *General Systems Theory*.

Within this framework, an approach to deal with processes of emergence *in a general setting* will be introduced on the basis of the concept of interaction, fundamental in *General Systems Theory* and in Physics. This allows the use of theoretical results obtained in Physics for introducing some possible approaches to model emergence in general.

Among the theoretical tools developed in the latter discipline, one of the most suitable for allowing a generalized use in General Systems Theory is the concept of ergodicity (Minati, 2002). This relates to the fact that every physical process (and *a fortiori* every behavior) can be endowed with two different aspects: one spatial (or configurational) and the other temporal. Regarding the latter aspect, L. Von Bertalanffy (Von Bertalanffy, 1952, p. 109) introduced a distinction between the different roles of time in studying systems. "In physical systems events are, in general, determined by the momentary conditions only. For example, for a falling body, it does not matter how it has arrived at its momentary conditions, for a chemical reaction it does not matter in what way the reacting compounds were produced. The past is, so to speak, effaced in physical systems. In contrast to this, organisms appear to be historical beings." The problem of time is also complicated by the role of the observer and by the conceptual interchangeability (in models) of causes and effects for some systems, as occurs, for example, in geomorphic systems (Schumm and Lichty, 1965).

Despite the apparent difference between spatial and temporal frameworks, adopted to describe behaviors, there could, nevertheless, be a sort of equivalence or trading-off between them. Such an equivalence, when present, is described using the concept of ergodicity. As we will see the interest in this concept stems from the fact that it allows one to deal not so much with changes over time in system behavior as with changes over time in its structure, together with the relationships and configurations of its components.

Let us start by recalling that the concept of ergodicity was introduced in various disciplines, such as Physics where the concept was invented, Theoretical Geomorphology, Population Dynamics and Economics.

6.1 Some definitions related to ergodicity

Introductory definitions of the concept of ergodicity, from various disciplinary fields will be presented. The concept was originally introduced in physics or, more precisely, in statistical mechanics.

6.1.1 Ergodic

In Webster's dictionary (Mish, 1987, p. 422) one finds the following definitions of this attribute:
1. of or relating to a process in which a sequence or sizable sample is equally representative of the whole (as in regard to a statistical parameter);
2. involving or relating to the probability that any state will recur, especially having zero probability that any state will never recur.

6.1.2 Disciplinary definitions

This subsection collects definitions and ways of using concepts related to ergodicity in various disciplines. This overview clarifies the different disciplinary applications of these concepts.

a) Physics

As stated above, the concept was originally introduced within statistical Mechanics. The word comes from the Greek: literally, *The road or pathway of energy*. Let us consider, for instance, using the kinetic theory of gases for a macroscopic system such as the content of a room full of air. The classical approach to the description of the system of molecules in the room consists of writing single motion equations for each particle of the gas, as if they were parts of a mechanical system. Statistical mechanics, however, teaches us that the system as a whole may be described by only a few macroscopic thermodynamic parameters. Turning to the microscopic description, the individual molecules can be considered as point-like objects moving in three-dimensional space. As the state of each particle at a given time is specified by six state variables, i.e. its three-dimensional position coordinates and the three components of its momentum, there is a six-dimensional phase space. In systems of N particles there will be, of course, a

6N-dimensional phase space. Within it, the instantaneous microscopic state of the whole system (the so-called *microstate*) can be represented as a single point. The trajectory of this point in time, governed by the equations of motion of gas particles, describes the behavior of the whole gas. Suitable constraints on the motion of this point in the 6N space may be introduced, for example, by requiring that the motion itself must occur over a hypersurface containing all points characterized by the same total energy. Often such a hypersurface is called an ergodic hypersurface. The attribute "ergodic" comes from the German expression: **"ergodenhypothese"**, originarily introduced by Ludwig Boltzmann (1844-1906). He coined the term *Ergode* for what Gibbs later called a "micro-canonical ensemble", characterized by a uniform distribution over a constant energy hypersurface within the phase space, starting from a modification of the term *monode* (again coined by Boltzmann). The latter was invented to denote a stationary statistical ensemble, from the Greek words *monos* (single) and *eidos* (appearance). As a consequence, the term *Ergode* should be interpreted as coming from the Greek words *ergon* (energy) and *eidos* and should mean "energy-like". This would lead to the translation of Boltzmann's *ergodenhypothese* as "hypothesis regarding the ergods". The attribute "ergodic" was never used by Boltzmann. It was introduced later by Paul and Tatiana Ehrenfest in their 1912 paper containing an exposition and a critical discussion of Boltzmann's work (see the more recent edition: Ehrenfest and Ehrenfest, 1990). In this paper (which later became a very popular book) the Ehrenfests introduced a very different meaning of the term *ergodic*, considering it as a derivation from the Greek: **ergon + odos** (energy path), which seems very different from the one originarily introduced by Boltzmann. The terms *ergodenhypothese* and *Ergode* appeared in papers published by Boltzmann in 1871 (Boltzmann, 1871) and 1884 (Boltzmann 1884a, 1884b), whereas the discussion by Paul and Tatiana Ehrenfest appeared much later. More historical detail can be found in the literature (see,for example, Krylov, 1979; Walters, 1982).

The *ergodic hypothesis* states that, given an infinite time duration, the trajectory of the point representing the entire system in the phase space will pass through every point (or as *close* as you want to every point, in the *quasi-ergodic* hypothesis) lying on the energy hypersurface. The relationship, or *trade-off*, between time and space comes from the fact that an average value, for the location of the point representing the system, determined by following its successive positions over time, will be the same when the average value is calculated over an ensemble of different points, representing different systems, at a single instant of time, provided they lie on the same energy hypersurface. *Sampling at a single time instant across an ensemble of different copies of the same system is equivalent to sampling through time for a single system: that is the notion contained in the ergodic hypothesis.*

The theory behind classical statistical mechanics is firmly based upon the Ergodic Hypothesis and Boltzmann statistics. These principles underlie the statistical analysis of the mechanical behavior of large sets of particles. It should be recalled that Boltzmann introduced his "Ergodic Hypothesis" as a *justification* for macroscopic thermodynamics, which, in those times, appeared as a branch of Physics lying somewhat outside of the main Newtonian framework.

The Ergodic Hypothesis is equivalent to stating that, in the long run, a system of molecules (or, in general, of microscopic components) will assume **all** possible microstates compatible with the conservation of energy. In other words, sooner or later, a system of molecules will go **arbitrarily close** to every conceivable microstate.

Before an introduction to the *Gibbs Postulate* about *time evolution and ergodicity* it should be mentioned that the expression *microcanonical ensemble* denotes any ensemble of different copies of a closed system in which all microstates have the *same* total energy, total momentum, etc., so as to fulfill the well.known Conservation Principles holding in classical mechanics.

Introduced by the theoretical physicist J. W. Gibbs (1839-1903), the *Gibbs Postulate* states that, *in the phase space, all states* in the microcanonical ensemble are equivalent, in the sense that they have the same probability of occurrence.

The assumption behind the Gibbs postulate is that after a long time every system will "forget" its initial conditions. In other words, the *probability of each microstate does not depend upon initial conditions.*

In mathematical terms, the ergodic hypothesis states the equivalence between two different ways of *averaging* **in the phase space.** Consider a system of N particles described by $3N$ coordinates $q_1, q_2, ..., q_N$, and $3N$ momenta $p_1, p_2, ..., p_N$. in the $6N$ phase space (each symbol denoting a 3-dimensional vector). The first way of averaging is defined:

1. By observing a single system, associated with a single point in the *phase space,* over a long time *(time averaging):*

$$\langle f(p, q) \rangle_\tau = \lim_{\tau \to \infty} \frac{1}{\tau} \int_0^\tau f\big(p(t), q(t)\big) \, dt \qquad (6.1)$$

2. By taking many systems from the observed microcanonical ensemble (i.e., with same total energy E_0) and by observing them at one and the same time (*ensemble average*):

$$\langle f(p, q)\rangle_E = \frac{1}{\Omega} \int_{E=E_0} f(p, q) \, dp^N \, dq^N \qquad (6.2)$$

where

$$\Omega = \int_{E=E_0} dp^N \, dq^N \qquad (6.3)$$

As the Ergodicity hypothesis states that *ensemble average is equal to time average*, it will be equivalent to postulate the identity:

$$\langle f(p, q)\rangle_\tau = \langle f(p, q)\rangle_E \qquad (6.4)$$

How large should τ be in (6.1) for the limit to be practically workable? The typical time-scale τ is related to the processes under study. In Ergodic systems τ is small, as often occurs in gases, liquids, crystals. In non-ergodic systems τ is large, and the observer, having a different time/space observational scale, cannot wait for equilibrium to occur, such as, for instance, in supercooled liquids or glasses.

It is very important to note that the *same* system can be *both* ergodic and non-ergodic depending upon the time scale of the observer, such as, for instance, in polymers (Veytsman and Kotelyanskii, 1997).

The study of the conditions which determine the ergodicity of a given system became, starting from the 1912 paper by the Ehrenfests, the subject of a special branch of mathematical physics, known as "ergodic theory" (for a simple introduction see Lebowitz and Penrose, 1973; celebrated books on the subject include Arnold and Avez, 1968; Sinai, 1977; Petersen, 1989; Gray, 1990; Walters, 1982). Without entering into detail on ergodic theory, two fundamental results shoould be cited:

In the 1970's Kolmogorov, Arnold and Moser proved that a number of suitable dynamical systems are characterized by a lack of ergodicity; these systems include our solar system, a chain of anharmonic oscillators (whose lack of ergodicity had already been discovered through computer simulations by Fermi, Pasta and Ulam in 1954) and some kinds of chaotic dynamical systems (see the literature cited above for further details);

The class of systems endowed with the property of ergodicity can, in turn, be organized in a hierarchical way, according to the way in which these systems acquire that property. One can thus introduce, within the class of ergodic systems, further specifications which allow the definition of *mixing* systems, *K-systems*, *Bernoulli* systems, and so on. Once again more detail may be found in the literature cited above.

In this book, however, use will not be made of the results obtained within ergodic theory.

b) Theoretical Geomorphology

In geomorphology scientists have to deal with the problem that they do not have sufficient time for observations because the time-scale of the observer and the time-scale of the phenomena are very different. Geomorphology is interested in using different approaches to consider and represent time.

Within this context there are two critical issues that have been dealt with by using the concept of ergodicity:
1. The need to substitute space for time;
2. The conflict between short-term- and small-space-scale studies and long-term and large-space-scale studies of landform evolution.

In Theoretical Geomorphology (Anderson, 1988, Kirkby, 1994) it is possible to find the following definitions:

"The mean of observations of an individual made over time is equal to the mean of observations made of many individuals at a single moment in time over an area." (Thorn, 1988 page 47),

and the *equivalent*

"If 15% of the population is in a particular state at any moment in time, can assume that each individual in the population spends 15% of time in that state." (Thorn, 1988, page 48).

The problem of *space and time* in geomorphology has been discussed (Paine, 1985; Schumm and Lichty, 1965; Thornes and Brunsden, 1977; Thornes, 1983) with specific references to ergodicity.

c) Economics

The concept of ergodicity is applied in economics and specifically when studying the dynamics of prices (Domowitz and El-Gamal, 1993; 1997; 1999).

A simple model of ergodicity has been introduced in Economics (Landon-Lane and Quinn, 2000). Consider a finite number n of different classes. Typically a class in a socio-economic system may be identified with an income class. These classes may be described as follows:

Class k	1	2	$n + 1$
	$M_1 <$ Income $< M_2$	$M_2 <$ Income $< M_3$	$M_n <$ Income $< M_{n+1}$

Let π_{kt} be the portion of the total population G in class k at time t. The vector $\pi_t = (\pi_{1t}, \pi_{2t}, \ldots, \pi_{nt})$ defines the status of the world at time t. Consider now the probability P of transition of vector π_t from a distribution i in income classes at time t-1 to a different distribution j in income classes at time t:

$$P (\pi_t = j \mid \pi_{t-1} = i) = p_{ij} \tag{6.5}$$

The assumption that the sequence of such transitions is a first order Markov chain is equivalent to stating that the status of the world π_t depends only upon π_{t-1} :

$$P (\pi_t \mid \pi_{t-1}, \pi_{t-2}, \ldots \pi_{t-s} = P (\pi_t \mid \pi_{t-1}), \; \forall \, s = 2, 3, \ldots \tag{6.6}$$

The Markov process transition matrix is $P \equiv [P_{ij}]$. Thus $\pi'_t = \pi'_{t-1} P$. As a consequence:

$$\pi'_t = \pi'_0 P^t \; \text{(i.e. } P \text{ applied } t \text{ times).} \tag{6.7}$$

An invariant or **ergodic** distribution is any distribution such that:

$$\pi' = \pi' P \tag{6.8}$$

The distribution π' is unique if there is only one eigenvalue of $P \bmod = 1$. For further mathematical details the reader may refer to Landon-Lane and Quinn (Landon-Lane and Quinn, 2000).

The notion of ergodicity used in this context is nothing more than a reformulation of the notion of ergodicity introduced in statistical mechanics. Namely, assuming that, in a large population of economic agents, the interactions are so complex as to make the whole system of income distribution insensitive to initial conditions, there will be the possibility for the system itself to fall into a situation in which the ergodic hypothesis holds. That is, the distribution of income classes is independent, in the long run or over a large scale, from the transition probabilities ruling the evolution of the system from one state to another. What is different, here, is that the evolution towards an ergodic distribution does not always occur. Suitable conditions on the dynamical evolution of the system are required, conditions which are more difficult to find than in the mechanical case. After all, economics is not always reducible to classical mechanics!

d) Population dynamics

In the field of population studie*s* the concept of ergodicity has been introduced mainly to support a technique for reconstructing the past evolution of a population starting from actual data, known as *inverse projection*. This was introduced by Ronald Lee (Lee 1974; 1978; 1985), as a logical inversion of conventional projection techniques. Some of the more frequently used procedures for the reconstruction of population structure over time require a knowledge of its age structure at an initial time. Unfortunately, these data are not usually available, whereas the global population without distinction of age or sex is available. This problem is very important, especially in deterministic techniques where structures of populations related to the first interval of the considered time-span significantly depend upon initial data.

As stated by Wachter (Wachter, 1986) ordinary projection is a way of passing *from rates to counts*, "say from a sequence of age-specific rates of giving birth and dying, to count over time of total births and deaths" Correspondingly, *inverse projection* is a way of passing *from counts to rates* "from total counts of events like births and deaths to a plausibly responsible sequence of age-specific rate sets chosen from within a model family (from total counts and age-specific rates together, the changing age structure over time is reconstructed)". Therefore the technique can reconstruct the structure of a population over a specific time-span by the availability of historical series of births and deaths.

Authors introducing and using the technique refer to the relationship with **ergodic reasoning** (Wachter, 1986; McCaa and Pérez Brignoli, 1989; McCaa and Vaupel, 1992; McCaa, 1993; 2001).

The theoretical foundations of inverse projection lie in the 1920s with the work of H. T. J. Norton on the weak ergodicity theorem (Norton, 1928) and in the 1960s with its rediscovery by A. Lopez (Lopez, 1961) giving rise, in turn, to the *weak convergence theorem*, stating that different populations subject to identically varying birthrates would converge to the same age distribution. In the long run, birthrates, not initial states, determine what a population will look like (Wachter, 1986; 1997). *Population age structures* are quickly forgotten (like *initial conditions* in physics), even when punctuated by demographically critical events such as epidemics, wars, baby booms or busts (Lee, 1985). The meaning of the *weak ergodicity theorem* is that by starting from different age structures, populations having the same initial population size and same dynamics of migration, births, and deaths, will converge to the same age distribution. On the other hand population size is a very important factor which can produce, when only partially known, inappropriately represented or misused, possible distortions (McCaa and Vaupel, 1992).

6.2 Ergodicity and stationarity

This chapter presents an approach to the problem of *detecting* the processes of emergence related to Collective Behavior. Such an approach can help in the modelling and management of emergence.

The importance of a well known distinction, holding within the theory of stochastic processes, between the concepts of *stationarity* and of *ergodicity* must first be stressed. In standard textbooks, a stochastic process, which we will denote hereafter as $x(t)$, is said to be stationary (in the broadest sense) if its mean $\mu_x(t) = E[x(t)]$ is constant and, further, its autocorrelation function $R_{xx}(t_1, t_2) = E[x(t_1) x(t_2)]$ is a function only of the difference $\tau = t_2 - t_1$. Here, as customary, the symbol E denotes the expectation value computed over all possible realizations of the stochastic process considered,, such as, for example:

$$E[x(t)] = \int P(x)\, x\, dx \qquad\qquad (6.9)$$

where $P(x)$ denotes the probability distribution associated with the values of x. On the other hand, the property of ergodicity (in a weak sense) implies that both the mean and the autocorrelation function, as computed through the previous formulae, coincide with the time averages relative to the process itself. This is equivalent to stating that:

$$E[x(t)] = <x> \ , \ R_{xx}(\tau) = <x(t)\, x(t + \tau)> \qquad\qquad (6.10)$$

Here the usual short-hand notations are used to denote the time averages; this means, for instance, that:

$$\langle x \rangle = \lim_{T \to \infty} (1/T) \int_{-(T/2)}^{(T/2)} x(t) \, dt$$

From these definitions it follows that an ergodic process must also be a stationary process. The reverse, however, is generally not true. Let us now try to understand how a stationary process could not be ergodic. For the sake of simplicity, the arguments will be limited to a stationary Gaussian process, in which the probability distribution for the occurrence of possible values of x is given by the well-known Gaussian function:

$$G(x) = [1/(\sigma \sqrt{2\pi})] \exp [- (x - \mu)^2/2 \; \sigma^2] \tag{6.11}$$

In order to make such a process non-ergodic, one could, for instance, adopt the following strategy: constrain the process in such a way that its outcome at time t is zero if t is an irrational number and the outcome is x (with Gaussian distribution) otherwise. It is easy to see that the average value of the process is zero (owing to the fact that, within the context of real numbers, the set of rational numbers has a zero measure), whereas its expected value is μ. If μ is different from zero, the process is not ergodic. The objection that in this case the expected value cannot be computed according to the usual formula can be easily circumvented by resorting to a suitable redefinition of the expected value itself which, avoiding here some of the technicalities, should give rise to the same result as before.

In any case, too much attention should not be paid to the mathematical details of this example (an infinite number of different examples of this kind could be invented). It was introduced merely to stress the fact that, in order to make a stationary process non-ergodic, some form of long-range constraint (on a temporal scale, for instance) must be introduced on the process being considered. At this point, it must be taken into account that, whatever the constraint introduced on the system, that constraint results, conceptually, from some action exerted from outside that system. This argument leads to the conclusion that, if the behavior of a system is described as a stationary stochastic process, then, in the absence of any external perturbation or of special kinds of inbuilt long-range orderings, the process itself is ergodic.

Such a conclusion suggests the possibility of using ergodicity to build an indicator for the occurrence of an emergent process. A suitable measure of the degree of fulfilment of the property of ergodicity in the behaviour of a given system, described, as above, by a stochastic process can be introduced. Such a measure, of course, depends upon the particular form of the model adopted to describe the system itself but, in its essence, it can be easily derived from that which holds for systems described by a single state variable $x(t)$, for which the *degree of non-ergodicity* can be defined as being related to the values of the differences:

$$d_1 = E\,[x(t)] - <x> \;,\; d_2 = R_{xx}(\tau) - <x(t)\,x(t+\tau)>\,. \tag{6.12}$$

Of course, any practical definition of the degree of non-ergodicity must be formulated such that, for ergodic systems, the degree of non ergodicity be zero. Consider now a process of emergence, associated with a suitable transient phase, during which the system is under the influence of suitable external constraints. It is to be expected that, owing to the presence of these constraints, the statistical properties of the stochastic processes describing the system behaviour may deviate from complete ergodicity, so that the degree of non ergodicity will change from zero to a finite value. Obviously one can not exclude that, after the end of the transient phase, the degree of non ergodicity will remain different from zero (as could occur, for instance, in phase transitions leading from ordered states to special kinds of glassy states; an example is given by Crisanti (Crisanti *et al*, 1996). In any case, the change in the degree of non ergodicity acts as a very good indicator for the occurrence of an emergent process, whatever its nature.

6.3 Ergodicity in Collective Beings

The ideas outlined above will now be applied to the case of Collective Beings. Neglecting for the moment the process of emergence of a Collective Being, once the Being is established and operating, there is a sort of conceptual **interchangeability** (Tables 3.1 and 3.2 3) between interacting agents which take on the *same roles at different times and different roles at the same time* (ergodic behavior). This occurs because in a Collective Being each agent belongs simultaneously to different systems. Provided the behaviour of the system of agents is stationary, such an interchangeability can be considered as a manifestation of its ergodicity or of the fact that its degree of non ergodicity is zero. It is stressed that interchangeability, as well as ergodicity, is not *per se* an objective property of a Collective Being. Besides the existence of the Collective Being itself, it also depends upon the presence of an *observer*, equipped with suitable detection devices, suitable goals and cognitive schemata. It must also be taken into account, however, that such a circumstance could hold, not for all agents belonging to a

Collective Being, but only for a fraction of them. The *degrees of ergodicity* proposed below in Section 6.4 will be introduced in order to express the percentage of agents for which there is the conceptual interchangeability mentioned above. This percentage, in turn, could be viewed as a sort of *order parameter*, suitable for describing the degree of progress of the 'phase transition' leading to the formation of the Collective Being itself. The approach proposed here seems to be suitable when agents are assumed to be autonomous, each provided with a cognitive system. In such a case they may behave by *deciding* the cognitive model to use and use the same model at different times. Thus the existence of interchangeability can be viewed as the most peculiar clue for the presence of a Collective Being. As the existence of the latter, in turn, implies that each agent must adopt an approach based on a dynamical use of cognitive models (otherwise interchangeability would be impossible), this means that another measure of the ergodicity of the system is given by the degree of the dynamical usage of models. This is the reason why an index of the dynamical usage of cognitive models may be very helpful for detecting processes of emergence and the establishment of Collective Beings.

Regarding the conditions necessary for *collective behaviors to be ergodic*, the approach previously used to model an economic context can be applied to study the systemic properties of collective behaviors, such as those characterizing flocks and swarms.

Consider a population G of elements X_i and distance classes instead of income classes. Thus elements X_i may be clustered through suitable clustering procedures based on a finite number C of distance classes characterized by the fact that in class k ($k \leq C$) the distance d between two elements belonging to the same cluster is smaller than a suitable value of M_k. An example of such a situation is depicted in the following table, showing the condition to be fulfilled by the reciprocal distance between two elements $D(X_i, X_r)$ for them to belong to the same cluster, as a function of the distance class:

Class k	1	2	n
	$0 < D(X_i, X_r) < M_1$	$0 < D(X_i, X_r) < M_2$	$0 < D(X_i, X_r) < M_n$

At any given point in time any element X_i may:
1. belong only to a single cluster;
2. belong simultaneously to J ($J < C$) different clusters.

Elements X_i may, for example, be represented as points in an n-dimensional space. A situation which could occur in a 2-dimensional space could, for instance, be depicted as shown in Figure 6.1.

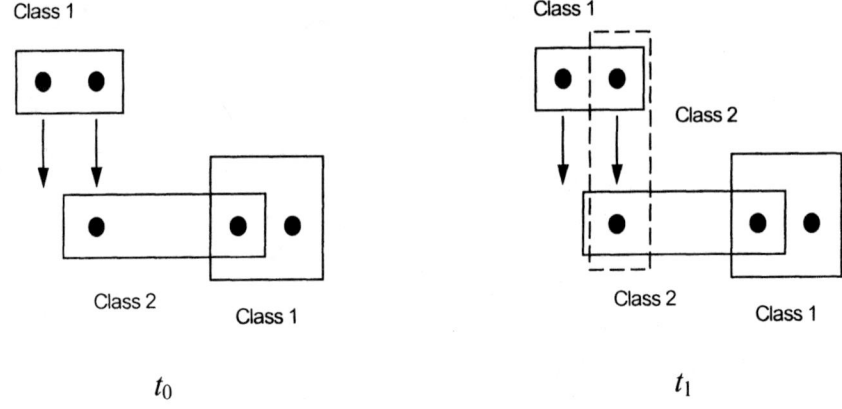

Figure 6-1. By considering only two classes of boids we have an example of spatial distribution at the time t.

As elements are supposed to be endowed with some motion (due, for example, to decisions taken by their cognitive systems), they can migrate from one cluster to another and thus clusters can appear, grow and disappear. And, of course, the reciprocal distances between the elements within each cluster can change with time.

Consider now the set of all possible distances between the different elements of G; they can belong simultaneously to different distance classes and, at any given point in time t, each distance class will be characterized by the number of distances falling within it. Thus, for any value of t, there is a distribution of the different distances within the different distance classes. Let π_{kt} denote the total number of reciprocal distances belonging to class k at time t. Then the vector $\pi_t = (\pi_{1t}, \pi_{2t}, \dots, \pi_{ct})$ defines the state of this distribution at time t. Of course, such a distribution will change over time, owing to the motion of the elements of G.

Let us suppose that this motion can only be described as stochastic processes. There are several reasons for making such a choice. First of all, a background stochasticity is intrinsic to every process taking place in the real world, owing to the unavoidable presence of noise. In the second place, it would be manifestly absurd to suppose that a system ruled by some cognitive system be equivalent to a mechanical system following completely deterministic rules of behavior, such as a falling apple or an oscillating pendulum. Even the most mechanistic psychologist or zoologist would reject such an extreme position. Moreover, the presence of an intrinsic source of stochasticity, even when superimposed upon deterministic rules of behaviour, automatically produces an overall stochastic behaviour, which can be described using the tools of the theory of stochastic processes.

This allows the introduction of the probability P of the transition of a distance from a class i at time $t1$ to a class j at time t, denoted as p_{ij}.

The first order Markov assumption (which turns out to be a very good approximation in most real cases) implies that the status of the world π_t depends only on π_{t-1} through Markov's transition matrix $[P_{ij}]$.

This implies that $\pi'_t = \pi'_{t1} P$.

Thus, an invariant or **ergodic** distribution is any distribution $\pi' = \pi' P$. In this case the distribution π' is also unique if there is only one eigenvalue of $P \bmod = 1$. It should be noted that, in this regard, two very different cases are possible: that of the uniqueness of the ergodic distribution and that of non-uniqueness. In the first case, if, in the absence of noise, there is only a single equilibrium state, it can be stated that this unique ergodic state is nothing but a stochastic version of the equilibrium state itself which, although perturbed by the presence of noise, does not change its nature. In the case of non-uniqueness, on the other hand, it can be stated that the noise has given rise to a new emergent structure, through a process of intrinsic emergence. That is, a structure which can not be predicted, owing to the stochastic nature itself of the system's evolution, during which one of the different ergodic distributions will be adopted. Such a situation recalls that of symmetry breaking already encountered in physics.

The interpretation of the above is that the ergodic state is invariant owing to the fact that agents (boids for instance) dynamically produce changes in the distributions of the classes over time, while keeping constant (or within suitable ranges) some parameters and/or suitable relationships between parameters, such as number of classes, frequency of class changing, speeds, directions, and so on.

More specifically, it can be assumed that the ergodicity manifested in Collective Behaviours emerging from agents endowed with a cognitive system is conserved due to the existence of suitable relationships between every agent and its neighbours (Minati *et al.*, 1997). Besides, such Collective Behaviour probably occurs due to limited variations in critical parameters such as speed, acceleration, altitude, and direction (Pessa *et al.*, 2004).

6.4 Emergence, Collective Beings and Ergodicity

Referring to the definitions of ergodicity introduced in Section 6.1.2 (Theoretical geomorphology) and with $x\%$ denoting the percentage of elements lying in a given state at any point in time and y% denoting the percentage of time spent by a single element in that state, the previous definitions can be reformulated as follows.

If $x\%$ of the population is in a particular state S at any moment in time and each individual in the population spends $y\%$ of time in that state, when the system is ergodic then $x\% = y\%$; a definition which is equivalent to: When the system is ergodic, the mean of observations on an individual made over a time y is equal to the mean of observations made on x individuals at a single moment in time over a domain.

By using these two definitions it is possible to define states S, characterized by a correspondence between x and y (or $x\%$ and $y\%$), like the one mentioned in a) and b). For instance, in a social system the states S could be those corresponding to buying a specific product, watching specific programs on TV, traveling by car, traveling by plane, etc. Of course, the property of being ergodic depends, for any given state, on the fact that experimental observations verify the fulfilment of a relationship such as $x = y$ (or $x\% = y\%$). Such a property for a given state, therefore, may or may not be present in any given situation or context. Thus there is the possibility of transitions from ergodicity to non-ergodicity and vice versa. Clearly, such a definition of ergodicity, related to a particular state, is somewhat different from the definition of ergodicity, related to dynamical evolution, which is usually adopted within Statistical Mechanics. This difference, on the other hand, does not necessarily imply an incompatibility. Here we claim that the definition of ergodicity related to the foregoing statements a) and b) is more suited to dealing with systems such as those occurring in the biological and social worlds (but even in physical situations which are uncontrollable, such as in geology). In such systems the dynamics is often poorly or totally unknown, or impossible to describe using a mathematical model. Thus, a set of suitable indices is required to detect in some way the occurrence of emergence even in the absence of mathematical models of possible behaviour or of algorithms capable of generating them.

For systems designers and systems modellers in particular contexts, such as those cited above, such information is of fundamental importance. The values of x and y are very important in order to assess the meaning and the relevance of ergodic behavior.

Consider the following situation. An observer O is watching a system emerging from a suitable set of interacting agents A_i. Let $\{B_i\}$ be the set of behavioural features of the single agent A_i, as detected by O, such as (keeping boids in mind) the speed, direction, acceleration, distance from neighbours, of a single boid. Each behavioural feature is assumed to be specified through suitable data relative to its time evolution.

Assume now that observer O detects the fact that the behavioral features of the single autonomous agents A_i are reciprocally coordinated in time and space as occurs, for example, during the flight of birds or insects. The problem is how to characterize the fact that system S of interacting agents A_i shows an emergent behaviour while the single agents assume asynchronous B_i behaviours.

To answer such a question, the arguments introduced in the previous Section lead to the conclusion that an answer is possible when and only when, for the same system, a situation was observed with no coordination of the individual behavioural features. That is, if a suitable measure of ergodicity is introduced, and we obtain, in the absence of coordination, a given value of this measure and a change in this value with the birth of coordination, then the observed coordination can be considered as emergent from the individual behaviors of the single agents A_i. The value of the measure itself is not important *per se*. What does matter is the fact that a *variation* is detected. This provides a precise tool for detecting emergent phenomena (and perhaps to foresee them), provided the behaviour of the system under study is carefully monitored.

Regarding the measure of ergodicity, there are an unlimited number of possible choices. However, within this section, one particular choice, between the infinite number available, will be illustrated.

To this end, let us assume that system monitoring involves a single, particular, behavioural feature F, which will be assumed to be associated with a finite number of different possible states F_i. For each of these states, let us assume that our monitoring (over a given observational time) of a system, containing a finite (and constant over time) number of elements, gave the average percentage of time spent by a single element in state F_i as $y\%_i$ and the average percentage of elements lying in the same state as $x\%_i$. A measure of the *degree of ergodicity* of that particular state is then given by:

$$E_i = 1/[1 + (x\%_i - y\%_i)^2] \tag{6.13}$$

Such a measure appears to be very useful from the mathematical point of view. Namely, when the state shows ergodicity, that is, when $x\%_i = y\%_i$, the index E_i assumes its maximum value of 1. On the contrary, when this condition is not fulfilled, the index will take a value (always greater than zero) between 1 and 0, irrespective of the particular values of $x\%_i$ and $y\%_i$. Such an index, of course, needs to be generalized so that the ergodicity of a system relates to the fact that all its possible states are ergodic (in the sense defined above). This can be done by introducing a new index, denoted as *global ergodicity degree* and defined by:

$$G = (\Sigma_i E_i)/N \qquad (6.14)$$

where N denotes the total number of different possible states F_i. It is easy to see that, when all values of E_i simultaneously reach their maximum values, index G will, in turn, reach its maximum value of 1, characterizing complete ergodicity. Variations in the value of G can then be used to detect transitions, within a given system, associated with the emergence, for instance, of collective behaviours.

Therefore, through the use of a suitable index of ergodicity, such as that introduced above, a rich phenomenology of possible situations can be examined. For instance, by using the definition of *collective behaviour* introduced in Section 3.2.2, with the definition of emergence proposed by Baas and co-workers (Baas, 1994; Baas and Emmeche, 1997; Mayer and Rasmussen, 1998) introduced in Section 3.1, it would be possible for an observer to detect the emergence of collective behavior at the lower observational level S^1 through the use of the index of ergodicity, without observing directly the emergent properties at the higher observational level S^2. This shows how useful such an index could be in frequently occurring situations where the observer is an integral part of the system under study and, as such, overall observation of the whole system is precluded to the observer.

Such a framework is particularly useful for studying both the emergence of collective behaviour, and especially the transitions from non-collective to collective behavior and vice versa. Such transitions are associated with variations in the ergodicity index.

Within this context, it is worth introducing some comments about circumstances which are known from statistical mechanics to be associated with non-ergodic behaviours. Although the definition of ergodicity introduced above does not coincide with the rigorous definition used in statistical mechanics, a knowledge of these circumstances can help to focus attention on some situations which could signal or anticipate the occurrence of a transition. First, the behavior shown by the components of a given system must not be too *complex*, in order to support an ergodic behaviour. Namely, if the components have very complex behaviours, as in *deterministic chaos*, then it is impossible, by definition, to find ergodicity.

Second, we must not forget the crucial role of the observer in detecting both ergodicity and emergence. Namely non-ergodicity may be the result of an insufficient time span allowed to the observer for detecting ergodicity. Ergodic behavior may disappear because of changes in the time span, although it may also disappear because the system components take on too complex individual behaviours.

One example is the situation in which boids are observed over a very short time scale, unsuitable for detecting the emergence of a flock. On the other hand, a collective behaviour of boids could disappear, that is lose its coherence, because of individual behaviours which become too complex, due to the fact, for instance, that continuously changing behavioral rules could give rise to a chaotic deterministic behaviour.

6.5 Further considerations

The process underlying changes in the degree of ergodicity over time, for a set or a system, depends in turn on the behavior of system components **and** on the learning abilities and cognitive system of the observer. It is interesting to observe **how most systems may be assumed, after an adequate amount of time, to converge towards an ergodic behavior.**

Within the context of this book, the interest in studying processes inducing a change in the degree of ergodicity in a set or in a system is connected to the need for a methodology suited to study the processes of emergence occurring within Collective Beings. Such a methodology should provide indicators able to show when these processes are taking place. It should not only detect processes of emergence , but even manage them (for instance, slowing them down or speeding them up) by acting on the dynamics of their degree of ergodicity.

Let us consider now some specific issues related to ergodicity and emergence within Collective Beings.

a) *Each* individual versus sub-populations

From what has been stated above it is clear that our previous definitions have only a statistical meaning. Referring to definition a), in Section 6.4, it can be seen that the requirement that *e u c h* individual is spending $y\%$ of the time in a particular state is very strong[4]. It would be better to introduce a different and more general formulation such as:

"If x% of the population is in a particular state S at *any* moment in time, and *all* subpopulations spend y% of time in that state, when the system is ergodic then $x\% = y\%$."

This means that, with reference to the original strong definition introduced in theoretical geomorphology (Anderson, 1988; Kirkby, 1994; Thorn 1982; 1988), instead of "each individual" the expression "all sub-populations" may be used. The requirement (to spend y % of time in a particular state) does not apply for the behavior of *each* single agent, but for *all* "sub-populations" of them allowing them to have no longer constant but only statistically significant properties. The definitions of sub-population, of course, derive from the interests of the observer, the same observer who is expected to detect emergence.

Moreover, it is conceptually more correct to compare the percentage of a population vs. all sub-populations, rather than the percentage of a population vs. each individual. For instance: "If x % of a population watches TV in *any* moment in time, it can be assumed that *all* sub-populations (i.e. all families of the social system) spends y % of time watching TV". In this case there are no explicit rigid prescriptions for single components of the family: it is sufficient that it is the family to spend y % of time watching TV and not the single family components. With respect to the question of sub-populations versus populations, it is also possible to consider the following example: some individuals may be excluded from taking part in certain activities in a system on principle, for instance, men cannot give birth to children, but nevertheless the mixed society of men and women may be ergodic, with reference to certain activities.

This reformulation of the concept introduced in theoretical geomorphology seems to be more appropriate for generalizations to different kinds of systems.

b) Ergodicity and emergence of Collective Beings

The relationship between ergodicity and the emergence of Collective Beings occurs when considering a state S as corresponding to the adoption of one specific, cognitive model, among the many available, both by the agents and observer (such a model is, of course, used to understand the emergent agent behavior). It could be said that the goal of the observer's model is to

---

[4] Thanks to H. Haken for useful discussions on this point (G. Minati).

detect emergence at level S^2, introduced by Baas and co-workers as cited above. For instance, agents belonging to a social system may behave (in the sense of taking on the corresponding cognitive model) as buyers, workers, drivers or students. In turn, the observer may be interested in studying a market, a company, road traffic or a school.

The need for many different models, corresponding to different simultaneous behavioral states S', S'', etc., occurs when we deal with the emergence of a multiple-systems Collective Being and is related to the need for adopting the *DYSAM* approach, already introduced in Section 2.4.

A temporal *succession of degrees of ergodicity* as introduced above may be very useful to characterize various kinds of Collective Beings and some of the processes leading to their emergence.

c) A generalization of the previous definition of ergodicity

Regarding definition b), in Section 6.4, a generalization of the measure of the *degree of ergodicity* introduced above is possible, in order to take into account cases involving situations characterized simultaneously by many different features.

Starting from the fundamental role of the observer who, by applying a specific cognitive model, will be able to extract from a given observational situation, a global index characterizing the situation in its wholeness. Of course, the particular way of computing such an index from individual observational features will depend upon the cognitive model adopted by the observer (the reference, generally, will be to some suitable *mean* of the practical observations). Furthermore, it is assumed that even when dealing with a number of different observers, possibly equipped with different cognitive models, looking at the same phenomena, it will be possible to extract, in some way (whose details are irrelevant for current purposes) a global index characterizing the observed phenomena (for instance, through a consensus among the observers themselves). The following formal notations can then be introduced:

let O_x be the global index from observations made by x individuals at a single moment in time over an area (a global picture in a single moment of a changing area is observed by x observers);

let O_y be the global index from observations made by an individual for a time y (a sequence of details considered by a single observer for a time y);

let F be a two-argument *similarity* function, such that $F(w,z) = u$, with $0 < u < 1$, where 0 corresponds to total dissimilarity and 1 to maximum similarity between w and z.

Then a general measure of the degree of ergodicity will be given by $E = F(O_x, O_y)$. Of course, the practical computation of E will require a suitable specification both of the explicit form chosen for F and of the process giving rise to the global indices associated with the different observational situations.

The condition $E = 1$ means that there is ergodicity. It is natural, when dealing with a system evolving in time, to consider the *temporal dynamics* of the values of E.

It should be noted that, when the value of the degree of ergodicity approaches 1, this means that the various ways of carrying out and collating the observations (corresponding to O_x and O_y) are similar, *compatible* for the observers, but not *identical* from a cognitive point of view, i.e. they are described in different ways, and lead to different interpretations and actions. When, on the other hand, the value **is** exactly equal to 1, they are *identical* from a cognitive point of view, that is they are described in the same way, leading to the same interpretations and actions. It should be stressed that this sort of equivalence occurs inasmuch as the observer (or the observers) adopts a specific *level of description* (for instance, ant-nests, flocks of birds are considered *equal*, that is having the same meaning).

Even in this case the *time history* of the degrees E is important because it shows the time evolution of coherence of the Collective Being, related in this case to the multiple-system of detections made by the various observers. When this *time history* is a sequence of values far from 1, then there is no ergodicity. Meanings of observations are no longer similar, and collective behaviors (i.e. observations in this case) are considered as *incompatible*. To characterize these situations in a more general way, the degree of ergodicity previously introduced may be further generalized through a *global ergodicity degree* defined by:

$$H = (\Sigma_i E_i)/N \tag{6.15}$$

where N denotes the total number of different possible observations and E_i is the degree of ergodicity associated with the i-th observational situation. In a similar way it can be seen that, when all E_i simultaneously reach their maximum values, the index H will, in turn, reach its maximum value of 1, characteristic of complete ergodicity.

Variations in the value of H can thus be used to detect transitions, within a given system of observations, associated with the emergence, for instance, of collective behavior. It should be noted that as the value of H depends, on the one hand, upon the practical observational results and, on the other, upon the cognitive system of the observer (and thus on the ensuing interpretations and actions), the global ergodicity degree has a double valence. Namely, it can be viewed as a sort of tool for detecting the existence of a transition within a Collective Being (as when the role of the observer is weak and unchanging). But it can also be viewed as a detector of a transition in the cognitive systems of the observers (not necessarily associated with a real change in the observed system), when the role of the observer is of capital importance and rapidly changing (as in the cases of learning, designing, and so on).

6.6 Some remarks and possible lines of research

The concept of ergodicity may help to better understand, model, simulate and control the *emergence of Collective Behaviors* from agents possessing a cognitive system. In the following subsections some *summarized* points are presented together with some open questions and simulation programs to be developed within future research activities on this topic.

1. Can a set (of interacting or non-interacting components) have different and simultaneous ergodic behaviors?

It depends on the level of description. A set of agents may have non-ergodic or different levels of ergodic behavior depending upon the observer, i.e. on time-scale, space scale, cognitive model adopted, etc. In turn the agents may belong to different systems emerging from different ergodic behaviors. The answer is negative if we use only one level of description: for instance, boids cannot be supposed to simultaneously belong to different flocks. This circumstance would be conceptually contradictory because it would generate incompatible behaviors.

2. Why do living systems take on ergodic behavior to give rise to a collective behavior?

What is the interest of ergodic behaviors for Nature? Living systems may behave ergodically to make emergent systems which are capable of doing something that they individually cannot as introduced in Section 3.2.4. Another possible answer is that by temporarily behaving so as to ensure high levels of ergodicity (even though always with different behavior compared with the situation taking place *when all the agents have identical behavior*) **the global system seems to behave similarly to its single components.** In living systems, for example, the system established by the emergence of collective behavior may, to a predator, seem similar to a single component (the prey), **but much bigger**. This may be the reason why small fish (Anderson, 1980) when detecting the predator not only establish a collective behavior (the fish school), by assuming an ergodic behavior, but also position themselves so as to reflect light towards the predator, inducing in it the perception of a larger animal (Vabo and Nottestad, 1997).

Living systems may assume ergodic behavior in order to make the behavior of the system emerging from collective behavior similar to one of the single agents. Identity holds only for artificially organized behaviors, as for platoons.

3. Different life duration.

Distinction must be made between systems having, *from the point of view of the observer's model*,
- *indistinguishable* elements, that is elements considered as *anonymous* particles, i.e., members of the crowd (the observer is interested in the global emergent system looking, for instance, for statistical aspects), grains of sand (the observer is interested in the shape of the beach), elements of a species (the observer is interested in population dynamics), elements in a chemical reaction;
- *distinguishable* elements, that is elements having differences distinguishable by the observer, i.e. members of a family (the observer is a family therapist), computers in a network (the network manager must distinguish between nodes, for instance, for security reasons), students in a classroom to set an educational program (the observer is the teacher).

In the former case the ergodic behavior refers to the behavior of existing, indistinguishable elements. In the latter *life duration* is important because the behavior to be ergodically reproduced may refer to elements which are no longer active (*living* or belonging to social systems). In such a case elements may ergodically reproduce the behavior of elements *no longer present* in the systems. An ergodic system is able to produce ergodic behavior on elements by referring to the behavior of elements which are no longer active.

4. Simulating the emergence of the collective behavior of a flock of Boids (i.e. a set of dynamic points having ergodic behavior).

In this subsection we will list some crucial issues to be dealt with in designing programs for simulating the emergence of collective behavior. Existing simulation programs follow different strategies, sometimes implicitly assumed when stating initial conditions and constraints.

a) *Minimum number of elements vs. the number of behavioral rules.* This issue relates to the need to balance the number of different behaviors that elements may assume and the allowable number of elements. Regarding the possibility of assuming ergodic behavior there are three extremes. In the first case, the number of elements is lower than the number of possible behavioral rules in such a way that at any time there is some behavioral rule which cannot be actually carried out. In the second case, the number of elements is equal or just greater than the number of allowable behavioral rules in such a way that at any time there is the possibility that a behavioral rule be performed by one or more elements. In the third case, the number of elements is much greater than the number of possible behavioral rules so that in time every behavioral rule may be simultaneously performed by a large number of elements (Gilbert and Troitzsch, 1999).

b) *Can two (or more) Collective Behaviors "merge"?* First of all we need to clarify what we mean, in this case, by the term "merge". One possible meaning is that two interacting Collective Beings (such as different ecosystems) are assumed to merge when the coherence and rules of one are taken on by the other. Another possible meaning is that from interacting Collective Beings a new, different one emerges. An element, for instance, may join or resign from an established Collective Being. Transitions between non-collective and collective, incoherent and coherent, non-ergodic and ergodic properties are possible. The issue relates to the possibility of the emergence of a Collective Being (associated with coherence, collective behavior, and ergodicity) from the interaction of two or more Collective Beings associated with the same properties.

c) *Initial conditions and continued coherence.* The issue relates to the sensitivity of a Collective Being to initial conditions. Are initial conditions irrelevant for a process of emergence establishing coherence? Processes of emergence of coherence are related, as mentioned in Section 6.4, to transitions between ergodicity and non-ergodicity. Section 6.1.2 also mentions how the *same* system can be *both* ergodic and non-ergodic, depending upon the time-scale of the observer. There is the issue relating to how *catastrophic* is the reduction of ergodicity for a Collective Being, not inasmuch as it results in a loss of coherence, but because it can entail a loss of the possibility of continued coherence (on continued coherence there is an extensive literature, ranging from classical models of dissipative structures to quantum models of phase transitions; to cite but a few, see, for example, Nitzan and Ortoleva, 1980; Umezawa, 1993;

Rivers *et al.*, 2002). Continued coherence is a process allowing or rather using a loss of coherence, like *using noise as a resource* (Mikhailov and Calenbuhr, 2002, pp. 94-102).

d) *Does a min/max number of elements exist for a Collective Being?* This relates to the possibility of theoretically identifying the min and max number of elements for a Collective Being, with reference, of course, to the number of behavioral rules, the time-scale and other constraints such as the space available and different life duration (previous point 3).

e) *The process through which an external element starts to belong to a Collective Being.* This question is different from the one dealt with when studying the transitions between non-ergodicity and ergodicity. The process leading to an element becoming part of a Collective Being may take place both when the element is performing one of the behavioral rules used by the system or when the elements already belonging to the system decide to perform the behavioral rule introduced by the new element. For the process of belonging to become *stable* the behavioral rules must be ergodically performed.

f) *The process through which the belonging of an element to a Collective Being ends.* On the one hand, the end of an element belonging to a Collective Being may be related to a *decision* or to a behavioral feature of an autonomous agent. On the other, it may be caused by the fact that the other elements do not perform the behavioral rule(s) performed by the autonomous agent, in such a way as to exclude it from ergodic behavior. In the latter case the behavioral rule(s) *proposed* by the autonomous agent are *no longer* reproduced by any other element, causing its expulsion to avoid the generation of incoherence.

g) *Perturbations able to "destroy" the ergodicity of a Collective Being.* Section 6.2 describes how the ergodicity of a system may disappear because of *too much complexity*. This issue refers to the fact that *external* perturbations could destroy an ergodic behavior. This could occur owing to an inference able to modify the ergodic dynamics in performing behavioral rules. An example of this circumstance, although a very simple one, is given by what occurs when the cognitive systems of the elements of a Collective Being must suddenly process urgent, unexpected information, such as the arrival of a predator, or other natural events like strong wind or waves.

5. Ergodicity for modeling emergence.

The study of the theoretical relationships between ergodicity and emergence may provide interesting contributions to a general, systemic theory of emergence. This is because such a study covers many interconnected problems as outlined above. In both contexts the role of the observer is crucial. Ergodicity may be studied as a *necessary condition* for establishing processes of emergence. It is *not a sufficient condition* because the observer is also required to be provided with a cognitive model such as to allow the detection of new, unexpected properties of the system, as occurs for collective behaviors. It is as if ergodicity can deal with a *necessary* aspect of the phenomena, while emergence deals even with cognitive aspects related to realizing new, unexpected properties. In any case, the crucial relationships between ergodicity and emergence are not only those just mentioned; moreover, it should also be noted that **ergodicity itself is an emergent property**.

Will the *combination* of the two approaches lead to the formulation of a *general theory of emergence*?

References

Anderson, J. J., 1980, A stochastic model for the size of fish schools, *Fish Bulletin* **79**:315-323.

Anderson, M. G., (ed.), 1988, *Modeling geomorphologic systems*. Wiley, New York.

Arnold, V. I., and Avez, A., 1968, *Ergodic problems of statistical mechanics*. Benjamin, New York.

Baas, N. A., 1994, Emergence, hierarchies and hyperstructures. In: *Alife III, Santa Fe Studies in the Science of Complexity*, Proc. Volume XVII, (C. G. Langton, ed.), Addison-Wesley, Redwood City, CA, pp. 515-537.

Baas, N. A., Emmeche, C., 1997, On Emergence and Explanation, *Intellectica* **25**:67-83.

Belousov, B. P., 1959, A periodic chemical reaction and its mechanism. *Sbornik Referatoo po Radiatsionnoi Meditsine*, Medgiz, Moscow, 145-147.

Boltzmann, L., 1884a, Über die Möglichkeit der Begründung einer kinetischen Gastheorie auf anziehende Kräfte allein, *Wiener Berichte* **89**:714-722.

Boltzmann, L., 1884b, Über eine von Hrn. Bartoli entdeckte Beziehung der Wärmestrahlung zum zweiten Hauptsatze. *Wiedemann's Annalen für Physik und Chemie* **22**:31-39.

Boltzmann, L, 1871, Einige allgemeine Sätze über Wärmegleichgewicht, *Wiener Berichte* **63**:679-711.

Crisanti, A., Falcioni, M., and Vulpiani, A., 1996, Broken Ergodicity and Glassy Behavior in a Deterministic Chaotic Map, *Physical Review Letters* **74**:612-615.

Domowitz, I., and El-Gamal, M., 1993, A Consistent Test of Stationary Ergodicity, *Econometric Theory* **9**:589-601.

Ehrenfest, P. and Ehrenfest, T., 1990, *The Conceptual Foundations of the Statistical Approach in Mechanics*. Dover, NewYork,. [Translated by M. J. Moravcsik, from

"Begriffliche Grundlagen der statistischen Auffassung in der Mechanik." *Encyklopaedie der mathematische Wissenschaften*, Vol. 4, 1912].

Gilbert, N., and Troitzsch, K. G., 1999, *Simulation for the Social Scientist.* Open University Press, Buckingham, UK.

Gray, R. M., 1990, *Probability, Random Processes, and Ergodic Properties.* Springer, Berlin-Heidelberg-New York.

Kirkby, M. J., (ed.), 1994, *Process models and theoretical geomorphology.* Wiley, New York.

Krylov, N., 1979, *Works on the foundations of statistical mechanics.* Princeton University Press, Princeton, NJ.

Lebowitz, J. L., and Penrose, O., 1973, Modern ergodic theory, *Physics Today* **26**:23-29.

Lee, R. D., 1974, Estimating Series of Vital Rates and Age Structures from Baptisms and Burials: A New Technique with Applications to Pre-industrial England, *Population Studies* **28**:495-512.

Lee, R. D., 1978, *Econometric Studies of Topics in Demographic History.* Arno Press, New York.

Lee, R. D., 1985, Inverse Projection and Back Projection: A Critical Appraisal and Comparative Results for England, 1539-1871, *Population Studies* **39**:233-248.

Lopez, A., 1961, *Problems in Stable Population Theory.* Office of Population Research, Princeton, N. J.

Mayer, B., and Rasmussen, S., 1998, Self-reproduction of a dynamical hierarchy in a chemical system. In: *Artificial Life VI*, (C. Adami, R. Belew, H. Kitano and C. Taylor, eds.), MIT Press, Cambridge, MA, pp. 123-129.

McCaa, R., 1993, Benchmarks for a New Inverse Population Projection Program: England, Sweden, and a Standard Demographic Transition. In: *Old and New Methods in Historical Demography*, (D. S. Reher and R. Schofield, eds.), Clarendon Press, Oxford, UK, pp. 40-56.

McCaa, R., and Pérez Brignoli, H., 1989, Populate: From Births and Deaths to the Demography of the Past, Present, and Future. Humphrey Institute Center for Population Analysis and Policy, University of Minnesota, Minneapolis, MN, working paper 89-06-02.

McCaa, R., and Vaupel, J. W.,1992, Comment la projection inverse se comporte-t-elle sur des données simulées? In: *Modèles de la démographie historique*, (A. Blum, N. Bonneuil and D. Blanchet, eds.), Institut National d'Etudes Démographiques, Paris, pp. 129-146.

Mikhailov, A. S., and Calenbuhr, V., 2002, *From Cells to Societies. Models of Complex Coherent Action.* Springer-Verlag, Berlin-Heidelberg-New York

Minati, G., 2002, Emergence and Ergodicity: a line of research. In: *Emergence in Complex Cognitive, Social and Biological Systems* (G. Minati and E. Pessa, eds.), Kluwer, New York, pp. 85-102

Minati, G., Penna, M. P., and Pessa, E., 1997, A conceptual framework for self-organization and merging processes in social systems. In: *Systems For Sustainability: People, Organisations and Environment*, (F. A. Stowell, R. Ison, R., Armson, J. Holloway, S. Jackson and S., McRobb, eds.), Plenum, New York, pp. 271-275.

Mish, F. C., (ed.), 1987, *Webster's Ninth New Collegiate Dictionary*, Merriam-Webster Inc., Springfield, MA.

Nitzan, A., and Ortoleva, P., 1980, Scaling and Ginzburg criteria for critical bifurcations in nonequilibrium reacting systems, *Physical Review* A **21**:1735-1755.

Norton, H. T. J., 1928, Natural selection and Mendelian variation. *Proceedings of the London Mathematical Society* **28**:1-45.

Paine, A. D. M., 1985, Ergodic reasoning in geomorphology: time for a review of the term?, *Progress in Physical Geography* **9**:1-15.

Pessa, E., and Penna, M. P., and Minati, G., 2004, Collective Phenomena in Living Systems and in Social organization, *Chaos & Complexity Letters*, pp.173-184.

Petersen, K. E., 1989, *Ergodic theory*. Cambridge University Press, Cambridge, UK.

Rivers, R. J., Lombardo, F. C., Mazzitelli, F. D., 2002, Classical Behavior After a Phase Transition: II. The Formation of Classical Defects, *International Journal of Theoretical Physics* **41**:2145-2160.

Schumm, S., and Lichty, R., 1965, Time, space and causality in geomorphology. *American Journal of Science* **263**: 110-119.

Sinai, Ya. G., 1977, *Lectures in ergodic theory*. Princeton University Press, Princeton, NJ.

Thorn, C. E., (ed.), 1982, *Space and time in geomorphology*. Allen & Unwin, Boston.

Thorn, C. E., 1988, *An introduction to theoretical geomorphology*. Unwin Hyman, Boston, MA.

Thornes, J. B., 1983, Evolutionary geomorphology, *Geography* **68**:225-235.

Thornes, J. B., and Brunsden, D., 1977, *Geomorphology and time*. Wiley, New York.

Umezawa, H., 1993, *Advanced Field Theory. Micro, Macro, and Thermal Physics*. American Institute of Physics, New York.

Vabo, R., and Nottestad, L., 1997, An individual based model of fish school reactions: predicting anti-predator behavior as observed in nature, *Fisheries Oceanography* **6**:155-171.

Von Bertalanffy, L., 1952, *Problems of Life. An Evaluation of Modern Biological and Scientific Thought*. Harper & Brothers, New York.

Von Bertalanffy, L., 1968, *General System Theory. Development, Applications*. George Braziller, New York.

Wachter, K. W., 1986, Ergodicity and Inverse Projection, *Population Studies* **40**:275-287.

Wachter, K. W., 1997, Reconstitution and Inverse Projection. In: *English Population History from Family Reconstitution 1580-1837*, (E. A. Wrigley, R. S. Davies, J. E. Oeppen, R. S. Schofield), Cambridge University Press, Cambridge, UK, pp. 515-544.

Walters, P., 1982, *An introduction to Ergodic Theory*. Springer, New York

Winfree, A. T.,1984, The prehistory of the Belousov-Zhabotinsky oscillator, *Journal of Chemical Education* **61**:661-663.

Zhabotinsky, A. M., 1964, Periodic liquid phase reactions, In: Proceedings of the Academy of Sciences, USSR, **157**: 392-395.

Zaikin, A. N., and Zhabotinsky, A. M. , 1970, Concentration wave propagation in two-dimensional liquid phase self oscillating system, *Nature* **225**:535-537.

Web Resources

Domowitz, I., and El-Gamal, M., 1999, A Consistent Nonparametric Test of Ergodicity for Time Series with Applications,
http://papers.ssrn.com/sol3/papers.cfm?abstract_id=179912,
Pennsylvania State University and Rice University - Department of Economics,

Domowitz, I., and El-Gamal, M., 1997, Financial Market Structure and the Ergodicity of Prices, http://www.ssc.wisc.edu/econ/archive/wp9719.pdf

Landon-Lane, J. S., and Quinn, J. A., 2000, Growth and Ergodicity: Has the world converged? The Econometric Society,
http://www.econometricsociety.org/meetings/wc00/pdf/0146.pdf

McCaa, R., 2001, An Essay on Inverse Projection,
http://www.hist.umn.edu/~rmccaa/populate/ipessay.htm

Chapter 7

APPLICATIONS TO SOCIAL SYSTEMS (1)
Growth, development, sustainable development and ethics

This chapter and the following one focus upon how the concepts of Collective Being and DYSAM can support processes of *control and management* of social systems and of the associated processes of emergence.

The concept of Collective Being was introduced in Chapter 3 as a particular multiple-system emerging from the Collective Behavior of agents equipped with a cognitive system and *simultaneously* belonging to different systems or *dynamically* belonging to different systems over time. In social systems this refers to agents *simultaneously* or *dynamically* interacting using the same cognitive models.

Regarding the control and management of a social system a *linear* way of influencing system behavior can be achieved by:
- acting upon networked elements taking into account their interdependence;
- acting upon specific interactions taking place amongst elements; this is done by influencing the interaction processes themselves through modifications of their environment and constraints;
- influencing system input.

In this case the influences act in a *direct* way upon the system under consideration. They can therefore be considered as *causes* producing *effects* (modifications of system behavior), expected to be directly proportional, in some sense, to the causes themselves. Whence our use of the term 'linear'. However, when dealing with Collective Beings it is possible to introduce a *non-linear* way of influencing them. This can be done by :

- influencing the behavior of autonomous agents in a system, taking into account the *non-linear* (due to processing through cognitive models) effects on other systems to which the agent belongs;
- influencing the multiple interactions which agents have to deal with;
- influencing the adoption of specific cognitive models.

In this case the process of emergence is influenced by acting upon the processing of multiple interactions, and thus we are dealing with a *non-direct* influence, which can be qualified as *non-linear*, owing to the impossibility of clearly recognizing causes and effects.

The particular multiple belonging of elements of Collective Beings to different systems makes it clear how processes of influencing and managing single elements while belonging to a specific system is also an easy way of influencing other systems to which the element belongs. Influencing the cognitive processing in one of the multiple systems effects the processing in others which are difficult to manipulate. The focus upon these aspects is one of the great advantages of using the level of description introduced for Collective Beings.

Before dealing in more detail with the processes of control and management of social systems we must remember that usually the latter are considered at two levels: *micro* (i.e., individuals) and *macro* (organizations, i.e., corporations, institutions). This distinction is analogous to the well-known one in Economics, between Macroeconomics and Microeconomics (see, for example, Mankiw, 2002; Pindyck and Rubinfeld, 2000).

In all social systems we have behavioral rules at both levels, with juridical, corporate, educational and military rules and regulations existing for (1) single individual behaviors and (2) organizational behaviors, i.e., the behaviors of individuals belonging to an organization and the behaviors of entire organizations. Violation of such rules and regulations may be punished in different ways. Within the linear approach the control and management of social systems is possible simply because they are structured by such rules. The upper level, the *macro* level, relates to the *general rules of the game*, the economic and political system, and the general scenario. Even this level, within the most complex social systems, is structured through international rules (De George, 1993) disciplining trade, financial relationships, copyright and patents, tourism, labour forces, military activities, etc. In short, within this approach the general conceptual framework adopted is based upon both the idea of controlling and managing through rules and upon the expectation of *linear* consequences of these rules, moving from simpler to more complex levels. By using the concept of Collective Being, however, a more sophisticated approach to control and management, based upon the adoption of DYSAM, is possible.

The discussion of this approach begins by dealing with the concepts of growth, development and sustainable development. The latter are considered crucial when designing tools to assess the effectiveness, especially at the macro level, of control and management strategies within a social system. The linear approach considers this effectiveness merely as a consequence of the rules holding within the system under consideration. We show here, however, the limits of this approach and how new perspectives arise when the same concepts are considered from a systemic point of view, using a new framework based upon emergence, Collective Beings and DYSAM.

7.1 Growth, Development and Sustainable Development

In this section we start from the difference between *growth* and *development*. According to the linear approach the two concepts coincide, with growth being directly proportional to the effectiveness of the rules imposed upon the social system considered. We believe, on the contrary, that the two concepts are profoundly different. The concept of growth can be applied to any system, whereas the concept of development can only be introduced within a context characterized by a number of different processes, often occurring simultaneously. This is the case, for instance, for a private company or a public organization, which are typically structured in subunits, departments, offices, and so on, each having different roles, goals, and *modus operandi..* The key problem in such kinds of social systems are the relationships (or coordination) between the various processes taking place in different parts of the system. A useful concept here is that of *harmonicity*: in general a development process regarding the whole organization is *harmonic* when it involves to the same or comparable extent all parts of that organization. Moreover, it is possible to introduce different levels of development having an increasingly systemic content:

1. development as a networked process of growth; the attribute 'networked' means that the growth itself is due to a coordination between different processes linking them into a sort of network;
2. development as a networked process of growth which is, however, considered as harmonic from the point of view of an observer located within the social system under consideration;

3. development as an emergent property of a system, resulting from a networked process of growth;
4. *sustainable development*, defined in more detail below, refers to an observer outside the system under consideration and to different points in time (e.g., future generations). How can one build *sustainable* social systems (Brown, 1981) and how can one induce and maintain *emergent sustainable development*?

It should also be noted that the social space of rules and limits, later grouped under the label 'ethics' in Section 7.2, within which Collective Beings behave, is expected to induce processes of growth and interactions among the Collective Beings themselves. Such rules and standards refer to the roles of both:

- *stakeholders* (people affected by corporate or Institutional behavior, such as consumers, etc., see Freeman, 1984) - this subject refers especially to *business ethics* (Nasi, 1995; Clarkson, 1998; Bowie, 1986) and
- *shareholders* (owners of company shares).

These two roles are not static nor rigidly well-defined and the boundaries between the two roles is fuzzy, changing as a consequence of interactions, in a simultaneous or dynamic way, as in Collective Beings.

7.1.1 Representing growth

In economics the term *growth* usually refers to the process of *increasing* some quantity, such as a country's Gross Domestic Product (GDP), a company's output or a population. Two main conceptual frameworks are generally used when this process is studied which are based upon the assumption of :

– the possibility of unlimited growth, described, for instance, by an exponential or a linear function;
– limited growth (Meadows *et al.*, 1972), often described by a *logistic function* (Figure 1.3), as introduced above in Section 1.4.4. This is used to study growths which may be limited by a limited amount of available resources, by environmental constraints or by physical size (Marchetti, 1981). This framework is used for designing scenarios for a future which is not driven by endless growth (Meadows *et al.*, 1993).

When considering the latter framework, it can be seen that within the logistic curve a shift from an initial stage of **increasing growth** to a second one of **decreasing growth** can be recognized. By limiting our considerations to the context of economic processes, clearly the logistic curve (Cramer, 2003) is very suitable for describing the whole evolutionary life of a process or of a good number of different circumstances such as:

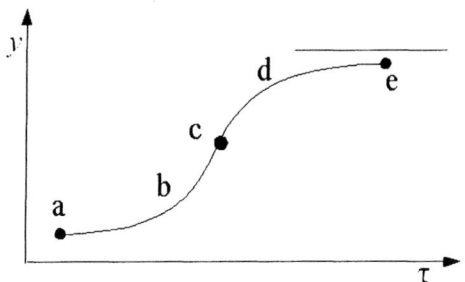

Figure 7-1. . Typical logistic curve with a-e representing the points at which the corresponding events are assumed to occur.

a – doing old things in a new way;
When the goal is to produce more efficiently using more advanced organizational approaches and/or technologies which are already available;
b – doing new things using new solutions;
When the goal is to produce new products or offer new services using new organizational approaches and/or new technologies;
c – doing new things using well-known technologies;
When established production systems and/or organizational approaches are used to produce new products and/or offer new services i.e., innovation using what is already available;
d – using new things in old ways;
When new products and new services are used without taking full advantage of their potential: traditional activities but using new technologies;
e – using old things in old ways;
Massive use of well-established technologies, looking only for high levels of production and mass markets.

7.1.2 Development

A development process may be described in a graphical way by resorting to the logistic curve introduced above. More precisely, development occurs when, after a first phase of increasing growth followed by a second phase of decreasing growth, we have a jump, denoted by **f** in Figure 7.2, to the initial part of another logistic curve due, for instance, to innovation. In electronics, for example, there was the jump from vacuum tubes to solid state electronics, then from solid state discrete component (transistor) electronics to the integrated circuits of microelectronics followed more recently by the transition from microelectronics to nanotechnologies and molecular-level devices)

The point **f,** where the shift from one logistic curve to another takes place, may be related to an event such as those mentioned in connection with Figure 7.1: innovation may occur on the product (starting at point a) or on the production process itself (starting at b), even if they occur together. After jump **f** a new growth curve begins. Subsequent linked processes of jumping from one logistic curve to another may be considered as a most suitable framework for representing development processes.

Good examples of development processes are found within *processes of technological evolution*. Computers, for instance, arose from a combination of expertise and technologies from both Information Science and Electronics. During their development technologies related to Mechanics and Electro-mechanics were able to support only their established applications. Nowadays, Electronics, after undergoing a growth process, lies, in turn, in a situation similar to that of Mechanics and Electro-mechanics at the dawn of the computer era. Namely, there is already the perspective of a combination of Electronics with Biology in order to generate new devices having much greater computational power, based on emergent computation (Forrest, 1990), biocomputing (Lundh *et al.,* 1997) or nanocomputers (Drexler, 1986; 1992). In the same way we have witnessed the development in mechanics from steam engines to internal combustion engines, arriving at the jet engine.

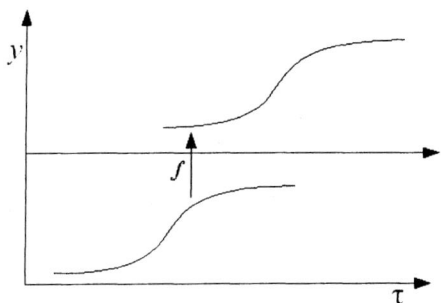

Figure 7-2. Process of development as a jump from one growth curve (logistic) to another made possible through scientific discoveries and technological innovations.

The scheme in Figure 7.2 can be further generalized. A developmental process, based on a sequence of innovations during growth processes, may in fact be represented by a new, more global, process of growth, fragmented into different processes, each having its own logistic description. This can be represented by a 2D-graph and plotting a more global variable such as, for instance, a suitable measure of the wealth produced along the y-axis (Figure 7.3).

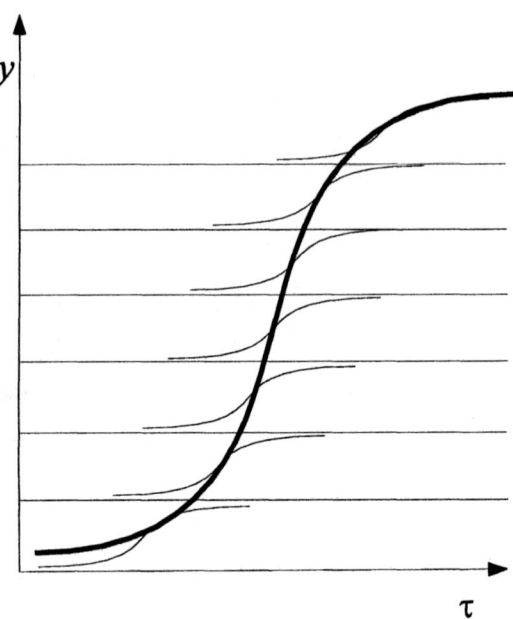

Figure 7-3. Representation of development through a superposition of different growth processes.

This kind of representation is also used in this book to describe **development processes** both:

–   as a sequence of linked processes of growth which depend, for example, upon successive innovations, and
–   as a set of linked, networked processes of growth considered as *harmonic* from the point of view of an observer within the social system where such processes take place (*harmonic* with reference to a plan, to a development project, to stakeholders or as a function of the adopted approach).

Harmony amongst growth processes is to be understood as being consistent with a development plan, a design for the social system under consideration based upon socio-economic considerations. In this case the concept of harmony can acquire a clear, well-specified meaning based, for instance, upon suitable relationships between the growth or the growth rates of the various components. One often used requirement for the development to be harmonic is, for example, that the component parts vary by reciprocal percentages.

In this case *harmony is identified with proportionality*, as observed in some biological systems (functional harmony of living bodies) and in corporations (balance between different divisions).

From the cases discussed above it emerges that *growth* seems to be a *necessary*, but not a *sufficient*, condition for development (Georgescu-Roegen, 1971; 1976).

Processes of development may also sometimes require a balance between the various growth processes, which may imply temporarily *negative* growth (for instance, in conditions of overproduction) or *zero* growth (perhaps to limit production to maintain price levels).

In any case, the concept of *development* can be reduced neither to single growth processes nor to a set of reciprocal harmonic, proportional growth processes. This is merely a simplification which is, however, useful for drawing the framework to describe the process in a more complete manner.

The concept of development must also take into account the ethical rule, accepted nowadays, that one can not design what is supposed to be a development for someone else (Banathy, 1996), because the concept of development is not an *objective* one.

However, a further generalization of the concept of development is possible. Consider the case where a number of different, mutually interacting growth processes occur. Within this more general context, development may be identified with an *emergent property* of the system of interacting growth processes. From this perspective, development is different from the set, or the sum of local growths. In a company, for instance, it may emerge from the interactions between marketing, production, delivery systems, human resources, finance, R&D and so on. This approach towards understanding development allows the use of systemic methods for managing processes of development. This kind of framework is now well established in modeling global phenomena in economics including those related to globalization.

Once a systemic view of developmental processes has been introduced, it can be used to focus upon a specific kind of development: so-called sustainable development. The original definition of this concept was introduced in the so-called «Brundtland Report». In 1987, the World Commission on Environment and Development (WCED), set up in 1983, published a report entitled «Our common future». The document came to be known as the «Brundtland Report» after the Commission's chairwoman, Gro Harlem Brundtland. (Tomlinson, 1987). This report , commissioned by the United Nations Organization, states that development is intended as *sustainable* when it is based upon a rate of resource consumption (by the system of growth processes allowing for the emergence of development itself) such as to leave to the system the time necessary to replace the resources employed.

Actually, some traditional views of sustainable development do take into account harmony and emergence amongst growth processes, but only:

- *within* the system, with reference only to its shareholders, and
- currently, without reference to the *configuration* of constraints and resources passed on to future generations, and thus influencing any future system.

This last point contrasts strongly with the original definition of *sustainable development* introduced by the Brundtland Report. It should be stressed that, within a systemic view of this concept, the harmony and emergent development to be considered refer not only:

- (in space) to growth processes regarding the context of the system under investigation, but also to *all* other systems interacting with it;
- (in time) but also to the produced output to be processed by systems that future generations will have to design for themselves.

The link between the concept of sustainable development introduced above and concrete problems such as climate change, desertification, acid rain, deforestation, air pollution, biological diversity, limited resources, greenhouse gas emissions, ozone depletion, population growth, renewable resources, night-sky pollution (light emission from artificial sources), sustainable levels of consumption, sustainable energy systems, have been dealt with in specific conferences and reports and are implicitly represented through the concept of *harmonic* development with reference to the stakeholders (represented by real people, future generations and nature itself - *the only supplier assumed to go unpaid in a normal commercial context*) both inside and outside the considered system. As a consequence *Sustainable Development* refers not only to a system of harmonic growth processes but also to the emergence of new ways of using resources (technology), to the production of new ones, as **knowledge** is produced in post-industrial society (see Section 7.2). In other terms, the kind of input currently being prepared for the systems to be designed by future generations is taken into consideration.

We stress that the nature of this input imposes very strong constraints upon system design and affects it in a significant way. Future systems can no longer be designed on the assumption of unlimited availability of growth processes and resources.

It now has to be taken into account that, by adopting a systemic approach to the concept of development, sustainability has to refer to growth processes which *maintain the emergence of development*. This in turn implies that, because the processes of emergence are not linearly dependent upon local features of the constituent parts, sustainable development is not linearly related to the sustainability of the single growth processes from which it emerges. In other words, the **sustainability of growth processes is neither a sufficient nor a necessary condition for the sustainability of the development which emerges from them.**

Now, within such a framework, the main questions are: how can the sustainability of a process of development, that is, of a system of growth processes, leave aside the sustainability of the latter? Does the non-sustainability of development imply the non-sustainability of the various growth processes involved?

To answer these questions, we begin by recalling that, when speaking of *development*, its characteristics are different from those of growth processes. Namely, a growth process is reversible, as that growth may stop and be followed by a decrease, taking a system back to its initial situation before the growth, whereas a development process gives rise to an entirely new situation in an irreversible way. For example, a company may undergo a growth process, consisting of an increase in its income, due to a favourable market situation. However, when its products lose their appeal for consumers, growth is transformed into a reduction in sales, which may lead to the disappearance of the company itself, that is to the same situation existing before the firm was founded. On the contrary, when a firm undergoes development due, for instance, to technological innovation, the situation changes in an irreversible way and can no longer return to that existing before the introduction of this innovation. Once the innovation has been introduced, merely the *knowledge* of its existence is enough to change the attitude both of shareholders and stakeholders of that company, even in the distant future. As a consequence of this, irreversibility is considered a very important feature of non-sustainability. Another concept related to sustainability is the *evolution of evolvability* based upon the possibility of a self-guided/conscious evolution (Banathy, 1989; 2000; 2003).

In some ways this recalls the novel approach followed by Prigogine (see Chapters 1 and 4) when introducing the thermodynamics of irreversible processes. Before his work, physicists focused upon equilibrium states and reversible processes, considering irreversible processes as phenomena of degeneration. Prigogine introduced a totally new approach by stressing the creative role of irreversibility. This led him to propose the theory of *dissipative structures*, a totally new way of considering such phenomena, thus opening the way for novel approaches in Systemics, even with reference to the physics of living matter.

Usually, sustainable development carries a positive connotation. It should, however, be recalled that this concept is, *per se*, devoid of any connotation, positive or negative. After all, the positivity and the negativity are not objective features, but only observer-dependent. For instance, within some systems the effects of single growth processes are very strong, affecting the features of the emerging system in a way which could be considered quite negative, as in ecosystems where development produces the extinction of a species. In these cases the non-sustainability of some processes induces the sustainability of an unwanted development, resulting in degenerative processes. This could be the case for mankind within the planet earth's environment.

The sustainability of particular systemic aspects of development processes, such as the production of knowledge and values, should be considered as being linked to finding a balance, to compensating, to synergic effects, to the creative role of irreversibility, mainly within social systems. In any case, the sustainability of the processes involved does not guarantee the sustainability of the overall emergent development process nor their suitability for the emergent system. Namely, an emergent development may structurally change the very process from which it emerges. One example is the development of science and technology (see, for instance, Mensch, 1975), which are inexhaustible sources and generators of resources but, on the other hand, give rise to pollution, waste, consumption of limited resources, irreversible modifications of the Earth and of life on the planet. In other words, *solutions to existing problems often produce new, often unexpected and usually more complex problems*.

We have already discussed development as emerging from interactions between growth processes. **Sustainability is therefore a systemic, emergent property of the system of growth processes generating development**. On this point, it should be stressed:

- sustainable processes of growth do not necessarily lead to the emergence of sustainable development;
- in the same way, a system of non-sustainable growth processes may give rise to sustainable development.

Both cases could occur for two main reasons: 1) the emergence of a coherent whole is a typically non-linear phenomenon where there is a feedback action of the emergent process on its sub-component processes; 2) this sort of emergence can only occur in *open* systems, interacting with their environment.

By resorting once again to the example of a company, situations may occur where the growth processes of its various departments are sustainable, whereas this is not true for the development of whole company because of negative market conditions. On the other hand, non-sustainable growth processes of its departments may lead to the emergence of a sustainable process of development of the company as a whole because of the usage, for instance, of new substitutive resources deriving from, in turn, new knowledge and cooperative effects induced by the environment. The concept of sustainability must therefore be referred, in space and time, to the global context.

Given that non-sustainability can be obtained and applied, would it not be possible to design sustainability as well? In reality the question is: *how to make sustainable development emergent, that is, a systemic property.* The answer is: by acting upon the *space of rules,* on the *cognitive space* where interactions take place, combining organizational constraints at the social level and emergence at the level of collective behaviors. Therefore making development *sustainable* may be matter of adjusting, balancing, compensating and correcting during the process rather than using pre-established rules. This recalls to mind the double-loop versus the single loop of second order cybernetics (see Section 2.1): in a single-loop the system is expected to be controlled by regulating the input as a function of the output; whereas in a double-loop the system is expected to redesign the rules and not just to apply them. That is, the *difference between playing a game and designing one.* **Designing sustainable development is related to designing emergence.**

When referring to the concept of development, one can not avoid mentioning a possible quantitative approach to this topic: that based on the methods of *Econophysics.* This term, as is well known, denotes a new discipline which investigates dynamic behaviors occurring within economic systems by applying the methods of Physics and, in particular, those of Statistical Physics (see, for example, Kertesz and Kondor, 1998; Mantegna and Stanley, 1999; Bouchaud and Potters, 2000; De Liso and Filatrella, 2002). This approach is, in a sense, a natural one, as any economic system can be described in two different ways: macroscopic (when dealing with the system as a whole) and microscopic (when dealing with the behaviors of its individual component parts). Of course, an economic system is not strictly equivalent to most physical systems, as the latter often consist of identical elements, whereas the individuals of an economic system can differ profoundly from one another with respect to their preferences, inclinations, cognitive systems, behavioural features, personalities and so on. However, although this might prevent a direct transposition of results already obtained for physical systems to economic systems (for a review, see Burda *et al.,* 2003), one can take advantage of more recent results from the Physics of *disordered systems* (*cfr.* Mézard *et al.,* 1987; Amit, 1989; Dotsenko, 1994; Newman, 1997; Mikhailov and Calenbuhr, 2002, Chap.8) to deal in a quantitative way with the complexities of economic dynamics. The topics investigated by Econophysicists include the dynamics and the statistical mechanics of markets, the analysis of financial time series, the emergence of collective behaviors amongst a set of agents playing competitive games, the dynamics of networks of interacting agents, the spreading of opinion, wealth distribution and many others. Of course, the development of companies and other organizations is included within these topics (see, for example, Ausloos *et al.*, 2004), as well as the role of ergodicity in complex economic systems (Lillo and Mantegna, 2000; 2001). The most important contributions of Econophysics to the study of development processes include the discovery that the features of such processes are associated not only with the form of growth laws (as shown in the previous discussion), but also to more subtle statistical features, such as the power laws relating the size of extinctions to their frequency or the unit cost of a product to the number of items produced. Such features, rather than a detailed forecasting of the time evolution of a particular organization, provide an understanding of the kind of economic *scenario* in which individual developmental processes occur. And, as previously illustrated, what is required for the sustainability of a given development is not a constraint acting upon the individual process but rather a constraint on the whole scenario within which the processes themselves take place.

We also mention the very interesting issues introduced by N. Georgescu-Roegen in reference to the exclusive attention given by economists to the topic of energy and ignoring the importance of *matter* (Georgescu-Roegen, 1971; 1977b; 1979).

7.1.3 Managing for development

The concept of development introduced above entails the possibility that growth processes may be suitable for describing the evolution of single (and even simple) systems whereas the focus on development may be more suitable when dealing with systems of growth processes, eventually occurring within the same system. Clearly, a very good conceptual framework for describing such systems of growth processes is that of Collective Beings. This is because individual agents in Collective Beings have multiple roles and emergence deriving from single systems of growth processes is facilitated by the common involvement of agents in different processes.

When dealing with Collective Beings, of course, the approach used is very different from traditional ones because within a Collective Being it is not possible to manage one individual system whilst ignoring the others.

Such a situation has, in fact, been known for a long time by people dealing with practical management issues. Namely, the reductionist approach based upon considering one system at a time does not work, even in small organisations. A similar lack of effectiveness has also been widely experienced in *planning* development. Once again, one has to take into account that the **development of a Collective Being is emergent.** It is possible to manage single growth processes and local development by *dynamically using different models* (DYSAM) for such a multi-dimensional process. This calls, in turn, for a new kind of leadership in organizations (Argyris, 1976; Magliocca, 2002).

Among the most typical contexts to which the concept of development outlined above can be fruitfully applied is the evolution of global world status; a topic of increasing interest because of its obvious impact on any national or local aspect (see, for instance, the series of annual publications of the Worldwatch Institute, "State of the world", W.W. Norton & Co., New York at http://www.worldwatch.org/). Many international conferences have been organized over the course of time, with the general goal of improving the global situation through the establishment of rules, designed to discipline relationships and interactions on given topics between different countries. The general idea was to set generally accepted standards, thus obtaining a global consensus. In particular, many international conferences have been organized on subjects related to development. The first of these may be considered as that which took place in Stockholm in 1972: the First International Conference on the Human Environment which produced *The United Nations Declaration on the Human Environment.*

Other international conferences, amongst many, which have attempted to harmonize, to find a general consensus on national and international policies for aspects of social and environmental impact, include:

- The Conference in Rio de Janeiro, Brazil, in 1992, which produced the *Declaration on Environment and Development*;
- *United Nations International Conference on Population and Development (ICPD)*, Cairo, Egypt, 1994;
- *Third Conference of the Parties* (official website http://unfccc.int/cop3/), Kyoto, Japan, 1997, whose purpose was to establish generally accepted rules to limit greenhouse gas emissions;
- Conferences of the *World Trade Organization (WTO)*;
- *World Summit on Sustainable Development*, Johannesburg, South Africa, 2002.

These conferences, as well as many others, highlighted the *non-sustainability* of development processes based on growth processes *requiring,* i.e., *grounded upon*, other growth processes having **necessarily**

permanent zero or negative growth rates. This is the paradox of the development of economic systems based upon the underdevelopment of other systems: the logic of exploitation.

However, despite the fact that all these conferences proposed new rules to improve the global state of the world and to avoid exploitation, these proposals have never been very effective. This can be easily understood by resorting to our approach based upon Collective Beings. Namely, within a Collective Being, every attempt to influence it by using only rules and/or rule changes is doomed to failure, owing to the fact that, within it, the individual agents have multiple roles. As a key role is played by the cognitive models used by those agents to govern their behaviour, at any level and in any context, the best way to influence a Collective Being is to act on the cognitive system of the agents composing it. Thus, the DYSAM approach would be suitable for managing such multiple cognitive modeling for social systems. Within this context, there is a fundamental resource necessary for designing and making emergent sets of rules and cognitive models, for managing other kinds of resources (such as knowledge in post-industrial society) and which will be used for designing any future for human systems: **ethics**.

7.2 Ethics

Ethics is fundamental resource for *deciding and managing*. Optimizing, looking for greater effectiveness, is a rational approach, which represents a *type* of ethics. The term *ethics* comes from the Greek *ethos,* behavior, and was introduced in philosophy by Aristotle. Usually it refers to that part of philosophy which studies human behavior, the criteria for its evaluation and making choices.

The term *ethics* is associated with a number of different meanings (for their classification, see, for example, Mautner, 1997, pp. 180-181).

o *Religious Ethics.* This is probably the most common way of understanding ethics. It is intended as a *collection of values*, of beliefs, or a doctrine, as presented by specific religious institutions, concerning what is right and wrong, good and bad, with reference to personal conduct.

o *Analytical Ethics or Metaethics.* This expression denotes a philosophical inquiry into the descriptions of behavioural concepts and related values, as good and bad, right and wrong, virtue, rights, and so on. This leaves aside the specific content of ethical norms.

o *Descriptive or Moral Ethics.* The study of behavioral uses, practices and beliefs adopted by a social system. It is also called *ethnological ethics* and belongs to the social sciences.

o *Positive morality.* Code or doctrine, generally adhered to by a group of individual members of a community, other social groups or those practising a given profession, concerning what is acceptable and what is not, what is right and wrong or good and bad.

o *Normative or Prescriptive Ethics.* The rational inquiry into not how people behave but how they ought to behave (Kagan, 1998; Postow, 1999). Normative ethics bridges the gap between metaethics and applied ethics, such as Bioethics, Business Ethics, Environmental Ethics, and Medical Ethics. It relates philosophy to moral philosophy. Examples of theories generated from this approach are Kantian ethics and utilitarian ethics.

o *Applied Ethics.* Medical ethics and business ethics (De George, 1997) are specific examples of applied ethics. It attempts to establish practical moral standards and behavioral rules.

Cicero was probably the first to use the Latin word *moralis* as the equivalent of the Greek *ethicos*. Much later, Hegel introduced a further distinction between the concepts of ethics and morality:

• *Morality*, introduced by Socrates and reinforced by Christian thinking, should refer to the concern of autonomous individuals, that is to the subjective aspects of behavior;

• *Ethics* should refer to a set of moral values actually carried out in history (e.g., within the context of a State or of social institutions such as the family).

The two terms are used as synonyms in some contexts, whereas in others they are used in a different or opposite sense. In this book the term *ethics* is intended as *normative ethics* for at least four main reasons:

• this kind of ethics may be discussed and investigated by anybody, without religious, political or ideological restrictions;

• it can be studied even by resorting to scientific concepts, results and principles;

• it relates to a **transformation of values into concrete rules**;

• it is suitable for dealing with *quality issues* rather than being taken for an extension of *quality*.

The crucial systemic feature of this kind of ethics is that the global ethics emerging from interacting agents, even when the latter have an identical morality and profess and practice the *same ethics*, is not the same as that shown by the individual agents. Moreover, the key point is that this kind of ethics, whatever its level of application, relates to behavioral rules and not to indirect, systemically induced effects. **Ethics simply does not apply to non-linear, long range effects**. For instance, social systems made up of components having ethics which prohibit killing may globally behave in such a way as to adopt an emerging ethics based on very different principles. For instance, such principles could allow the exploitation of weaker social systems which, in turn, could induce difficult and precarious living conditions for their components because of a lack of economic development, unregulated use of resources, pollution or health problems.

Both the law and ethics produce changes in the environment (see Stone, 1975) in which human agents interact by applying their cognitive models. The *space of rules* of the environment is no longer given only by *natural* laws. This space *self-develops*, increasing its own complexity through human agents by the addition of limits and rules to be *used* by their cognitive models. This occurs because human agents are able to design and change limits and rules. The environment in which human agents interact is thus a combination of (1) a *natural* part shared with any other agent provided with a cognitive system and (2) a *self-designed* one based on limits and rules established by law and ethics. This latter aspect is the *social space* shared by single individual agents and by multiple-systems emerging from the collective behavior of agents which give rise to Collective Beings. This *social space* constituted by rules within which agents and emergent Collective Beings (Steels, 1990, 1991) behave gives rise to new processes affecting both the natural environment and the new social one. This *social space* is based upon the limits and rules established by laws and ethics (Minati, 2002).

The topic of ethics is intimately connected with system design and planning. In fact, the activity of human beings is often based upon using what is available at the moment before having any comprehensive plan of usage. This refers to any kind of resource and is well described by Herbert Simon's theory of *Bounded Rationality* (Simon, 1955).

This is even true for technologies and scientific results, which are always used when they become available, in the absence of any planning which takes into account the possibility of further innovations. A more advanced approach to system design, however, should allow for:

1. new technologies and scientific results becoming available *during* the design process itself;
2. the use of new, more appropriate, conceptual tools for developing a provisional comprehensive plan for exploiting a given technology, avoiding the use of more traditional approaches suitable for lower technological levels.

In such a design process many transformations and re-definitions can occur. The constant element is the need for social control related to detection of the emergence of systemic properties. This is true for any kind of social system, from the microscopic to the macroscopic level.

Within this context, the difference between *planning* (rational use of resources to reach goals, such as *growth*) and *designing* (inventing new systems, as in *systems engineering*, and methodologies for *systems management*) becomes clear. *Strategic designing* refers to long-term evolutionary perspectives considered at the current level of knowledge, but even with reference to the evolution of knowledge itself.

Both are steps in what may be called *systemic designing for emergence* based upon the consideration of dynamical, emerging scenarios where it is possible to play local roles by influencing local variables, to plan where appropriate, to induce processes (by making or removing constraints, by activating, slowing down or accelerating processes, by making resources available or not), to influence *social software* (i.e., cognitive models) used by autonomous agents, for instance, through their *ethics*. Examples of this kind of higher-level designing for long-term scenarios took place in history through crucial decisions such as those adopted by some social systems when abolishing slavery or imposing the usage of a single language. Examples of this approach to designing in social systems at a local level include the introduction of new media for communication, such as satellites or Internet, or of a new currency, as recently happened for the Euro.

The concept of *Collective Being* allows a representation of the context for *Systemic designing for emergence* and provides a description of how this process works. **Designing is based upon the use of multiple roles of agents at the same and/or different times in establishing processes of emergence. This strategy can influence the cognitive models of agents belonging to a given system producing significant effects, especially in systems where direct influence upon the system as a whole is more difficult.**

Here, we introduce a distinction between *systems design* and *systemic designing for emergence*. The former is a *methodology* appropriate for dealing with the complexity of the categories of local processes such as management (Jackson, 2000) and education (Banathy, 1996). The latter is not a methodology, but a process of dynamically understanding and using emergent properties, as well as traditional systems management. It uses, *dynamically*, various models (DYSAM), strategies, tactics, *systems design* and planning. It is related to *strategic management* (Freeman, 1984) for emergent socio-economic processes based upon evolutionary views and not only upon local goals.

7.2.1 Ethics and Quality

In post-industrial societies, where knowledge is a primary resource (see Section 7.2), the focus is more and more on the issue of *quality* as introduced by the International Organization for Standardization (ISO), a worldwide federation of 148 national standards bodies, established in 1947. The ISO international regulation 8402 defines *quality* as the set of properties and characteristics conferring on a product or service the capacity to satisfy customer needs. The power of this definition for social systems is evident when considering the quality of services such as health, education, the environment, transport, food production and conservation. It is difficult to find an issue where *quality* does not apply.

Quality is now a well-established and accepted concept in socio-economic contexts. Certification standards have been introduced and are now almost universally recognized. Generally speaking, two different approaches have been introduced. In the first, processes of quality certification refer to production, distribution, preservation, design and maintenance regarding industrial processes and services. In this case *objective* measurement criteria are clearly identifiable. In the second approach, processes of quality certification refer to *social accountability* (Romm, 2001; Zadeck *et al.*, 1997), *social auditing*, focusing on systemic, synergic, induced effects. Within this approach, objective, certifiable measurements and evaluations to be submitted to stakeholders are very difficult to identify. As discussed below the main reason for such a difference is that the first approach the issues dealt with may be effectively considered within the conceptual framework of *closed systems* or of systems for which it is possible to assume approaches based upon *conceptual closeness* (see Appendix 1, point 23). In this case quality can be assessed with reference to what has to be measured *only* in an observer-independent way: namely the observer is not a user. Standardization is thus considered to be sufficient for obtaining a reasonable level of objectivity.

In the second case, on the other hand, one has to assume conceptual approaches based upon the concept of *systemic openness*. Thus, the measurement of quality is not *only* an engineering problem, and standardization is not appropriate. It becomes a matter of user profiling and learning processes.

The relationship between *quality and ethics* can be described through the following four contexts:

- **Quality for *closed* systems**. In this case reference is made to engineering characteristics, to pre-established *functionalities* of the product or service rather than to user requirements. Measurements are objective and follow engineering approaches. In this case a good example of closeness is given by the way some instruction manuals for products and services have been designed. It is assumed that they must be precise, accurate, complete, up-to-date and consistent. But they may not be very useful because their design did not take into account the real needs of users, clients, and stakeholders. This occurs because, despite their having the desirable characteristics listed above, they may be hard to understand, exhaustive but too complicated, having too many details, up-to-date but difficult to use when the user has specific urgent needs, and so on. Technically, this may be considered as a *human-machine interface problem both for devices and services.*

- **Quality for *compatible* systems.** In this case the context is taken into account at different levels and then modelled, stereotyped and standardized: various possibilities are considered, but their nature is given in advance and their number is limited. The reference is to other products and services available in the given context, but tailored to stereotyped user needs. It becomes, in fact, a *marketing problem.* The interest is not in the real stakeholders, but in the prototype of the generic consumer: the more generic the users, the larger the market. Standards are adopted and used, for instance, more for compatibility among products and services than for meeting specific user needs.
- **Quality and Normative Ethics for Social Systems.** In this case quality and normative ethics overlap. Quality standards and rules are established not only with reference to products and services, keeping in mind marketing issues, with reference to stakeholders belonging to a well-defined social system. Here reference is made to generic stakeholders, distributed over time and space including future generations or other, *remote*, systems which may become merely consumers or manufacturers (as in the case of electronic equipment assembled in developing countries at very low cost).

Within a systemic approach, not only must we take into account the network of relationships and interactions amongst all the social agents involved (such as stakeholders, resources, environment and activated processes), but also accept the crucial role played by the awareness of emergence of new aspects within the global system. Moreover, **it becomes clear why standards for quality and normative ethics are adjacent.** Beyond *quality* there is, as already mentioned, *ethics* as a *system of values* to be transformed into rules, that is, *normative ethics.*

7.2.2 Effectiveness and advantages of Ethics

The effectiveness of any given ethics is related to an observer (usually belonging to the system adopting them) and must be evaluated by considering the purposes of the designers of the social space, the associated emerging processes and their outcomes. Generally speaking, the effectiveness of ethics, as a consequence of the arguments presented above in relation to the concept of development, may be assumed to correspond to their ability to sustain, over time, social systems, avoiding social disintegration, while making emergent sustainable processes and not just maintaining currently existing ones. Different ethics may have different *powers of aggregation* among the components of a given social system. One can also introduce the concept of the *power of ethics* with reference to their ability **to activate, induce and sustain emergent social processes**. The reference is to processes such as economic growth and development, population growth, increasing quality of life, peace, etc.

When dealing with multiple-systems it is possible not only to have *multi-ethical systems*, i.e., systems composed of different subsystems each having different ethics, often corresponding to different communities, different native languages, or different religions, but also multiple-systems composed of agents, each of which use, over time, different ethics, either simultaneously or changing them dynamically, depending upon the context Besides, a given system of values can allow different prescriptive implementations (*normative ethics*), depending upon the context. Thus, for instance, it is possible to profess the value of life while, at one and the same time, and without seeing any contradiction, belong to a social system which accepts the death penalty, the sale of products containing cancer-causing substances, and whose economy is based upon exploitation and non-sustainability. This can occur because within a Collective Being we can have multiple roles which, however, are not managed on the basis of criteria of *coherence*, but by using the same cognitive resources (with the minimal possible effort) in different contexts, a circumstance which requires the application of different cognitive models.

In this section the focus is mostly on micro social systems such as corporations and institutions. We attempt to introduce economically meaningful reasons why a company obtains advantages through caring, managing, and deciding its ethics. The general idea is that ethics is also *locally* effective and convenient, not only over the long term.

Therefore, the main question is: why is it interesting and worthwhile to examine the issue of ethics for corporate applications? The answer is: to manage *protective* and *evolutionary* issues. More specifically, the reasons why ethics pay can be summarized as follows:

- As mentioned above, *every* company uses (explicitly or implicitly) ethical standards. It is a matter of identifying them on the basis of their effectiveness, profitability, strategic importance and consistency with sustainability;
- The application of given ethics could allow entrepreneurs to *sustain* their companies, through their development and innovation, and not only to start them up and look for *growth*;
- To increase their strategic role in the hosting global social system by acquiring the ability to understand strategic scenarios, enabling them to design *development*;
- To protect them from unexpected costs, often disastrous when it regards a question of corporate behavior considered to be highly unethical (in a general sense with reference to values) by the global hosting system.
- To gain competitive advantages and to increase the added value to products or services.

What kind of businesses are not interested in ethics? Short-term, non-strategic and non-sustainable ones. This is the case, for instance, of businesses based upon technologies rapidly becoming obsolete or saturating markets, upon still *unregulated* markets (where it is possible to act as *raiders*) or using in a non-sustainable way non-replaceable resources. Ethics in these cases are reduced, in short, to exploiting a sudden opportunity, to making as much money as fast as possible: the "fast-buck" approach. The proceeds may then be invested in other short-term, very profitable, or longer-term, more serious and sustainable, business activities. At the extreme it amounts to robbery from the hosting system. These are the business ethics of the Mafia and similar organizations.

7.2.3 Ethics for quality and effectiveness

The above discussion on the relationship between ethics and quality leads us to a problem, a crucial one in modern society, which is a source both of ineffectiveness and poor quality.

This problem stems from the fact that the combination of science and technology has usually been activated by focusing on scientific questions considered *by themselves*, isolated from any wider context. Thus, written problems have been formulated and solutions found. *Problems have been on paper*. That is, it has not been taken into account that problems are not only abstract (except, perhaps, initially), but also have to be extended and integrated by taking into account **who** has the problem and **who** is going to use the scientific knowledge generated and the technological solutions found. The user and uses have not been considered as an integral part of the problem. A well-known example is given by human-machine interfaces, whose design was considered for many years merely as an extension of a specific engineering problem. Such an attitude was overcome only very recently, when learning, adaptive systems, requiring user profiling (Stewart and Davies, 1997; Kobsa, 1993; McTear, 1993), were introduced in many fields. We recall here, that the need to take into account the role of the user is related to *systemic openness* and especially to systemic *logical openness*, described in Appendix 1. The latter implies that the receiving system is not only expected to receive information, but to process it at *different dynamic levels of complexity*.

For a long time the development of science and technology has often been a matter of simplification, of reduction: this approach allows people to focus on abstract, context-free problems, avoiding any consideration of users and usability. This strategy has been very effective for the design and development of machines and tools. It was, of course, not only adopted for the sake of simplification, but also because most scientists and technocrats believed in reductionism, in the possibility of reducing complex problems to simple component parts. In any case, even not being aware of having selected such a strategy gives rise to further, various kinds of problems, even of a scientific nature. Namely, within a systemic view, as well as within modern scientific thinking, the role of the observer, of the user, can not be considered, as discussed in Chapter 2, just as a concession to the role of the human being, due to a kind of benevolence of human nature. It has to be recognized as an integral part of the scientific problem itself and of its solution (see also Chapter 3 on *emergence*). Therefore an approach based on neglecting the observer's role can only be considered as an *approximation*.

Unfortunately, in the past, attention has been devoted only to designing technological, theoretical, and social devices able, in the abstract, to solve problems independently of their expected use and application by users. **This led to the consequence that technological solutions were often designed more for the problems than for the people *having* those problems**. A list of representative examples includes:

- hospitals designed more for medical staff than for patients;
- corporations and services designed more for shareholders and workers than for customers;
- schools designed more for teachers than for students;

- control and regulation equipment designed more for the device to be controlled and regulated than its users;
- user manuals written to explain functions, with emphasis on the device itself rather than user requirements;
- software designed more to execute certain functions rather than to interact with users expected to benefit from it and to use the output produced.

A typical case where such an approach led to the failure of a whole line of research is given by *automatic translation* projects, which were based on the illusion of mechanizing the translation process, independently of who is translating. This is equivalent to forgetting that the observer can be considered neither as a generator of noise, nor of relativism, but as a generator of existence, as in the generation of meaning associated with the translation process. As a matter of fact, from the same text in a language L1 it is possible to produce different translations into a language L2, all of them being original, with their copyright and well-identified authors. It is to be underlined that the concept of the **centrality of the human being as active observer,** not based upon anthropocentrism or ethnocentrism, but having a deeply ecological meaning (Capra, 1996), shows first an awareness of and then the responsibility of the human role without giving it the meaning of a presumptuous omnipotence.

The prevailing reductionist attitude in the design of any device, theoretical or practical, has rarely taken into account the question of its **usability** by the targeted user. The latter was not expected to be involved in the design of the device itself. In recent years, however, this aspect has become more important. This has occurred mainly for marketing reasons: the availability of user-friendly, easy-to-use interfaces, expands the market for such devices. Of course, the attention toward usability issues carries with it novel theoretical problems in designing interfaces. It is important to underline that in Computer Science the purpose is not just the availability of easy-to-use and easy-to-understand interfaces. Thanks to the availability of user models it is now possible to design *adaptive interfaces*, able to learn from interactions with users. *In this way it is no longer the user adapting to the interface, but rather the interface adapting to the user* (Allen, 1990; Carroll, 1989).

The real challenge for future IT applications in social systems is related to the availability in the market, *at the user-level,* of information systems:

- requiring a user profile (when a system is switched on it needs to know WHO is the user);
- with an adaptive human-machine interface;
- with learning abilities;

- acting as a support system rather than just deterministically performing a given function.

We believe that the strategy required must pass from "function to usage" (Wiener, 1950).

The robustness, reliability and availability of traditional function-based systems is limited by their complexity. The introduction of self-learning abilities, the combination of deterministic approaches and usage may be the key to the control and management of complexity. In the machine age, the attention towards, and the emphasis placed upon the ability of the user was a point of weakness in mass consumer markets based upon deterministic, algorithmic approaches. Now, on entering the systemic age, an age of complexity and of learning, adaptive, self-organizing systems, the crucial role of the user is no longer a cause of weakness, but of robustness.

7.2.4 Ethics and Globalization

The so-called process of globalisation may be intended as taking place between two extremes:
- *Openness*, i.e., free trade, reduced borders, global communications, merging of international corporations, easier international financial operations, easier and clearer alignment between various national institutions (degrees, professions, law), and so on. At this extreme there is processing of the input, there are processes of integration.
- *Losing identity*, i.e., loss of national languages, ways of life (food, dress, life style), local religions, and so on. At this extreme the input is not given by processes, but just fitted into things as they are, as a new, forced element. There is no input, but transposition.

This issue can be approached by resorting to the concepts of Collective Being and DYSAM. In the case of globalisation, the classical approach is based upon **reducing** Collective Beings to just one of the component systems, by imposing the rules and dynamics of that one individual, dominant system to all the others. To avoid such reduction one has to deal with issues such as the protection of diversity, considering it as a resource (maintaining social multiplicity, emergence of Collective Beings) and not as a problem.

We therefore need to find indicators of emergence, as already introduced in Chapter 6, i.e., tools to manage continuous processes of emergence, providing us with the ability to design without focusing only upon the behaviour of individual components. This implies the design of *social software* and of spaces of rules by establishing processes of continuous interaction and not only of regulation. This activity is conceptually related to the framework of *Evolutionary Game Theory* introduced in Chapter 2.

However, as we all know, there is currently a heated debate about the consequences of globalisation. Namely, on one hand, thanks to the spread of new technologies, people around the globe are more closely connected than ever before. People, goods, information and money flow quickly around the world. Goods, services, ideas (and *problems*) produced in one part of the world are increasingly available and producing effects everywhere else. On the other hand, there is an economic globalisation, associated with looking at economic and financial issues, instruments and portfolios from a worldwide rather than from a single-country viewpoint.

There are several problems with this process, related to the two extremes mentioned above. The main one regards the risks of standardization, of no respect for diversity (corresponding to a possible *imperialistic* use of language, currency, standards), of societies consisting of systems of non-autonomous agents, where stronger agents (such as very large corporations with their own worldwide operational networks) decide and design for individuals and even whole communities. A global society may seem to be a place where peace is the acceptance of such a new status rather than a process, an emergent property. Instead of a system of autonomous agents building emergent social systems through the processing of new *inputs*, there is a global market of *transposed* goods and services. There is an inherent risk that various societies around the world will become little more than markets from which to extract wealth with people living merely to consume, as buyers in an age where marketing technologies are also used for *consent manipulation* (Evans, 1995; Herman and Chomsky, 1988; Minati, 2004).

Thus, the systemic approach and the concepts introduced regarding ethics may help to provide a deeper understanding of the various aspects of globalisation. More specifically, within the scientific literature dedicated to the principles of Systemics it emerges that fundamental aspects of ethical behavior exclude:

1. Designing a system without involving the targeted users, future stakeholders, the clients. This happens by assuming that what is good for the designer *must* also be good for the clients of the system (Banathy, 1991; 1996; 2000);
2. Forcing someone else to do something, not only against his/her will, but assuming that what the designer wants is the best choice. This derives from the assumption that the cognitive models, values and goals of the designer are the best ones possible and so they must be *imposed* upon

others. *This is merely designing and not facilitating or inducing emergence.*

Much more suitable, systemic concepts for dealing with such issues are those of emergence and *openness*. It is very helpful to have a clear idea of how openness between systems, and between systems and their environment does not mean lack of boundaries, lack of distinction. *Systemic openness* (Minati *et al.,* 1996; 1998), may be considered as more than the permeability of system boundaries to matter and energy, as introduced in early definitions. Namely, the extended concept of openness, the *logical openness* introduced in Appendix 1, point 23, refers to the **interaction** between autonomous systems through the mutual use of different levels of models. *Ethical considerations emerge when one subject acts with the intention, the plan to change, to modify the other and its context.* **The ethics of such an interaction between systems** regards the possibility of mutually influencing their respective behaviors, while respecting the systems' *autopoietic* (see Appendix 1, point 5) processes, i.e., *system identity* (on autopoiesis, see Varela *et al.,* 1974; Maturana and Varela, 1980). To design in the place of others is equivalent to destroying their identity, that is, their ability to reproduce themselves. **An ethical process of interaction must be based on co-creation, co-designing whilst respecting the autopoiesis of the system.** Interacting systems are not assumed to simply establish a set of systems or lead to one dominating or enclosing the other. **Systems interact making emergent new multiple-systems, as in Collective Beings, and not just a new system.**

Processes of globalisation having no respect for autopoietic and emergence processes, have **objectivism** as their philosophical background, because they assume that "the best" exists, is unique and, what is worse, is what the stronger system (winner) does. On the contrary, the fact of sharing common infrastructures (i.e., telecommunications, railway systems, standards, currency) and of facilitating the crossing of boundaries, as is currently happening in Europe, should not mean forcing different systems to identify with others, but rather to facilitate the process of emergence of an overall system, which can only exist **while** autonomous systems continue to interact.

Another concept related to the subject of globalisation is that of *sustainability*, already introduced above. The current global stability, far from equilibrium, of the system "world" appears not to be sustainable because the stability of the system is based on such an unbalanced configuration of interacting subsystems. To keep stability between unbalanced subsystems requires the *consumption* of energy and resources which may be greater than those available and produced by the global system: *global system becomes poorer while one of the unbalanced subsystems appears to become richer.*

With reference to the subsystems of an *unbalanced* configuration and particularly to those under the greatest stress, the emerging system may react to such an unbalanced, unstable and very large energy-consumption situation with *internal* and *external* strong, destructive reactions. To avoid the waste of energy and resources to keep an unbalanced system stable, the subsystems must be harmonized, reaching a balance. There is no way out. Linear reactions to disharmonies may only increase the complexity of the unbalanced system, by emphasizing the disharmonies. Linear reactions may be *locally*, in time and space, necessary, but they are not strategic in nature.

Non-ethical behavior, non-ethical interactions, non-sustainable economic processes may be very profitable for the individual and in the short run. Ethics is strategic and profitable in the long run. In economics, this consideration applies when the different systems interact searching only to maximize their particular interest without any overall ethical considerations about the global effects of their actions over space and time. The crucial question is: who is expected to pay, to forego short-term profits, in exchange for more ethical interactions? Who has to accept reduced profit margins in order to allow for sustainability?

References

Allen, R. B., 1990, User model: theory, methods, and practice. *International Journal of man-machine Studies* **32**:511-543.

Amit, D. J., 1989, *Modeling Brain Function. The world of Attractor Neural Networks.* Cambridge University Press, Cambridge, UK.

Argyris, C., 1976, *Increasing Leadership Effectiveness.* Wiley, New York.

Ausloos, M., Clippe, P., and Pekalski, A., 2004, Model of macroeconomic evolution in stable regionally dependent economic fields. Physica A 337:269-287.

Banathy, B. H., 1989, The design of evolutionary systems, *Systems Research and Behavioral Science* **6**:193-212.

Banathy, B. H., 1991, *Systems Design of Education. A Journey to Create the Future.* Educational Technology Publications, Englewood Cliffs, NJ.

Banathy, B. H., 1996, *Designing Social Systems in a Changing World.* Kluwer, New York.

Banathy, B. H., 2000, *Guided Evolution of Society: A Systems View.* Kluwer, New York.

Banathy, B. H., 2003, Self-Guided/Conscious Evolution, *Systems Research and Behavioral Science* **20**:309-321.

Bouchaud, J. P., and Potters, M., 2000, *Theory of financial risks. From Statistical Physics to risk management.* Cambridge University Press, Cambridge, UK.

Bowie, N., 1986, Business Ethics. In: *New Directions in Ethics*, (J.B. De Marco and R. Fox, eds.), Routledge & Kegan Paul, London, pp. 158-172.

Brown, L. R., 1981, *Building a Sustainable Society*. Norton, New York.

Burda, Z., Jurkiewicz, J. and Novak, M. A., 2003, Is Econophysics a solid science? *Acta Physica Polonica B*, **34**: 87-112.

Capra, F., 1996, *The Web of life*. Anchor/Doubleday, New York.

Carroll, J. M., (ed.), 1989, *Interfacing thought: Cognitive Aspects of Human-Computer Interactions*. MIT Press, Cambridge, MA.

Clarkson, M. B. E., (ed.), 1998, *The Corporate and its Stakeholders: Classic and Contemporary Readings* , University of Toronto Press, Toronto, Canada.

Cramer, J. S., 2003, The Origins and Development of the Logit Model. In: *Logit Models in Economics and Other Fields* (Chapter 9), Cambridge University Press, Cambridge, UK, pp: 149-157.

De George, R. T., 1993, *Competing with Integrity in International Business*. Oxford University Press, Oxford, UK.

De George, R. T., 1997, *Business Ethics*. Macmillan, New York.

De Liso, N. and Filatrella, G., 2002, Econophysics: the emergence of a new field? *Economia Politica*, **19**: 297-332.

Dotsenko, V., 1994, *An introduction to the theory of Spin Glasses and Neural Networks*. World Scientific, Singapore.

Drexler, K. E., 1986, *Molecular engineering: An approach to the development of general capabilities for molecular manipulation*. Anchor Press, New York.

Drexler, K. E., 1992, *Nanosystems*. Wiley, New York.

Evans, R. G., 1995, Manufacturing consensus, marketing truth: guidelines for economic evaluation, *Annals of Internal Medicine* **122**:59-60.

Forrest, S., 1990, *Emergent computation*. MIT Press, Cambridge, MA.

Freeman, E. R., 1984, *Strategic management - a Stakeholder Approach*. Pitman, Boston, MA.

Georgescu-Roegen, N., 1971, *The Entropy Law and the Economic Process*. Harvard University Press, Cambridge, MA.

Georgescu-Roegen, N., 1976, *Energy and Economic Myths: Institutional and Analytical Economic Essays*. Pergamon Press, NewYork.

Georgescu-Roegen, N., 1977a, Bioeconomics: A new look at the nature of the economic activity. In: *The Political Economy of Food and Energy* (L. Junker, ed.), University of Michigan, Ann Arbor, MI, pp. 105-134.

Georgescu-Roegen, N., 1977b, Matter matters, too. In: *Prospects for Growth: Changing Expectations for the Future* (K. D. Wilson, ed.), Praeger, New York, pp. 293-313.

Georgescu-Roegen, N., 1979, The Role of Matter in the Substitution of Energies. In: *Energy: International Cooperation on Crisis* (A Ayoub, ed.), Press de l' Université Laval, Québec, pp. 95-105.

Herman, E. S., and Chomsky, N., 1988, *Manufacturing Consent*. Pantheon Books, New York.

Jackson, M. C., 2000, *Systems approaches to management*. Kluwer, New York.

Kagan, S., 1998, *Normative Ethics*. Westview Press, Boulder, CO.

Kertesz, J., and Kondor, I., (eds.), 1998. *Econophysics: an emerging science*. Kluwer, Dordrecht.

Kobsa, A., 1993, User Modeling: Recent work, prospects and hazards. In: *Adaptive User Interfaces: Principles and Practice* (M. Schneider-Hufschmidt, T. Kuhme, and U. Malinowski, eds.), Elsevier Science Publishers B.V., Amsterdam, pp. 111-128.

Lillo, F., and Mantegna, R. N., 2000, Variety and volatility in financial markets. *Physical Review E*, **62**: 6126-6134.

Lillo, F., and Mantegna, R. N., 2001, Empirical properties of the variety of a financial portfolio and the single-index model. *European Physics Journal B*, **20**: 503-509.

Lundh, D., Olsson B., and Narayanan, A., (eds.), 1997, *Proceedings of BCEC97, BioComputing and Emergent Computation.* World Scientific, Singapore.

Magliocca, L. A., 2002, Transforming leadership through coalitions: building the ethics of sustainable development in globalization. In: *Emergence in complex cognitive, social and biological systems* (G. Minati, E. Pessa, eds.), Kluwer, New York, pp. 235-250.

Mankiw, G. N., 2002, *Macroeconomics.* Worth Publishing, New York.

Mantegna, R. N. and Stanley, H. E., 1999, *An introduction to Econophysics: Correlations and complexity in finance.* Cambridge University Press, Cambridge, UK.

Marchetti, C., 1981, Society as a learning system: Invention, and Innovation Cycles Revisited, *Technological Forecasting and Social Change* **18**:267-282.

Maturana, H. R., and Varela, F., 1980, (eds.) *Autopoiesis and Cognition: The Realization of the Living.* Reidel, Dordrecht.

Mautner, P., (ed.), 1997, *Dictionary of Philosophy.* Penguin, London.

McTear, M., 1993, User modeling for adaptive computer systems: A survey of recent developments. *Artificial Intelligence Review* **7**:157-184.

Meadows, D. H., Meadows, D. L., and Randers, J., 1993, *Beyond the Limits: Confronting Global Collapse, Envisioning a Sustainable Future.* Chelsea Green Publishing Company, White River Junction, VT.

Meadows, D. H., Meadows, D. L., Randers, J., and Behrens III, W. W., 1972, *The limits to growth: a report for The Club of Rome's project on the predicament of mankind.* Universe Books, New York.

Mensch, G., 1975, *Stalemate in Technology: Innovations Overcome the Depression.* Ballinger Publishing Company, Cambridge, MA

Mézard, M., Parisi, G., and Virasoro, M. A., 1987, *Spin Glass Theory and Beyond.* World Scientific, Singapore.

Mikhailov, A. S., and Calenbuhr, V., 2002, *From Cells to Societies. Models of Complex Coherent Action.* Springer, Berlin-Heidelberg-New York.

Minati, G., 2004, Buying consent in the "free markets": the end of democracy?, *World Futures* **60**:29-37.

Minati, G., Penna, M. P., and Pessa, E., 1996, Towards a general theory of logically open systems. In: *3rd Systems Science European Congress*, (E. Pessa, M. P. Penna, A. Montesanto, eds.), Edizioni Kappa, Rome, Italy, pp. 957-960.

Minati, G., Penna, M. P., and Pessa, E., 1998, Thermodynamic and Logical Openness in General Systems, *Systems Research and Behavioral Science* **15**:131-145.

Nasi, J., (ed.), 1995, *Understanding Stakeholder Thinking*, LSR Publications, Helsinki.

Newman, C. M., 1997, *Topics in disordered systems.* Birkhäuser, Basel, CH.

Pindyck, R. S., and Rubinfeld, D. L., 2000, *Microeconomics.* Prentice Hall, Upper Saddle River, NJ.

Postow, B., 1999, *Reasons for Action: Toward a Normative Theory and Meta-Level Criteria.* Kluwer, Dordrecht.

Romm, N. R., A., 2001, *Accountability in Social Research: Issues and Debates.* Kluwer, New York.

Simon, H. A., 1955, A behavioral model of rational choice. *The Quarterly Journal of Economics* **69**: 99-118.

Steels, L., 1990, Cooperation between distributed agents through self-organization. In: *Decentralized AI* (Y. Demazeau and J.-P. Muller, eds.), North-Holland, Amsterdam, pp. 175-196.

Steels, L., 1991, Toward a theory of emergent functionality. In: *Simulation of Adaptive Behavior: From Animals to Animats*, (J. A. Meyer and S.W. Wilson, eds.), MIT Press, Cambridge, MA, pp. 451-461.

Stone, C., 1975, *Where the Law Ends: The Social Control of Corporate Behavior*. Harper & Row, New York.

Tomlinson, C.,1987, *Our Common Future*. Oxford Paperbacks, UK.

Varela, F., Maturana, H., and Uribe, R., 1974, Autopoiesis: The Organization of the living systems, its characterization and a model, *BioSystems* **5**:187-196.

Wiener, N., 1950, *The Human Use of Human Beings*. Houghton Mifflin, New York.

Zadeck, S., Pruzan, P. M., and Evans, R., (eds.), 1997, *Building Corporate Accountability: The emerging Practice of Social and Ethical Accounting, Auditing & Reporting*. Earthscan/James & James, London.

Web Resources

Minati, G., 2002, Ethics as emergent property of the behavior of living systems. In: *Encyclopedia of Life Support Systems (EOLSS)*, Vol. 1, Physical Sciences Engineering and Technology Resources, Systems Science and Cybernetics: The Long Road to World Sociosystemicity, (Parra-Luna F. ed.), EOLSS Publishers, Oxford, UK. http://www.eolss.net

Stewart, S., and Davies, J., 1997, User Profiling Techniques: A Critical Review, In: *Information Retrieval Research*, (J. Furner and D. Harper, eds.), Springer Electronic Workshops in Computing. http://www.springer.co.uk/ewic/workshops/IRR97/

Chapter 8

APPLICATIONS TO SOCIAL SYSTEMS (2):
systems archetypes, virtual systems, knowledge management, organizational learning, industrial districts

In this second chapter devoted to social applications of the concepts of Collective Beings and DYSAM we introduce systems archetypes and virtual systems. We also discuss other issues such as knowledge management, organizational learning and industrial districts.

8.1 Systems Archetypes and Collective Beings

Systems Archetypes (Kim, 1994; 1995; Kim and Anderson, 1998; Braun, 2002) describe common behavioral patterns of organizations. They relate to recurring families of problems in management. They are used in System Dynamics (Senge, 1990a) to represent and tackle managerial problems and also as diagnostic tools (Goodman and Kleiner, 1993; Lane, 1996; Wolstenholme, 2003). Their introduction is rooted in the early work of Forrester (e.g., Forrester, 1961; 1968), which was, in turn, inspired by models of the evolution of social and economic systems based upon differential equations. As already illustrated in previous Chapters, these models are often difficult to study and, in most cases, even their analytical solutions, provided we could find them, would be useless to anyone using the model for decision-making in a complex environment. Namely, they have always been built by relying on high levels of abstraction and by neglecting more specific features which, however, could be very important in real-world problems. For this reason, from Poincaré onwards, both pure mathematicians and applied scientists as well as engineers and managers have tried to focus only on general features of the behavior of these models, neglecting the details of the evolution of particular instances of them. Such an approach underlies, for example, all the work done so far within Dynamical Systems Theory, illustrated above in Chapter 4. Behind it, there is the belief that the general features outlined above are very few or, at least, groupable within a small number of categories. These categories are the archetypes, developed by Jay Forrester, and still currently used within System Dynamics.

Most of the time, however, this belief lacked any serious mathematical proof of its validity. In the 1960s and 1970s, however, a number of mathematicians showed how, in the case of systems with a small number of dependent variables and parameters, this belief was well grounded. More precisely, they were able to prove that the patterns of structural behavioral change (so-called *bifurcations*, as introduced in Chapter 4) occurring in these systems could be subdivided into a small number of prototypes, so-called *catastrophes*. Thus, *catastrophe theory* (see the celebrated book by Thom, 1975; other useful and more recent books are Poston and Stewart, 1978; Arnold, 1992; Gilmore, 1993) seemed to offer a sound basis for understanding and classifying, in a simple way, all patterns of change occurring in the biological and socio-economic world. In this sense, the concept of systems archetypes appeared to be well grounded, even mathematically.

Further research, however, began to cast serious doubts upon the possibility of applying catastrophe theory to most real problems. It was shown that, on increasing the number of variables and parameters, it became impossible to group the patterns of change into a small number of categories (e.g., Arnold *et al.*, 1999). This circumstance, already present in previous and celebrated theorems such as that of Smale on *structural stability* (see, for example, Smale, 1966; Palis and de Melo, 1982; Arnold, 1988), practically dominates the world of chaotic phenomena and of partial differential equations. The introduction of more realistic features within models of biological and socio-economic behaviors implies an increasingly closer relationship with models of *disordered systems*, where individuality is more important than general laws. Thus, systems archetypes remain merely as a framework which is useful only in the absence of significant fluctuations or perturbations, a situation often encountered in many domains, such as that of small economic systems, such as industrial organizations, which manage to survive within a relatively stable environment. We believe that the concept of Collective Being could extend the range of usable systems archetypes, with particular reference to interactive systems (see Senge *et al.,* 1994).

Several kinds of archetypes have been introduced in the literature. Grouped according to type, some are listed below (those in italics are then briefly described):

a) Underachievement
    1. *limits to growth*
    2. tragedy of the commons
    3. *growth and underinvestment*
b) Out-of-control
    4. *shifting the burden*
    5. *fixes that fail*
    6. accidental adversaries
c) Relative achievement
    7. success to the successful
    8. attractiveness principle
d) Relative control
    9. *eroding goals*
    10. *escalation*

Systems archetypes introduce typical, invariant, systemic problems often for representing paradoxical, irresolvable situations, revealing the lack of effectiveness of many classical managerial approaches.

The idea introduced in this book is that systems archetypes should be applied not only to single systems, but even to Collective Beings. In this way System Dynamics models could refer to elements and processes having multiple meanings with reference to multiple-systems. It is possible to develop multiple, linked models which mutually interact. In this way components and interactions can be influenced by using different approaches. It should be stressed that, on the basis of the above, paradoxical situations related to systems archetypes may be *managed, not solved* by resorting to the DYnamic uSAge of Models (DYSAM).

Applying systems archetypes to multiple systems provides the flexibility needed to use DYSAM, that is, to invent and not just detect multiple belonging. By using some of the classical archetypes not in single open or closed systems nor in subsystems, but in the various simultaneous or dynamic systems of Collective Beings, it is possible to draw up different strategies and actions having synergic effects by operating at different levels and using simulation techniques (Dowling *et al.,* 1995).

Systems archetypes useful when dealing with Collective Beings through DYSAM include:

1) *Limits to growth.*

The situation dealt with by this archetype occurs when the process of growth is hindered by factors opposing it. A typical case, considered above in Chapter 7, is that of *saturation*, classically described by the *logistic curve*. Other cases will be discussed when dealing with a related archetype: *Growth and Underinvestment*.

In some businesses reaching the saturation point is planned and the entire process is considered on a short-term basis. The market is not only expected to have a limited time span, but this is actually planned. In this case the archetype does not give arise to any paradoxical or contradictory problem to be solved.

In other long-term business activities the formulation of the problem within a single system leads to having to find ways to extend the curve as much as possible, delaying arrival at the saturation point.

However, when dealing with this archetype within the context of a Collective Being it is also possible to find different usages of the same technologies which, in principle, could activate new logistic curves in different systems. It becomes not only possible to use the same technology in different ways, using DYSAM, but to activate different business activities related to that technology, such as selling know-how, the technology itself, allowing the use of specific expertise in different contexts. The recognition of the occurrence of this archetype, along with the effects of technological innovation, is the reason why companies sometimes change their focus on business activities.

3) *Growth and underinvestment.*

The problem introduced by this archetype relates to actions to be taken when system growth approaches a limit as in the *limits to growth* archetype above. The most important aspect managers have to face is deciding whether the limit is due to a systems problem or to external factors, as in the case of *saturation*. This problem is crucial for the survival of a company. If the limit is due to *saturation*, strategy should be based not on investing in improvements which will have no effects on a saturated market. If the limit is due to the inappropriate use of resources, technological resources, production facilities, distribution, pricing and so on, then the problem can be resolved through adequate investments. In the case of underinvestment, resources may be found by reducing performance standards and this action justifies further underinvestment.

By considering this archetype within the context of a Collective Being it becomes easier to detect an unbalanced situation between the different systems which are expected to support, through their synergy, the growth of the overall system. The roles of the various systems may not only be related to their functioning as subsystems, but also to the fact that they behave as autonomous systems emerging from the sharing of common resources. The need for investment, then, can not be decided by considering individual system performances, but by taking into account multiple roles, multiple effects of investments. The latter, in turn, support the idea in investors' minds that the value of the investments made is amplified and not strictly confined to a single subsystem. The multi-dimensional view introduced by considering multiple roles has beneficial effects by focusing more upon synergic effects than on linear, functional effects acting upon single systems.

4) *Shifting the burden.*

Typically, this situation occurs when focus is placed upon relieving the effects of a problem rather than facing the problem. Within a Collective Being one possible strategy is to separate the problem in such a way as to deal with it within two different systems considered to be *autonomous*, that is, having different reasons for being dealt with. In this way the symptomatic solution works in one system while the search for a fundamental solution works in another. By adopting DYSAM the two problems are related at a higher level, but not at the level of single systems. The manager adopting DYSAM will find reasons, independent of the symptoms, for coping with the fundamental problem. In particular, a specific system may be invented, having its own trade-offs and economics, just to reach the goal of dealing with the fundamental problem.

5) *Fixes that Fail.*

Actions taken to fix a problem may have unexpected effects making the original problem even worse. Unexpected effects may come from actions taken after focusing only upon the individual problem, without paying attention to effects deriving from systemic interactions. Actions introduced may unbalance the system. This often highlights the lack of effectiveness of the reductionist approach.

How can the multi-dimensional, multi-systemic view adopted in Collective Beings be helpful? The real problem with this archetype is its general lack of a systemic view. Problems may be dealt with through

DYSAM in a multi-dimensional way by considering not only their interaction, as in the same system or in subsystems, but in a context where the same elements have multiple roles.

9) *Eroding goals.*

The inability to reach a goal and the consequent formulation of an achievable, similar but more limited goal is the core of this archetype. If the process of reaching a goal and of formulating the goal are considered within the same system this archetype usually emerges.

The proposal is to consider the processes of formulating a goal and reaching it, from a complex and more complete managerial point of view, within the context of different systems belonging to a Collective Being. One of them may be the operative framework of the company. Others may be at the Training level or in the Research & Development subsystem. The same final goal may be expressed in multi-dimensional ways, by describing its meaning from different corporate points of view. In Collective Beings, the fact of using DYSAM and of acting upon different aspects of a goal is not a kind of reductionism, but a way to design the emergence of the overall goal from the same interacting components having different roles.

10) *Escalation.*

The loop triggered by two competitive behaviors is not sustainable. The conclusion is, at best, the collapse of one competitor and a less and less profitable market for the survivor. It takes place when mutual actions are considered within the *same* system. A possible way to escape the establishment of such a deleterious loop is to consider the respective actions in *different* systems, that is, to manage a Collective Being which shows this kind of effect only when it *degenerates* into a single system. Actions taken by a competitor should be counteracted at different levels. For instance, price reductions should be counteracted by changes in functionality, product characteristics, by innovating existing products or by introducing new ones for the same market, produced using the same technology and facilities. In the trap of the downwards spiral, when it occurs within a single system, the reaction to a price reduction is a further price reduction, in turn implying further reductions in quality, performance, after-sales services, reliability and so on. DYSAM allows usage of the same resources to face the same problems-opportunities but from different simultaneous approaches.

The other systems archetypes listed above, will not be described here as they acquire meaning only within the context of single, and often closed, systems. Therefore they disappear when using the framework of Collective Beings, in which we placed ourselves from the beginning.

8.2 Virtual systems

This section describes the concept of *virtual systems* with particular reference to their usage in **post-industrial societies** and the possibility of managing **virtual systems** in multi-dimensional ways, allowing business activities to be based upon *information* and *knowledge management*. *Virtual systems* are based on resources **effectively** usable *as if, instead of* because they use one resource as if it were functionally endowed with the characteristics of another different resource. This is possible because the systems are able to perform simulations, dynamically and simultaneously using different resources. The systemic aspect *emerges* through usage, through interaction with the users. Virtual systems are also *temporarily* structured sets, modelled by an observer in order to view any kind of resource *organized* in a suitable way to perform given tasks. For example, a virtual company *really* exists only as a temporary way of using resources belonging to other companies (Skyrme, 1999).

In post-industrial societies the fundamental resources are knowledge and information. Both allow for the establishment and processing of virtual systems. This has led to the birth of suitable terminology, based on expressions such as New and Old Economy, which are widely used even in everyday language. They were introduced to mean that the *New Economy* is no longer based upon traditional industrial production and the processing of material goods as in the *Old Economy*. The *New Economy* is, on the contrary, based upon the immateriality of virtualized goods, i.e., **upon the processing of information related to real goods and to itself** due to the primary importance of technologies such as communications, simulation and the Internet. The *two* economies are mutually related. They interact with each other and the New Economy can actually pull forward the Old one, having not just a complementary role.

The circumstances associated with to such a change will now be considered in more detail. Starting from the observation that the post-industrial or knowledge society focuses upon the production and usage of knowledge and then referring to the evolution of socio-economic systems and to concepts already introduced in the literature (Bell, 1973; Drucker 1968, 1970, 1989) a possible succession of stages of development of a socio-economic system can be outlined:

Stage 1: the economy is **pre-industrial**. In this phase social systems are based upon physical power as the main resource in order to *directly* process (using only very simple tools as an extension of the human body) objects and phenomena;

Stage 2: the economy is **industrial**. In this phase social systems are still based upon physical power as the main resource, *but amplified by machines*;

Stage 3: the economy is **post-industrial**. In this phase, also referred to as the information or knowledge society, social systems are based upon scientific knowledge as the main resource, as it has a greater potential for profitability. In the current phase, also controversially named *post-business society* (Drucker, 1989), the main processes refer to information regarding physical goods (i.e., electronic or virtual trading, financial services and packages, software, telecommunications, etc.).

In the post-industrial stage goods are produced and processed by applying technical solutions having the ability to not only optimize such processes (making them more economical, more reliable, more effective and then more sensitive and flexible to scientific and technological developments), but also to make them *information-based processes*. In this way, on the one hand, production processes in the New Economy take advantage of computer and telecommunication technologies while, on the other, the information regarding the goods produced and traded becomes subject of electronic marketing and trading. And they become, *recursively*, in their turn, subjects of the New Economy. For instance, companies doing business on the Internet in the electronic trading of goods produced within the framework of the Old Economy may themselves become the object of electronic transactions.

It has been pointed out in the literature that **Industrial Society** was not born from science, i.e., from the application of theoretical knowledge. Industrial societies originated from the activity of people inventing and applying technologies in the nineteenth century, processing physical objects, their properties and effects. Applications regarded more the *discovered effects* rather than *scientific knowledge*, by proceeding with the combination of experiences, with conjectures, empirically applied, more than with theories (Drucker, 1968; Drucker, 1970).

Basic "technologies" of pre-industrial society were used for objects and effects having such physical dimensions as to be *directly* manageable by human beings, without interfacing, by hand, without sophisticated technological interfaces which, in the future would allow human beings to act on infinite and infinitesimal dimensional scales. Thus, the scales actually used to measure force, spatial dimensions, time, temperature, pressure and so on, are of the same order as those used to measure the physical characteristics of human beings.

On the contrary, processes and objects of the Industrial Society, as well as fundamental inventions, were managed by human beings mostly as extensions of themselves, using intermediate devices, particularly in the material use and application of inventions. As a consequence, Industrial Society technologies allowed human beings to extend their physical abilities. Technological solutions had the purpose of reproducing the discovered effects in order to use them and then transform them into products.

Some particular inventions contributed in a crucial way to designing the actual form of Industrial Society and certain specific industries, still active in the marketplace although using production processes updated by new scientific knowledge. They include: the electric motor, the elevator, the electric light bulb, the internal combustion engine, the motor car, the phonograph, the telegraph and the telephone. The development of the Industrial Society was also based upon other much older inventions such as, for instance, iron and steel production, glass, the compass, gunpowder, printing. A systematic, introductory treatment of this aspect may be found in Mensch (Mensch, 1975).

One way of measuring the extension of physical capabilities through the application of Industrial Society technologies is to consider the per-capita availability of power, in kWatt per capita. In a similar fashion, during the early stages of computer availability, corporations used to consider the availability of computing power in Millions of Instructions Per Second (MIPS) per-capita. This is characteristic of the previous age where *Computer* was the name of a job and not of a *device*.

In one of his most important contributions, Peter Drucker (Drucker, 1968) pointed out that in the 1970s social systems entered into a **discontinuity age**, the attribute denoting shifts which took place:

- *from* societies and industries based on practical experience, and on applications of effects and empirical discoveries which are **well-known, but not necessarily theoretically explained** (and therefore not usable as scientific knowledge for further sophistication and applications),

- *to* industries based on the application of knowledge, with problems and solutions, as well as generated technologies, far from the everyday experience typical of previous societies.

A new stage began where science and technology influence each other: *the problems and solutions of one are also the problems and solutions of the other.* Examples of new knowledge-based industries include electronics in general, robotics, the computer and software industry, telecommunications, aerospace, chemistry, pharmacology and biotechnology. The pressure driving development towards the Industrial Society decreased following the establishment of in*dustries* based upon the application of effects rather than scientific knowledge. "*An economist of 1913 could, therefore, have forecast* *the industry structure of the 1960s with reasonable accuracy . . . Most industrial technology is an extension and modification of the inventions and technologies of that remarkable half-century before World War I . . . This continuity, in turn, has made for stable industry structure. Every one of the great nineteenth-century inventions gave birth, almost overnight, to a new major industry and to new big business. These are still the major industries and big businesses of today"* (Drucker, 1968, page 7).

Discontinuity occurred because of the application of knowledge as the main resource. The complexity, unpredictability and non-linearity of post-industrial society are produced by such discontinuity. The expression *knowledge industry* was used for the first time in "Production and Distribution of Knowledge in the U.S." by F. Machlup, an economist at Princeton, in 1962 (Machlup, 1962). It should be stressed that **the extremely powerful effects activated by the knowledge industry come not only from applying its results, but also from applying them to itself.**

We close this section by making reference to the origins of the Internet, introduced above as the most effective device available for making goods and services *virtual*, a fundamental process for the New Economy. The ancestor of the Internet came from the Department of Defense Advanced Research Projects Agency (DARPA) which developed a packet-switching telecommunications network connecting, in 1974, 36 sites and 40 computers.

8.2.1 Being Virtual: Philosophy, Physics, Computer Science and Economics

Once the need for dealing with virtual systems in the socio-economic world has been justified (see previous Section), we can briefly analyze the meaning of the attribute 'virtual' (Lévy, 1995) within Economics. This attribute is widely used within a range of disciplines, such as Philosophy, Physics, Computer Science. Its common origin stems from the Latin attribute *virtualis*, deriving, in turn, from the word *virtus*, meaning 'power' or 'strength'. According to the philosopher Saint Thomas Aquinas (1225-1274) an effect is said to be "*formally contained*" in its cause if the nature of the effect is already present in its cause. Moreover, an effect is said to be *virtually* present in its cause if the cause may induce it without having the nature of the effect itself.

In economics this classical philosophical meaning remains more or less unchanged . Namely, within the context of this discipline the attribute *virtual* refers to the possibility of using something that is simulated, of interacting with it as if it were really available. Therefore, the effects of such a use and of these interactions have a nature of a concrete kind, different from that of the entity simulated or assumed to exist but which is merely virtual. By extension, this attribute may also refer to designing new *uses* for what is already available and originally designed for a single specific function. One example is a magnetic card which can be used to access a protected location, but also to buy something, as if it corresponded to real money.

The purpose of the so-called *New Economy* is **to make resources as virtual as possible** in order to allow such *uses* to be as suitable as possible for market innovations. Take, for example, financial markets (which use *virtual money*) where the development of financial products is limited only by fantasy (and law). They use and sell in a new way a resource which was already available: money. Another example is the establishment of virtual goods market, including money, stocks, shares and securities. This includes the purchase of credit, linking time with risk.

New uses or better, innovative usage of products and services are already potentially available, but somebody has yet to *invent* them, bring them to life, carry them out, combining and configuring, linking them by relationships and interactions.

8.2.2 The Opposite Process: Making Virtual the Current Reality

Reference has been made so far to the conceptual process that goes from virtual to real. Why would it be interesting to focus upon the **reverse process**? Because, when dealing with socio-economic processes, **making virtual a configuration of material resources** such as goods and services means to infinitely amplify its potential. Why? The answer lies in the strict relationship still existing between most products and their specific usage, for which they were originally designed. This relationship prevents the creation or discovery of new possible uses of them and thus limits their applicative power. By making them virtual we can, instead, break this relationship and open the way for the design of new uses. How this can be done? The best way here appears to be the introduction of a suitable *parametrization* of products, so as to leave the user with the freedom of choice of parameter values, those best suited to implement the usage created by the user according to his/her goals. Another strategy is to transfer all the information processing relating to the products into a suitable virtual medium, such as the Internet. An example of the first strategy is given by a software house, producing initially a software package for specific fixed purposes, not modifiable by the users themselves, and then transforming the same software into a general *shell*, with a modifiable interface through which the user can adapt the software by modifying its parameters as a function of his/her needs. Examples of the second strategy are:

- by making it possible to *visualize* the products (for instance on TV, Internet, etc,) not only as pictures, but also by inducing in the observer the emergence of complete scenarios; in this case, more than just the physical *transmission* of pictures or movies, there is the possibility of *generation* of images in the observer, suitable for use as *models*; this can also be implemented by allowing for a simulated interactive usage; such kinds of effects are possible when suitable technological resources are available to the set of possible users, such as telephones, mobile phones, computers, HI-FI devices, video, CD, DVD players and recorders, radio-TV devices, Internet connections, etc.; in brief, where there is *accumulation* of electronics having information processing capabilities (Krueger, 1991);
- by allowing the possibility of financial transactions using virtual money;
- by allowing the possibility of looking for products and services in which users are interested, using devices (i.e., Internet browsers) to search for them *virtually* over the entire world, that is, wherever there are centers linked to the Internet offering the products required.

The introduction of virtuality means to add novel "valences of use" which, from an economic point of view, enormously increase the **added value** of the products themselves.

It should be noted that being *non-material* is in no way synonymous with *virtual*. Virtual means to assume the existence of networks of relationships/interactions among components which allow the establishment of *actualisation* processes. Being *non-material*, on the other hand, makes reference more to the nature of an element which can then be handled through suitable processes. In classical economics non-material goods were identified with services as well as professional, financial and artistic activities, etc. Information is a good example of a non-material resource. It is actualized by its diffusion, when communicated. It is materialized when printed and recorded. The same can be said of knowledge. It is actualised when applied, when used. It becomes materialized when it gives rise to patents, to publications.

8.2.3 Virtual Corporations

The most suitable context for discussing the role of virtuality within Economics is that of *Virtual Corporations*. First of all, however, the differences between the concepts of *enterprise* and *corporation*(Malhotra, 2000) must be clarified. From the classical point of view an *enterprise* is intended as a fundamental organ of economic activity, an organization of capital and work, whose purpose is to produce goods and services to be sold. The term *Corporation* in Civil Law, on the other hand, denotes a complex of goods organized to operate an enterprise. **A corporation is therefore intended as a patrimonial projection of an enterprise.** When speaking of virtual corporations, reference will be made to complexes of available but *not yet actualized* resources even though they are *usable for actualization*.

The expression *virtual corporation* was probably introduced for the first time in the report entitled "21st Century Manufacturing Enterprise Strategy", published by the Iacocca Institute, Lehigh University, in 1991 (Nagel and Dove, 1991). In this report the expression is used to identify a corporation established by selecting organizational resources from several corporations and synthesizing them into a single electronic company. In the so-called New Economy a new and different meaning exists; according to which a *virtual corporation* (Davidow and Malone, 1992):

- is composed of resources (such as companies, professionals, consultants and technologies) configured in real-time according to the user's needs; the user instead of **selecting** from a pre-established menu (in a world of companies intended as *closed* social systems there are products, solutions *searching for user requirements, for problems and not vice versa*) **orders, co-designs** solutions, in collaboration with the virtual corporation, becoming himself an integral part of the company value-added generation process;

- *uses* the abilities of traditional corporations, giving them structure, within the framework of an actualization design;

- is no longer based upon contracts of employment *only* with workers, but mostly upon *de facto* **agreements**, because its own legal form is no longer traditional; contracts, legal devices, are **used** as instruments in the dynamics of a virtual corporation's configuration and for its business, but does not coincide with nor is it reducible to them;

- is based on a role of *capital* different from the traditional one: it changes *from being an instrument for property and control to become merely one resource amongst others*; in this way, situations emerge where a corporation may have a positive market value (as quoted in the Stock Exchange) even though it runs at a loss; this can occur due to the controlled market share and to its strategic role in economic projects (such as buying Internet domains to take advantage of future markets).

It is important to make a distinction when referring, for instance, to a corporation, between the attributes *flexible* and *virtual*. In the first case it is a matter of suitableness to market needs, for instance, by adapting rigidities in production to turbulence, by using financial tools to manage delay in producing value or by using organizational solutions in order to manage personnel depending upon the company's requirements. In the second case the reference is to the availability of virtualized resources and virtual usages. The strategy consists of nothing more than taking on the client as an active agent in corporate processes themselves, actualizing virtual resources (co-designing).

It should be noted that a virtual corporation may be emergent from usages, as well as being designed, established by a network of roles organized on demand.

Referring to the previously introduced concept of *system*, when dealing with a virtual corporation there is a shift from the conceptual use of *closed* systems to that of *open* systems, because:

- the client is, *in principle,* part of the ideation, design and even, in virtual corporations, of the production process itself;
- *stakeholders*, by referring to social systems, not only make evaluations and judgments, but induce, produce shifts *from growth to development* as introduced above; they actuate, produce projects, the *future* (Ackoff, 1981), as the observer is fundamental for carrying out reality (Berger and Luckmann, 1966);
- the observer of physical phenomena produces existence, as for emergence (see Chapter 3);
- in technological systems it is the observer who decides whether a device is *working* or not and how to use it.

It is interesting to note how the virtual level of behavior of a corporate system may be indicated by the distance between:

- the time limit required by clients to obtain goods or a service **produced in a non-standard way** (otherwise it would be sufficient to have it in storage and the time would coincide with delivery time)
- and
- the time used in reality to make it available.

The more goods or a service come from the actualization of virtual resources (i.e., generated by a virtual corporation), the smaller the difference between the two times.

One must distinguish between:

- the use of virtual resources, to be actualized in a design carried out **with** or even **by** the user-client himself, minimising the distance between the two times mentioned above or even leading to their coincidence, and
- the general availability of technologies able to immediately produce services and products required by the users; such technologies allow acting upon the assumption that the standard **client** looks for *immediate* availability (in real time, depending upon the temporal parameters of the process considered) of the product-service: it is assumed that the difference between the two times is zero, making product-service *immediately* available; examples are:
  - reproduction of original documents photocopied and bound immediately;
  - products introduced by Polaroid for instant photographic developing;
  - instantaneous picture and movie production with digital video cameras;
  - telecommunications allowing immediate information exchange using modems, fax machines and Internet technology;
  - electronic money transfer systems reducing paper money circulation through the use of credit cards, currency transactions and the virtual point in time from which interest begins to accrue (distinction between legal starting date and transaction date), telecommunication networks;
  - automatic carwashes and laundrettes;
  - rapid delivery services;
  - fast food restaurants;
  - instantaneous and even simultaneous attendance at meetings and conferences in different locations, using technologies such as teleconferencing.

The concept of *virtual corporation* introduced is thus applicable as an innovation both within the New Economy and within contexts where the economic convenience to actively collaborate *in a network of roles as part of an ideal virtual corporation* emerges case by case.

Clearly, the focus is placed upon *added value production*.

Establishing a virtual corporation thus means to go beyond *alliances* among companies which usually means that a suitable synchronization and collaboration are established with the aim of reducing costs. On the contrary, establishing a virtual corporation **means to establish synergic relationships, a potential system among companies which can be actualized on demand.**

In a *virtual corporation,* resources, solutions and problems of single companies are nodes in a *web* of relationships and interactions different from those traditionally established in single corporations or companies belonging to the same holding company. In such a traditional case relations and interactions are coincident with hierarchical, functional, economic and contractual links, etc., given by the underlying *structure*. A virtual corporation, on the other hand, may exist only for the time necessary to comply with an order, to produce a product or a service. It is the client, the market, making decisions concerning its life, not the shareholders.

In order to be able to offer constant, qualified and excellent ways for carrying out professional activities based on the actualization of virtualized resources, the *alliance*, the *synchronization* among companies no longer suffices. A virtual corporation emerging from a web of relationships and interactions among single companies is required which gives rise to a complex system characterized by its ability to actualize virtual resources.

Due to the multiple, dynamical roles of elements, virtual corporations may be effectively considered as Collective Beings emerging from the simultaneous or dynamical use of resources which can be managed with a DYSAM-based approach. The ability to design multiple roles for resources must be coupled with the ability to organize multiple levels of management.

We emphasize how this approach is particularly necessary in Post-Industrial societies where the central resource is knowledge expressed in its many dynamical, disciplinary aspects. The non-material aspect of this resource is extremely suitable for multiple approaches and roles in its management.

This section can be summarised by saying that the focus of our discussion lies upon the change from economic activities dedicated to specific products and services to applying knowledge to the design and management of products and services.

As a matter of fact, corporations dedicated to knowledge production and management, virtually and simultaneously redesign themselves for dealing with different projects often corresponding to different, often induced and designed, market needs. A DYnamical uSAge of Models is necessary for coping with simultaneous and different kinds of problems.

For instance, the design of Information Systems is, on the one hand, based upon theoretical and applicative interdisciplinary knowledge dealing with engineering, computer science, telecommunications, economics, laws and so on. On the other, applicative needs (banking, finance, industry, hospitals and so on) may be such as to require simultaneous application of different approaches to the same data, i.e., different representations, by allowing confidentiality and availability for processing, by having different roles for *Decision Support Systems* (DSS) (Turban *et al.*, 2004) and *Management Information Systems* (MIS) (Laudon and Laudon, 2004).

8.2.4 Money

Various kinds of abstractions have been introduced in the history of human beings which have profoundly affected the formation, evolution and preservation of collective phenomena, as well as the establishment of social systems. Disciplines studying these processes include, for instance, anthropology, ethnology, sociology, economics and history.

This section focuses on one of the abstractions which have been decisive in the development of social systems: the ability to calculate and handle very simple correspondences, such as the *bi-unique* ones. Without needing to know how to carry out sophisticated computations, the technique of "each pebble corresponds to a sheep and *vice versa*" was used: primitive human beings were then able to control the quantity of some elements while being unable to perform higher-level arithmetic operations.

A more sophisticated abstraction, based both upon the ability to count and the concept of worth, is *money*, materialized first through symbolic objects, such as coins also having an intrinsic value (made from precious metals such as gold and silver) and then through more conventional materials (such as more common metals and paper) and information.

In bartering, the exchange of goods takes place under the form of a permutation (having mutual usefulness for traders), whereas trade economy is based on money, representing the unit of measure of the **value of goods**. General consensus (imposed by the State, if not by free choice) among all involved traders on adopting money, changeable and shareable, allows one to identify two steps in the exchange process:

- assignment of property on payment;
- acquisition of goods by assignment of money.

In this way all trade relationships are expressed by using units of measure, that is **money units**, acting as a sort of common denominator of the relationships themselves. In this way, the trade ratios become prices. Moreover, by reducing the units of measure (for instance, litre, metre, kilogram) to those of money it becomes possible to compare prices.

A powerful innovation in the use of money was the invention of the possibility of considering a transaction effective *before* the date of the real transaction. The idea was to distinguish the real date of a financial transaction from the date of its effective validity. Moreover, the introduction of interest rates on loans and on non-contextual, delayed payments emphasizes the role of another element: the **time**. Another factor taken into account is **risk**, which comes into play when the reliability of a debtor is evaluated. This implies the coupling of the time factor entering into play when giving credit, *advancing* money in order to receive interest, (by applying a price, considering money also as virtual goods) accepting the risk of advancing money, with only a given probability of having it back. This idea opened a new market: not only the market of goods paid for by using money, *but the market of money itself.* Money, the unit of trade, is also considered as goods, especially due to its being coupled with time, thus becoming independent from contextual and effective availability. In this way many financial products, financial transactions and their related markets were also born.

Money changed from being virtual goods and from being a device to establish virtual goods, to being virtualised in its own right. The knowledge of the rules of these processes, and the power of establishing and modifying them, are the basis of well-established kinds of processes characterizing the so-called New Economy.

In the New Economy, the reason for buying a company no longer coincides with an interest in buying its ability to increase the value of processed goods in industrial processes. The reason could also be an interest in buying a market to handle it by applying different strategies. A company, for instance, could be bought to eliminate its product from the market and to substitute it with that of the buyer. The question then becomes: how much does it cost to *buy a market*?

When trying to answer such a question, the role played by information in this kind of process must be taken into account. Namely, the coupling of a certain amount of money (typically a price or stock) with goods or services, or any kind of activity, acts as *information*. Such a piece of information may take on the role of indicator of many aggregated characteristics such as quality, reliability and availability. Information, then, may be also be used for market distortion, for insider trading, for instance, or jobbery.

These considerations highlight the virtual nature of financial markets. The latter may be dealt with in an effective way by resorting to the concepts of Collective Beings and of DYSAM, that is by considering multiple, simultaneous and dynamical effects when acting upon multiple-systems generated by the same agents. This approach makes it clear how, in virtual systems, it is a poor strategy to consider only systems of interacting subsystems. The systems are virtually and dynamically redefined over the same elements, as for Collective Beings. For instance, financial actions not only generate effects on different social subsystems through their associated interactions, as claimed by the classical systemic view. Above all, they redefine the role of agents, which simultaneously and dynamically establish different systems.

An example of the classical dynamics originating from interacting systems is given by considering the cascade effect of price variations of primary goods and services (such as oil and transportation) as well as those of money itself. However, within the post-industrial society of the New

Economy we deal with virtuality, such as in banking which, by definition, is associated with two general simultaneous and dynamic basic roles: deposits and loans. A customer may simultaneously and dynamically have relationships with a bank and with the banking system through many different services relating to the two basic roles mentioned above. Banking cannot be managed like a company producing specific products or offering only a specific service. Banking is based upon the dynamic use of resources such as money, information and knowledge by inventing financial products to be simultaneously offered and managed.

8.3 Other applications

Many other applications of the concepts of Collective Beings and DYSAM to social systems are possible in various areas such as knowledge management, organizational learning and industrial districts. Some general considerations with reference to these are made below.

8.3.1 Knowledge management

What is *knowledge*? In the context of this book the focus will not be upon the philosophical aspects of the definition of knowledge, nor upon the very important topic of *knowledge representation*, critical for the effectiveness of its use. The focus here is upon knowledge as a *resource* in post-industrial societies and on related management problems. This requires two main categories of processes to be taken into account: *knowledge production* and *knowledge processing*.

When considering the former category one cannot avoid dealing with (perhaps) the most important of the many institutionalised social processes of knowledge production, that is science. Two different aspects of this need to be underlined:

1) within an objectivistic view, scientific research is intended as searching for the laws of nature, for understanding, that is *discovering* how nature *really* works; this view of scientific activity may be metaphorically grasped through the image of *mining*; from this point of view democracy for science is a very difficult subject; as a matter of fact this kind of science cannot be democratic at all: a theory is not "right" because the majority votes for it; it is therefore possible to consider

science as *independent* of democracy: it is not possible to discuss the *objectivity* of scientific results, but only, at the most, their usage; in short, according to this view *knowledge production* is intended as *knowledge discovery*;

2) the non-objectivistic approach to knowledge production makes reference to the active role of the observer, to having different ways of understanding a process, i.e., modelling it, and to removing the division between science, technology and applications; in this view the role of democracy is more evident when deciding the *direction* (many are possible) of what was assumed to be mining (looking for what) and how to use the results, the availability and the diffusion (political, economic, social and educational) of knowledge; the production of knowledge is implemented through a complex system of social activities leading to the processing of phenomena in effective ways by using specific disciplinary knowledge, as in chemistry, physics, biology and engineering; on the contrary, the objectivistic approach is based upon a rigid correspondence between problems and disciplinary knowledge, and between problems and tools while paying little attention to interdisciplinary, systemic knowledge.

When adopting the non-objectivistic approach, one has to acknowledge the occurrence of entirely new questions which, within the objectivistic approach, would have been impossible to formulate. Namely, within the former view, science is nothing but the result of a human and mainly social effort, whose effects strongly depend upon the goals, desires, cognitive schemata, beliefs and social values of the human beings participating in this enterprise. This implies, in turn, that in the *knowledge society*, not only is scientific research a business activity as well, but also every scientific statement is deeply rooted in the *theoretical centrality of the observer,* that is *the user* (Derek, 1977). The role of the latter in modern science has already been introduced above in Chapter 2. This paradigm shift has been very important to help us realize that we form part of what we are studying, as if *living matter became aware of itself.* One might ask "which *evolutionary* advantage exists for Nature, *loving mother for species and insensitive stepmother for individuals*, in knowing itself?" **A possible answer is that this knowledge makes possible changes which natural evolutionary processes are unable to produce. The new necessary resource is knowledge.** Some circumstances requiring this change include facing:
• processes associated with reaching the asymptotic level of logistic growth curves;

- problems produced by non-sustainable processes, established for any reason, such as resource consumption or overpopulation;
- the need for changing the context of human life, such as when we are forced to live in totally new physical conditions.

This approach recalls the view focusing upon the *evolution of consciousness and on conscious evolution* in social systems (Hubbard, 1998). A related topic, already mentioned in Chapter 6, is the concept of *evolution of evolvability* (Banathy, 1989; 2003).

The production and usage of knowledge is based upon different strategies in *using* the special device which, endowed with different levels of complexity, many living beings have: the *brain*. A number of studies on cognitive processing, some of which focusing upon the emergence of mind, or how mind studies itself, *its* brain and *its* body, showed that this kind of processing is based upon combined strategies of the *creation and discovery* of knowledge. In a sense, this could be viewed as a combination of an *objectivist* and a *non-objectivist* strategy.

When speaking of knowledge processing, one cannot avoid the reference to *collective knowledge*, a concept already introduced above. It can be exemplified by *swarm intelligence*, taking place when a swarm is able to do something that agents cannot (see Section 3.2.2). The concept of *collective knowledge* applies, of course, to social systems as in the case of a company. In this case, *corporate knowledge* (Beazley *et al.*, 2002; Holsapple and Whinston, 1988) emerges from the interactions between cognitive resources used by autonomous agents. These resources include:
- information (represented in a suitable way for processing);
- memory (referring to functional abilities, such as recording, finding, reconstructing);
- learning and adaptive abilities;
- ability to make inferences, like those based upon deduction, induction and abduction;
- ability to *generate and use* methodologies, strategies and tactics;
- ability to *use* errors as a source of information and not only to avoid them;
- ability to *invent* new uses of available resources;
- ability to *create* relationships and analogies;
- ability to *depict*, to see the *possible* (Barrow, 1998; Ceruti, 1994);
- ability to understand, use and apply single discipline knowledge in an integrated way, that is in a systemic way;
- ability to explain;
- ability to use theories and technologies;
- ability to process meaning.

Although many of the cognitive resources mentioned above, as well as their use, overlap, it is important to note how collective knowledge, such as corporate knowledge, emerges from their interaction.

The establishment of **knowledge** in post-industrial society, intended as introduced above, that is **as the main resource**, requires that social systems, and thus corporate systems, are able to manage it, that is to produce, to apply, diffuse, remember and learn it through collective processes. It is to be noted that, in the context of the **post-industrial society,** knowledge is intended both as a system of resources, as introduced above, *and* as referring to single discipline knowledge which:

• deteriorates and loses effectiveness if it is **only** preserved and hoarded;
• contrary to material resources, it **increases** when diffused (for instance, by teaching) and applied (due to competition, innovation, and processes of improvement).

A social system expected to manage knowledge is a very complex system, i.e., having a very dense network of non-linear relationships and interactions both among its component parts and among relationships/interactions themselves especially because *knowledge* is a non-material resource, existing only when processed and applied by agents and, collectively, by emergent systems. *Knowledge* is not *virtual* (that is, usable as if it were something else, see Section 8.2.1), but rather a *non-material resource*, like information.

The production and use of knowledge in social systems refers to emergence from multiple-systems. We therefore have to deal with the knowledge of Collective Beings and especially with the enormous amount of Collective Knowledge (Feldman and Massard, 2001) **continuously produced and to be managed.** The knowledge of a Collective Being is not simply the dynamic sum of disciplinary or interdisciplinary knowledge of single systems. Rather it emerges from that produced and used by single systems, even though this entails tremendous problems of knowledge Representation, Storage, Retrieval, Transmission, Composition, and Updating. The **emerging Collective Knowledge** is not only based upon shared, dynamic usage of single knowledge; it requires new, open, systemic approaches, as in education (Banathy, 1991). It emerges when autonomous agents **interactively** use their individual knowledge during dynamic usage of the same cognitive models. Examples of emergent collective knowledge are, for instance, that of a research team, of a detective squad, of a political leadership or that of an orchestra.

In some Collective Beings, the Collective Knowledge is used by agents which interact in a predefined way, by following rules. This is the case of the phenomenon of the existence of a *collective representation while individuals are unable to represent* (see Section 3.3), as occurring when ants go looking for food while leaving a chemical track (pheromone).

The availability of tools and methodologies can affect the emergence of knowledge and this is equivalent to having at our disposal very powerful tools to induce, control and manage the emergence of Collective Beings themselves. This is why Knowledge Management for corporations is such a critical strategic issue.

The ability to manage emergent collective knowledge is at one and the same time related to the ability to manage, i.e., to *model*, the multiple-systems environment (Chun, 1999) with its related processes of emergence.

This is closely related to the well-known topic of **organizational learning**.

8.3.2 Organizational learning

We approach the subject not from a managerial, business point of view, the most discussed and widespread in the literature because of its immediate, linear applications (Argyris and Schön, 1978; 1996; Fiol and Lyles, 1985; Levitt and March, 1988; Senge, 1990b). Here, the subject is introduced from a more theoretical point of view, focusing upon the systemic approach. We stress that, although some distinctions are made between learning organizations and organizational learning, the general approach refers to learning processes.

The latter can be considered from different points of view such as, for instance, that adopted by *machine learning* (Mitchell, 1997), or that based upon models of *systemic learning* in *education* (Banathy, 1991). In general, learning calls for high levels of openness, as introduced in Appendix 1, point 22 (*Open and Closed systems*). This is because the input must produce changes in an agent's cognitive models, in his/her way of processing information. In turn, this calls for an analysis of the problem, for instance, of *knowledge representation* (Sowa, 2000) and of the *cognitive architectures* supporting it (Anderson, 1993; Newell, 1990).

Another specific theoretical approach to learning is based upon considering *learning as a process of phase transition* for the system (Botta *et al.*, 2000, Giordana, 2000; Giordana *et al.*, 2000; Penna and Pessa, 1995; Saitta and Zucker, 2000). This approach could be useful for modelling both individual and organizational (collective) learning.

In this context, we focus upon the systemic aspect of the process, connected to the fact that learning established by organizations can not be considered as being linearly coincident with the sum of the learning processes regarding the individual agents belonging to them (Cowan, 1995). Learning organizations should rather be considered as Collective Beings, processing information through collective processing, by establishing processes of emergence, like those associated with *swarm intelligence* discussed in previous chapters.

As is well known, learning organizations are systems which can support organizational learning, that is, to learn how to do something which individuals cannot but which organizations can. As discussed previously, organizations (systems) of this kind may be established by applying organizational rules to an agent's behavior (e.g., an assembly line) or by inducing emergence from the interacting agents.

A learning organization is not supposed to learn what individuals are used to learning, such as information and knowledge. Rather it learns how to collectively use and process different individual learning processes. The business of a system is to convert single inputs into global effects (behaviors, rules) and to collectively react to them by establishing self-organized processes. For instance, if a part of a living body becomes stronger or weaker, other parts adapt to the new configuration by balancing and compensating. Moreover, that which is learned by learning organizations is *virtual*, that is, it becomes explicit, material, only when applied.

The main problem when dealing with learning organizations is how to induce and facilitate not only the learning process, but also the use, the communication, the storage, and the representation of the organizational knowledge learned . In this respect, the approach based on Collective Beings helps one to understand how the same knowledge is transformed in different ways when processed by different agents, as in translations of the same text performed by different translators. Of course, if the process is considered only from a perspective focused on performance, the result can be viewed merely as a set of non-homogeneous versions of the same knowledge, requiring standardization. The individual differences are considered as a source of error, or of confusion. If, on the other hand, we adopt a different perspective, such as that based upon DYSAM, then the differences can become the sources of new processes, of new forms of emergence. This recalls the role of observers, who not only just generate relativity.

Thus, by considering a collective usage of the different versions learnt of the same knowledge, that is, *knowledge management*, it is possible to realize its simultaneous or dynamical usage. Within this perspective, *standardizing* is a form of degenerated, reductionist knowledge management.

Organizational learning, of course, has been the subject of many investigations and intense modelling activity (for more detailed information and useful references, see Lomi and Larsen, 2001; Ouksel and Vyhmeister, 2001; Carley, 2002). One of its key features is the fact that, from the point of view of an outside observer, it manifests itself in the form of increased productivity as a firm gains experience in production. What is interesting is that the law governing such an increase appears to be *universal*, in the sense that its form appears to be independent of the kind of product or firm considered. The functional form of this law is:

$$y = y(1) x^{-b}$$

where y denotes the cost of production (measured in a suitable way, such as the number of hours of production required) of the x-th unit and b is a suitable exponent which varies across firms, but which typically has a value near 1/3 (Dutton and Thomas, 1984).

This requires explanation and interpretation, in the light of our previous arguments about Collective Beings, Collective Learning and DYSAM. At first sight, the universality of this law seems to require an explanation based upon a random graph model, which simply produces functions of this kind. Such a model has been concretely built by Huberman (Huberman, 2001) and is based upon the hypothesis that the decrease in production costs depends upon both the introduction of new shortcuts in the production process and on improvements in the effectiveness of decision-making procedures. However, the form of this law may provide further insight regarding the nature of processes taking place within a learning organization. Let us introduce a new variable, and call it *effectiveness* of the production process and denoted by E. Let us also assume that, for large values of y (and therefore small values of x), it depends upon y through the relationship:

$$E = A/y$$

where A is a suitable constant. By substituting the functional form of $y(x)$ previously introduced, we obtain:

$$E = [A/y(1)] x^{b}$$

This relationship appears to be very similar to that which gives the value of order parameter as a function of the control parameter near the critical point of a phase transition (incidentally, in the Landau theory of phase transitions the exponent b should have a value exactly equal to ½, whereas in our case the value is closer to 1/3). If this interpretation is correct, it implies that at the beginning of the production process, when the number of items produced is still small, the firm behaves as a system *after undergoing* a phase transition, which can be viewed as the event which gave rise to the production process itself, transforming a set of agents into a coherent whole devoted to organizing this production. It is the phase of *growth*, already introduced and discussed in the previous Chapter, where the system is still sensitive to the influence of perturbations of any kind, even very small ones.

What happens, however, in a situation where x is very large and therefore y very small? In this case, owing to the smallness of y, one can resort to an expression of E through a power series development in terms of y. By retaining only the first order term, we obtain a relationship such as:

$$E = K - B\,y$$

K and B being suitable constants. Substituting again the functional form of $y(x)$, we obtain directly the law:

$$E = K - B\,y(1)\,x^{-b}$$

describing a saturation process in which the growth tends to the asymptotic limit K. This is just the *decreasing growth* phase described in the previous Chapter when illustrating the role played by logistic curves in representing the development history of an organization. Of course, within this phase the sensitivity to disturbances has disappeared and the whole production process is more akin to a crystal, where structural changes are very difficult and organizational learning is progressively losing its effectiveness.

This analysis shows how, starting from the hypothesis that a firm operates as a Collective Being and that the establishment of a Collective Being is a consequence of an emergence process (that is, some kind of phase transition), this alone is sufficient for forecasting some general behavioral features of the production process and to identify the phase of production in which it is more convenient to act upon the organization itself in order to improve organizational learning. One consequence of this reasoning is that the introduction of a strategy based upon DYSAM provides more hope of success in the first growth phase than in the final one of decreasing growth. Of course, the role of technical innovations which, in many cases, transform the decreasing growth phase into a new initial growth phase have been neglected. But even this transformation can be viewed as analogous to a kind of phase transition which, in any case, can be triggered only by the adoption of a DYSAM-based strategy.

8.3.3 Industrial districts

To complete our discussion, we briefly dwell on some behavioral features of industry. Neoclassical economic theory focuses upon the analysis of industry and particularly upon the manner in which firms compete in the market. In a very basic way *cooperation* may be intended as establishing mutually advantageous business relationships. Cooperation may also be intended in terms of *collusion* between firms establishing oligopolistic structures. In this case the reason for cooperating is for the purpose of establishing "cartels", attempting to set prices at or near to monopolistic levels. But the kind of cooperation (Axelrod, 1984) we consider here is not explicit, designed, planned: it is *emergent*, as in the collective behaviors where cooperation emerges from single agent behaviors. It has been noticed, however, how business relationships are often simultaneously competitive *and* cooperative (Axelrod, 1997; Dei Ottati, 1994) as within *industrial districts* (Marshall, 1920), a circumstance requiring the introduction of the notion of "co-opetition" (Brandenburger and Nalebuff, 1997). Marshall, in fact, considered that *industrial districts enjoy the same economies of scale that only very large companies normally enjoy.* As Marshall put it, "The mysteries of the trade become no mysteries, but are as it were in the air." (cited in Surowiecki, 2000, p. 68).

Many recent publications investigate industrial districts in different countries (Best, 1990; Enright, 1998; Fujita and Hill, 1998; Herrigel, 1996; Porter, 1998a; 1998b; Pyke and Sengenberger, 1992; Schmitz, 1999). The concept of industrial district has been introduced in the literature to capture the success of agglomerations of small firms in specific areas of various countries. Because of their economic and political importance, industrial districts are under study in many economic scenarios. Amongst many others, Silicon Valley and Washington Beltway in the US are examples of districts in information, communication and biotechnology clusters. Examples in Italy include silk production in the area of Como, textiles in Prato, shoes in Naples amongst many others (Albertini, 1999).

According to the approaches to complexity of the Santa Fe Institute, an industrial district can be defined as an *evolving network of heterogeneous and localized interacting firms* (Anderson et al., 1988; Arthur et al., 1997; Squazzoni and Boero, 2002).

Industrial districts are good examples of emergence from interacting multiple-systems, competition and cooperation being the fundamental economic interactions. They are therefore also very good examples of Collective Beings. To model an industrial district (Auyang, 1998) many different aspects must be considered simultaneously: first of all that the behavior of all systems involved have continuous and simultaneous aspects of competition and collaboration. Single specific behaviors may interpreted both as competitive and/or cooperative depending upon timing, the specific systems considered, the observer, the cognitive models adopted (Belussi *et al.,* 2000). That is, the typical case of Collective Beings where the DYSAM approach becomes extremely useful.

The multiple-systems dimension of industrial districts may be considered from different points of view (Cowan *et al.,* 1994). For instance:

1. One point of view refers to the **legal, social, and economic** aspects of an emerging system which may be very well identifiable from a functional point of view by an observer; for instance, in some cases taxation can avoid the classical non-systemic approach in which each component company is taxed individually and a taxation policy taking into account the process of emergence adopted. Who, if anyone, is expected to represent an industrial district? Laws and rules for contexts hosting emergence in industrial districts must be oriented to induce and protect emergence more than to regulate or control. Social systems designers must learn how to induce, start, preserve, orient and measure processes of emergence in industrial districts and also establish appropriate fiscal policies.

2. Another refers to **knowledge**. If the reason for setting up a geographical economic cluster is competition based upon reduced prices-costs, then the emergence of an industrial district will be based upon its ability to learn how to innovate and use information and knowledge as a resource. It is not only matter of the use, management, production and acquisition of specific knowledge. It is also a matter of impact on the social hosting context such as education requirements and *vice versa.* In order to keep itself emergent, an industrial district must not only find the human resources necessary to keep the system as it is, but also to improve it and to support the competitive aspects. The impact on education induced by an industrial district is very important to ensure that the dynamics of the related social system continue to host specific processes of emergence. In this way, an industrial district may be considered as a learning system, collectively processing knowledge.

3. A third point of view refers to **ethics**. Does ethics have a role in the process of emergence of an industrial district? We believe that ethics and its relationship with quality (see Chapter 7) have a specific, crucial role. Basic values of industrial districts such as quality, knowledge and excellence are transformed into implicit behavioral rules. Often the heart of it is to do something special, and be proud of it. Then it becomes a business.

4. A fourth approach refers to **management**. A firm must be managed keeping in mind that the emergent system to which the firm belongs is, in

reality, more important than single businesses calling for different styles and strategies for using individual corporate resources. This holds for all corporate areas of management: human resources, finance, legal affairs, budgeting, production, marketing, distribution and suppliers. The management of a firm belonging to (that is, contributing to make emergent) an industrial district must simultaneously keep in mind the single, specific firm as well as the industrial district, giving and taking advantage of this situation (that is, dealing with Collective Beings).

5. One final aspect is the degree to which an industrial district may be considered a **virtual** system. Does an industrial district possess the ability to actualise? What is *virtual* in an industrial district? Probably that which is virtual, waiting to be actualised, in an industrial district is the cultural, craft, behavioral, environmental, scientific, technological and traditional background of a social system. From this point of view an industrial district is a *device* able to actualise such a background, transforming it into corporate abilities suitable to be applied by a collective multiple-system more than by a single, structured company.

Many different approaches have been introduced to model industrial districts (Conte *et al.,* 1997; Easton and Araujo, 1992), including those which resort to agent-based computational models. The possibility of using not only the DYSAM approach, but also *ergodicity* as a way of introducing indices and measurements to the process of emergence should be briefly mentioned. By measuring the economic and social aspects occurring per single agent per unit of time and per subset, then applying the **Ergodicity** hypothesis (i.e., *ensemble average is equal to time average)*, and assuming that when processes of emergence take place the indices of ergodicity may have particular dynamics (see Chapter 6) provides the possibility of obtaining new information which can be used in modelling the emergence process.

References

Ackoff, R., 1981, *Creating the corporate future*, Wiley, New York.

Albertini, S., 1999, Networking and Division of Labour. The Case of the Industrial Districts in the North-East of Italy, *Human Systems Management* **18**:107-115.

Anderson, J. R., 1993, *Rules of the Mind*, Erlbaum, Hillsdale, NJ.

Anderson, P. W., Arrow, K., and Pines D., (eds.), 1988, *The Economy as an Evolving Complex System*, Addison-Wesley, Redwood City, CA.

Argyris, C., and Schön, D. A., 1996, *Organizational Learning II: Theory, Method, and Practice.* Addison-Wesley, Reading, Mass.

Argyris, C., and Schön, D., 1978, *Organizational Learning.* Addison-Wesley, Reading, MA.

Arnold, V. I., 1988, *Geometrical methods in the theory of ordinary differential equations.* 2nd edition, Springer, New York.

Arnold, V. I., 1992, *Catastrophe Theory.* 3rd edition, Springer, Berlin.

Arnold, V. I., Afrajmovich, V. S., Ilyashenko, Yu. S. and Shilnikov, L. P., 1999, *Bifurcation Theory and Catastrophe Theory.* Springer, Berlin.

Arthur, B. W., Durlauf, S. N., and Lane, D., (eds.), 1997, *The Economy as an Evolving Complex System II.* Addison-Wesley, New York.

Auyang, S. Y., 1998, *Foundations of Complex- System Theories in Economics, Evolutionary Biology, and Statistical Physics.* Cambridge University Press, Cambridge, UK.

Axelrod, P., 1984, *The Evolution of Cooperation.* Basic Books, New York.

Axelrod, R., 1997, *The Complexity of Cooperation. Agent-Based Models of Competition and Cooperation.* Princeton University Press, Princeton, NJ.

Banathy, B. H., 1989, The design of evolutionary systems, *Systems Research and Behavioral Science* **6**:193-212.

Banathy, B. H., 1991, *Systems Design of Education. A Journey to Create the Future.* Educational Technology Publications, Englewood Cliffs, NJ.

Banathy, B. H., 2003, Self-Guided/Conscious Evolution, *Systems Research and Behavioral Science* **20**:309-321.

Barrow, J. D., 1998, *Impossibility. The limits of science and the science of limits.* Oxford University Press, Oxford, UK.

Beazley, H., Boenisch, J., and Harden, D., 2002, *Continuity Management: Preserving Corporate Knowledge and Productivity When Employees Leave.* Wiley, New York.

Bell, D., 1973, *The Coming of Post-Industrial Society. A Venture in Social Forecasting*, Basic Books, New York.

Belussi, F., and Gottardi, G., (eds.), 2000, *Evolutionary Patterns of Local Industrial Systems. Towards a Cognitive Approach to the Industrial District.* Ashgate, Aldershot, UK.

Berger, P. L., and Luckmann, T., 1966, *The Social Construction of Reality.* Penguin Books, New York.

Best, M. H., 1990, *The New Competition - Institutions of Industrial Restructuring.* Harvard University Press, Cambridge, MA.

Botta, M., Giordana, A., Saitta, L., and Sebag, M., 2000, Relational learning: Hard problems and Phase transitions. In: *AI'IA 99. Advances in Artificial Inteligence,* (E. Lamma, P. Mello, eds.), Springer, Berlin, pp. 178-189.

Brandenburger, A. M., and Nalebuff, B. J., 1997, *Co-Opetition. 1. A revolutionary mindset that combines competition and cooperation. 2. The Game Theory strategy that's changing the game of business.* Doubleday, New York.

Carley, K. M., 2002, Computational organizational science and organizational engineering, *Simulation Modelling Practice and Theory* **10**:253-269.

Ceruti, M., 1994, *Constraints and possibilities: The evolution of knowledge and knowledge of evolution*. Gordon & Breach, New York.

Chun, W. C., 1999, *Information Management for the Intelligent Organization. The Art of Scanning the Environment*. Information Today Inc., Medford, NJ.

Conte, R., Hegelmann, R., and Terna, P., (eds.), 1997, *Simulating Social Phenomena*. Springer, Berlin.

Cowan, D. A., 1995, Rhythms of Learning: Patterns That Bridge Individuals and Organizations, *Journal of Management Inquiry* **4**:222-246.

Cowan, G. A., Pines, D., and Meltzer, D., (eds.), 1994, *Complexity. Metaphors, Models, and Reality*, Addison-Wesley, Reading, MA.

Dei Ottati, G., 1994, Co-operation and Competition in the Industrial Districts as an Organizational Model, *European Planning Studies* **4**:463-483.

Davidow, W. H. and Malone, D. M. S., 1992, *The Virtual Corporation: Structuring and Revitalizing the Corporation for the 21st Century*. HarperCollins, New York

Derek, L. P., 1977, *Wittgenstein and the scientific knowledge*. Macmillan, London.

Dowling, A. M., MacDonald, H. R., and Richardson, G. P., 1995, Simulation of Systems Archetypes. In: *Proceedings of the 1995 System Dynamics Conference* (K. Saeed, ed.), Tokyo, Japan, pp. 454-463.

Drucker, P. F., 1968, *The Age of Discontinuity*. Heinemann, London.

Drucker, P. F., 1970, *Technology, Management & Society*. Harper & Row, New York

Drucker, P. F., 1989, *The new realities*. Harper & Row, New York.

Dutton, J. M. and Thomas, A, 1984, Treating progress functions as a managerial opportunity, *Academy of Management Review* **9**: 235-247.

Easton, G., and Araujo, L., 1992, Non-economic exchange in industrial networks. In: *Industrial Networks – A New View of Reality*, (B. Axelsson and G. Easton, eds.), Routledge, London, pp. 62-84.

Enright, M., 1998, Regional Clusters and Firm Strategy. In: *The Dynamic Firm – The Role of Technology, Strategy, Organization, and Regions*, (A. Chandler, P. Hagstrom, and O. Solvell, eds.), Oxford University Press, Oxford, UK, pp. 315-342.

Feldman, M. P., and Massard, N., (eds.), 2001, *Institutions and Systems in the Geography of Innovation*. Kluwer, New York.

Fiol, C. M., and Lyles, M., 1985, Organizational Learning, *Academy of Management Review* **10**:803-813.

Forrester, J. W., 1961, *Industrial Dynamics*. MIT Press, Cambridge, MA.

Forrester, J. W., 1968, *Principles of Systems*. Wright-Allen Press, Cambridge, MA.

Fujita, K., and Hill, R. C., 1998, Industrial Districts and Economic Development in Japan: The Case of Tokyo and Osaka, *Economic Development Quarterly* **6**:181-198.

Gilmore, R., 1993, *Catastrophe Theory for scientists and engineers*. Dover, New York.

Giordana, A., 2000, Phase transitions in Relational Learning, *Machine Learning* **41**:217-251.

Giordana, A., Saitta, L., Sebag, M., and Botta, M., 2000, Analyzing Relational Learning in the Phase Transition Framework. In: *Proceedings of the 17th International Conference on Machine Learning* (P. Langley, ed.), Morgan Kaufmann, Stanford, CA, pp. 112-120.

Goodman, M., and Kleiner, A., 1993, Using the Archetype Family Tree as a Diagnostic Tool, *The Systems Thinker* **4**:5-6.

Herrigel, G., 1996, *Industrial constructions: The sources of German industrial power*. Cambridge University Press, Cambridge, UK.

Holsapple, C. W., and Whinston, A. B., 1988, *The Information Jungle: A Quasi-Novel Approach to Managing Corporate Knowledge*, Dow Jones-Irwin, Homewood, IL.

Hubbard, B., 1998, *Conscious evolution: Awakening the Power of our Social Potential.* New World Library, Novato, CA.

Huberman, B. A., 2001, The Dynamics of Organizational Learning, *Computational & Mathematical Organization Theory* 7:145-153.

Kim, D. H., 1994, Predicting Behavior Using Systems Archetypes, *The Systems Thinker* 5:5-6.

Kim, D. H., 1995, Systems Archetypes as Dynamic Theories, *The Systems Thinker* 6:6-9.

Kim, D. H., and Anderson, V., 1998, *Systems archetype basics: From story to structure.* Pegasus Communications, Waltham, MA.

Krueger, M., 1991, *Artificial Reality II.* Addison-Wesley, Reading, MA.

Lane, D. C., 1996, Reinterpreting 'Generic Structure': Evolution, application and limitations of a concept. *System Dynamics Review* 12:87-120.

Laudon, K., and Laudon, J., 2004, *Essentials of Management Information Systems - Managing the Digital Firm.* 6th Edition, Pearson - Prentice Hall, Upper Saddle River, NJ..

Levitt, B., and March, J. G., 1988, Organizational learning, *Annual Review of Sociology* 14:319-340.

Lévy, P., 1995, *Qu'est-ce que le virtuel?* Editions La Découverte, Paris.

Lomi, A., and Larsen, E. R., (eds.), 2001, *Dynamics of Organizations: Computational Modeling and Organizational Theories.* MIT Press, Cambridge, MA.

Machlup, F., 1962, *The production and distribution of knowledge in the United States.* Princeton University Press, Princeton, NJ.

Malhotra, Y., (ed.), 2000, *Knowledge Management and Virtual Organizations.* Idea Group Publishing, London.

Marshall, A., 1920, *Industry and Trade.* Macmillan, London.

Mensch, G., 1975, *Stalemate in Technology: Innovations Overcome the Depression.* Ballinger Publishing Company, Cambridge, MA.

Mitchell, T. M, 1997, *Machine Learning*, McGraw-Hill, New York.

Nagel, R. N. and Dove, R., 1991, *21th Century manufacturing enterprise strategy: An industry led view* (vol. 1). DIANE Publishing Company, Collingdale, PA.

Newell, A., 1990, *Unified Theory of Cognition.* Harvard University Press, Cambridge, MA.

Ouksel, A., and Vyhmeister, R., 2001, Performance of Organizational Design Models and their Impact on Organization Learning, *Computational & Mathematical Organization Theory* 6:395-410.

Palis, J., and de Melo, W., 1982, *Geometric theory of differential systems.* Springer, New York.

Penna, M. P., and Pessa, E., 1995, Can learning process in neural networks be considered as a phase transition? In: *Neural Nets, Proceedings of the 7th Italian Workshop on Neural Nets*, (M. Marinaro and R. Tagliaferri, eds.), World Scientific, Singapore, pp. 123-129.

Porter, M. E., 1998a, *On Competition.* Harvard Business Review Book, Boston, MA.

Porter, M. E., 1998b, Clusters and the new economics of competition, *Harvard Business Review* November-December:77-90.

Poston, T., and Stewart, I., 1978, *Catastrophe Theory and its applications.* Pitman, London.

Pyke, F., and Sengenberger, W., (eds.), 1992, *Industrial districts and local economic regeneration*, International Institute for Labour Studies, Geneva.

Saitta, L., and Zucker, J. D., 2000, Abstraction and Phase Transition in Relational Learning. In: *Proceedings of the Symposium on Abstraction, Reformulations and Approximation. SARA 2000.* (Y. Choueiry and T. Walsh, eds.). Springer, Berlin, pp. 132-143.

Schmitz, H., 1999, Collective efficiency and increasing returns, *Cambridge Journal of Economics* **23**:465-483.

Senge, P. M., 1990a, *The Fifth Discipline: The Art and Practice of the Learning Organization*. Doubleday/Currency, New York.

Senge, P. M., 1990b, The Leader's New Work: Building Learning Organizations, *Sloan Management Review* **32**:7-23.

Senge, P. M., Kleiner, A., Roberts C., Smith, B., and Ross R., 1994, *The Fifth Discipline Fieldbook: Strategies and Tools for Building a Learning Organization*. Doubleday/Currency, New York.

Skyrme, D. J., 1999, Knowledge Networking: Creating the Collaborative Enterprise, Butterworth-Heinemann, Oxford, UK.

Smale, S., 1966, Structurally stable systems are not dense. *American Journal of Mathematics* **88**:491-496.

Sowa, J. F., 2000, *Knowledge Representation: Logical, Philosophical, and Computational Foundations*. Brooks Cole Publishing Co., Pacific Grove, CA.

Surowiecki, J., 2000, The Financial Page: Why Do Companies Like Company, *The New Yorker* April 24 and May 1, p. 68.

Thom, R. , 1975, *Structural stability and morphogenesis: An outline of a general theory of models*. Benjamin, Reading, MA.

Turban, E., Aronson , J. E., and, Liang T-P, 2004, Decision Support Systems and Intelligent Systems. Prentice Hall, Upper Saddle River, NJ

Wolstenholme. E. F., 2003, Toward the definition and use of a core set of archetypal structures. *System Dynamics Review* **19**:7-26.

Web Resources

Braun, W., 2002, *The system archetypes*. Available at the web site
http://www.uni-klu.ac.at/~gossimit/pap/sd/wb_sysarch.pdf.

Squazzoni, F., and Boero, R., 2002, Economic Performance, Inter-Firm Relations and Local Institutional Engineering in a Computational Prototype of Industrial Districts, *Journal of Artificial Societies and Social Simulation* vol. 5, no. 1
http://jasss.soc.surrey.ac.uk/5/1/1.html

Chapter 9

APPLICATIONS TO COGNITIVE SYSTEMS: BEYOND COMPUTATIONALISM

9.1 TRADITIONAL COGNITIVE SCIENCE

As often stressed in this book, in principle we cannot deal with Collective Beings without taking into account that the agents of which they consist are endowed with a cognitive system. The expression "cognitive system" will be used to denote not only the individual system responsible for cognitive processes in the usual sense, such as memory storage and recall, attention, perception, thinking and reasoning, but even the system responsible for emotional and affective processes. Namely, as is evident even in everyday situations, these two kinds of processes are strictly interconnected and it is practically impossible to draw a boundary separating them. Of course, it is easier to investigate cognitive processes in the usual sense, for obvious technical reasons, and this explains the label "cognitive" often adopted to denote the whole complex of mental processes.

Investigating the operation of cognitive systems, be they natural or artificial (provided they exist) is the fundamental task of Cognitive Science. This expression denotes, in a generic way, a transdisciplinary enterprise aiming at the scientific investigation of knowledge, of its transformations, of its acquisition and of its use in humans, animals and machines (standard textbooks are Barsalou, 1992; Stillings *et al.*, 1995). This study requires a high level of integration of different capabilities, such as those of psychologists, neurophysiologists, computer scientists, philosophers, mathematicians, physicists, and many others. Historically, Cognitive Science was born in 1977, when R. Schank, A. Collins and E. Charniak founded a new scientific journal, entitled "Cognitive Science", whose aim was initially stated in a paper by Collins (Collins, 1977). Later, in the first conference of the new Cognitive Science Society, held in La Jolla, California in 1979, D. A. Norman defined in a more precise way the investigational domains of the new discipline (see Norman, 1980). In its initial development phase Cognitive Science was strongly dominated by the framework then current in cognitive psychology and in Artificial Intelligence, well summarized by the Physical Symbol System Hypothesis of Newell and Simon (e.g., Newell and Simon, 1976), which claims that:

–   cognitive abilities and whence, in a broad sense, "intelligence" are possible only in the presence of a symbolic representation of events and situations, external as well as internal, and of the ability to manipulate the symbols constituting the representations themselves;
–   all cognitive systems share a common set of basic symbol processing abilities;
–   every model of a given cognitive process can always be cast in the form of a program, written in a suitable symbolic language, which, once implemented on a computer, produces exactly the same behavior observed in the human beings in which we assume that this same cognitive process is occurring.

It is to be remarked that 1), 2) and 3) imply automatically that the computer, on which every model of cognitive processing can be implemented, must be a *digital* computer, as an analogic computer is not suitable for manipulating discretized symbols (even though it could perform such a task in an approximate way, provided suitable conventions were to be adopted).

The framework described above characterizes so-called *traditional* Cognitive Science which has been very popular in the past among cognitive psychologists and is still popular among developmental psychologists (see, for instance, Spelke *et al.*, 1992; for a review of this topic see Laurence and Margolis, 2002). In a sense, it can be considered as a generalization of the usual mechanistic view, previously adopted in 19th century physics, based on the concept of *computation*, rather than on   *energy*, so as to account for behavioral phenomena. Of course, a computational process can, in principle, be endowed with systemic features. However, within traditional Cognitive Science, the concept of computation has often been used in a restricted way, so as to include only input-output processes performed by specialized *modules* belonging to a hierarchical architecture, like subroutines in a highly structured computer program. In short, this view implies that cognitive system operation can be described by a (hopefully sophisticated) general computer program, including specialized subroutines, each being devoted to a specific cognitive operation and accessible for separate investigation. Both cognitive psychology experiments and Artificial Intelligence simulations can help discover the instructions constituting both the general program and the individual subroutines, owing to the fact (tacitly assumed but never clearly proven) that the complexity of cognitive processes stems not from the complexity of the laws ruling them (the programs are simple!) but from the huge amount of memory required.

The principles underlying traditional Cognitive Science, sometimes labeled as the *symbolic computational approach* (illustrated in a clear and exhaustive way in Pylyshyn, 1984) led to the birth of several myths. The first consists of believing that traditional Cognitive Science describes cognitive processes as *information processing*, where the word 'information' has the meaning attributed to it in Information Theory. This myth even gave rise to a popular approach in cognitive psychology, known as the *Human Information Processing* (HIP) approach (see, for instance, Lindsay and Norman, 1972; Reed, 1988) but this myth was shown to be false (see, in this regard, the analysis contained in Palmer and Kimchi, 1986). Cognitive psychologists or, more in general, cognitive scientists never made use of the technical concepts of Information Theory which, so far, remain strongly connected to those of Thermodynamics (a domain rarely visited by most cognitive scientists).

The second myth holds that traditional Cognitive Science identifies cognitive processes with computations in the sense of Turing's mathematical theory of computation. In this regard, two remarks are in order: 1) cognitive psychologists and, more in general, cognitive scientists have never made practical use of the mathematical theory of computation (except in philosophical discussions); 2) Alan Turing, as well as other founding fathers of the mathematical theory of computation, did not introduce the abstract description of computation processes to account for the phenomenological features of computation processes (or, more in general, of cognitive processes) in human beings. As is well known by anyone with a minimum of competence in logic or in computer science, Turing's goal was only that of showing the seemingly paradoxical limitations to long-term predictions involved in *some simple* (and *seemingly innocuous*) *properties of computations ruled by a deterministic algorithm*. And the Halting Theorem shows that this goal has been reached in a highly successful way! However, a deeper philosophical analysis of the concept of 'computation' shows that it can be characterized by at least three aspects (see Sloman, 2002):

- *functional*, that is relevant to the goals underlying a computation and to information we could obtain by performing it;
- *syntactic*, that is relevant to the formal structure of the computation process itself;
- *implementational*, that is relevant to the physical structure of the device allowing execution of the computation.

Standard mathematical theory of computation deals only with the syntactic aspect and has so far neglected the other two. However, the latter are clearly essential to understand why and how human beings compute or design computing machines. This implies that we cannot reach any conclusion about the relationship between mental activity and computations by resorting only to the mathematical theory of computation. As a matter of fact, the latter has never been used for modeling human or animal cognitive processing or in performing experiments or computer simulations.

The above arguments clearly show how cognitive scientists started their investigations on knowledge and mental processes without a precise definition of the main concepts (such as 'information' and 'computation') underlying the framework they were adopting. This appears, at first sight, as a serious drawback, which could led to the failure of the whole enterprise of Cognitive Science. However, this is not so: the framework adopted by traditional Cognitive Science, on the one hand, is based on a precise philosophical theory, to be identified with *functionalism*, and, on the other, solves a number of conceptual problems, which were irresolvable within both dualistic and monistic approaches to the study of mental processes. Functionalists claim that mental states are nothing but *functional* states, that is characterized not by their content but rather by the causes generating them and by the physical effects they produce. This entails that identical mental states, that is characterized by the same functional roles, can be physically implemented in different ways. If this view is supplemented by the assumption that the relationships between different mental states are of a computational nature, a stance made popular by philosophers such as Jerry Fodor (see Fodor, 1981; 1983; 1987), we simply arrive at the conceptual basis of traditional Cognitive Science.

This approach solves a fundamental problem of all materialistic theories of mental processes, known as the *homunculus* problem. This arises from the fact that, in a functionalist theory of mental states, we need an external supervising mechanism recognizing these states and connecting them in the correct manner. This mechanism corresponds to the *homunculus*. Let us suppose, for instance, that we have the intention of grasping a glass and that, as a consequence, we move our arm toward the glass. We will then have two mental states, the one corresponding to the intention of grasping the glass, and the other corresponding to the control of movements of our arm to reach the glass. The latter mental state is, of course, caused, by the former mental state. However, the relationships between the two mental states holds only if we have a supervisor - the *homunculus* – which, when looking at the first mental state, identifies it as corresponding to the intention of grasping the glass and then activates precisely that mental state allowing for the control of our arm movement. At first sight this could not seem to be a problem: after all, many complex systems have a centralized control mechanism. However, once the existence of *homunculus* has been accepted, we are forced to account for its operation, and this needs an investigation of its own mental states, in turn supervised by a second (smaller) *homunculus* lying within the brain of the first one. But the mind of this second *homunculus* can work only if supervised by a third (even smaller) *homunculus*, and so on: in short, a *regressio ad infinitum*!

The computationalist view solves this problem very easily. Namely, if we consider a digital computer, we observe that the first cause for its "intelligent" behavior lies in the presence of suitable programs written in a given high-level programming language. However, other lower-level programs transform the high-level programs in longer sequences of simpler instructions. Moreover, further programs, at a still lower level, transform the latter into even more simple instructions, and this chain of transformations continues down to the lowest level, in which we have only simple physical processes, such as variations in an electric potential, corresponding to operations on single bits. Within this whole hierarchical architecture, however, there is no *homunculus*! Everything works and all levels are interconnected only through the correct arrangement of the architecture of different computational programs. Therefore, if the mind operates according to principles similar to those used in designing a digital computer, no *homunculus* is needed to build a theory of mental processes.

The flaw in this argument is that it neglects the fact that the "intelligent" operation of a digital computer has been made possible by the existence of a human *designer*. The only way to answer this objection consists of assuming that the designer of the human mind be identified with Nature, that is with natural evolution which, after millions of years of genetic mutations, produced the human brain and the human mind we know today. This kind of answer implies automatically an *innatist* view of the distinctive features of cognitive processes. As a matter of fact, this view pervaded the whole of traditional Cognitive Science (on this topic, see Piattelli-Palmarini *et al.*, 1980).

However, this circumstance was not directly responsible for the decline in traditional Cognitive Science. Namely, at first sight, many high-level cognitive processes, such as logical reasoning, mathematical computation, using language in a grammatically correct way, appear simply as computations, for which the most important task is to discover their rules, without worrying about their origin. On the contrary, the fall of traditional Cognitive Science began when people recognized that, on one hand, the essential feature of all cognitive processes, that is of being endowed with meaning, had absolutely nothing to do with digital computation, and that, on the other hand, the hierarchical modular architecture, required by the computational view of mind, was not evidenced by experimental findings.

Of course, we cannot summarize here, for lack of space, the intense debate on these topics, which lasted for many years. But with regard to the former aspect we mention only the significant influence exerted by the celebrated Searle's Chinese Room argument (Searle, 1980; 1999; Hauser, 1997; Preston and Bishop, 2002), evidencing how no symbolic computational procedure can produce the understanding of the meaning of manipulated symbols. The latter aspect is connected to the fact that experimental and theoretical studies on cognitive systems, as well as on the biological brain, were unable to find convincing proof for the existence of a hierarchical organization of specialized cognitive modules predicted by traditional Cognitive Science.

We recall that the only tool so far available for detecting such modules is the so-called *method of dissociations*, also named, in other contexts, as the *method of decompositions* (see Bechtel and Richardson, 1993). This can be used in two different ways:

-       *top-down*, when applied to cognitive processes, as observed at a macroscopic level in psychological experiments;
-       *bottom-up*, when applied to experimental and clinical data coming from Neuroscience.

The top-down version of the decomposition method includes, in turn, two different sub-cases:

d.1) *model-based decomposition*;

d.2) *task decomposition*.

Method d.1) has been widely used within Artificial Intelligence. It consists of a decomposition of the procedure used to perform a given cognitive task, on the basis of purely *logical* arguments, into smaller sub-procedures, down to a level in which each sub-procedure can be implemented in an easy and intuitive way by a single module. This method allows for the building of efficient software programs able to simulate the performance of the cognitive task under study. Unfortunately, even when the method works, this fact does not automatically mean that the human cognitive system is working in the same way as the computer program simulating its operation.

Method d.2) was proposed by Tulving (see Tulving, 1983) and is based on the consideration of pairs of different cognitive tasks, sharing a given independent variable. If the manipulation of this variable produces an effect on the performance in one of the two tasks, but not on the performance in the other, then this circumstance is considered as evidence that the two tasks are performed by two different modules. The problem with this method is that, even if we assume its validity (which is far from being proven), it allows the detection of modules whose nature is very different from that expected on the basis of the Physical Symbol System Hypothesis. Namely, these modules are associated with functions of a very general kind and are linked by horizontal interconnections, whereas the modules expected within the computationalist approach should be highly specific and linked mostly by vertical interconnections. The most celebrated example of the modules detected by the application of method d.2) is given by the Tulving distinction between episodic and semantic memory and it illustrates how these general-purpose memory systems cannot be considered as proof for the existence of a modular mind architecture.

Let us now briefly discuss the bottom-up version of the decomposition method, usually known as the method of dissociations (see, for instance, Gazzaniga, 2004). It can only be applied to subjects characterized by a severe impairment in one specific cognitive ability. If this impairment is associated with a lesion in a specific brain area, then we can identify this area as the seat of the module responsible for the considered ability. This method, implemented in association with the use of different kinds of brain imaging techniques, has recently led to the identification of a number of brain modules devoted to specific cognitive tasks such as, for instance, face recognition, number manipulation, language understanding, reading. Even this method, however, is plagued by a number of shortcomings, such as:

s.1) the number of subjects to which the method can be applied is very small; namely it is very rare that a subject be characterized by the impairment of only one specific cognitive ability or by a single localized brain lesion; often the impairment regards a number of different abilities and the lesion is widespread instead of being localized; moreover, the reduced number of available subjects automatically implies a loss of statistical validity of the results obtained;

s.2) the number of modules which can be detected, if the method is applied without suitable criticism, is very high; this appears as somewhat unrealistic because, if the brain were characterized by such a complex modular architecture, it would be far too vulnerable with respect to noise, errors, disturbances, as is the case for a digital computer;

s.3) the application of the method itself could be biased by the belief in the existence of a modular cognitive architecture.

To conclude, we can state that none of the methods described above seems able to provide a reliable solution to the problem of demonstrating the hierarchical modular architecture of mind. The only way out is to acknowledge the intrinsic incompleteness of the symbolic computational approach and to search for a different, or more general, approach for the investigation of cognitive systems.

9.2 Is Cognition equivalent to Computation?

The points raised in the previous section, while describing the drawbacks of traditional Cognitive Science, left open the way for different possible remedies. We outline here approaches constituting an alternative to the symbolic computational one, grouping them into two different classes:
-        approaches relying on a *logical* theory of computation more powerful than that of Turing;
-        approaches based on non-digital forms of computation.

Approaches belonging to the former class are collectively denoted as *hypercomputationalism* (Siegelmann, 1999; MacLennan, 2001; Stannett, 2001). It received strong support from many philosophers who, severely criticizing traditional Cognitive Science, announced the new era of *post-computationalism* (including Bringsjord, 1992; Fetzer, 2001; Bringsjord and Zenzen, 2003). The thesis of hypercomputationalism claims that the inability of the symbolic computational approach to account for mental processing features stems from the fact that the operation of the mind is not reducible to a Turing-like computation, but rather corresponds to a sort of hypercomputation, in principle far more powerful than the usual Turing one. The problem with this approach is that of offering a practical example of hypercomputation (except mental processes themselves). In this regard most supporters of hypercomputationalism claim that all (or, at least, most) analogic computations are examples of hypercomputations. This amounts to stating that, for instance, computing the analytical solution of a differential equation is a kind of hypercomputation and entails that all physical and biological devices (including brain neurons) working in an analogic way are performing hypercomputations.

There are two problems with this view, one philosophical and the other technical. The philosophical difficulty stems from the fact that, practically, the introduction of hypercomputation does not produce an appreciable change in the symbolic computational approach. Namely, both hypercomputationalists and traditional cognitive scientists agree on the fact that mental processes are computations, the only point of disagreement being whether they are more powerful than Turing's or not. While recognizing that analogic computations can offer new interesting modeling tools, we must acknowledge that substituting hypercomputation to standard Turing computation does not lead to any progress in solving problems, such as that of meaning, or in answering the Searle Chinese Room argument. As a matter of fact, taking limits or performing integrals or derivatives does not lead to the emergence of meaning any more than performing logical conjunctions or using truth tables.

The technical difficulty is connected to the fact that, in most cases, we have no mathematical proof that analogic computations are more powerful than those performed by a Turing machine. However, it has to be taken into account that it is very difficult to obtain such proofs and that the literature on this topic is very scarce (fundamental references are Pour-El and Richards, 1983; 1989). A further source of difficulty stems from the fact that, from an experimental point of view, we have no means of distinguishing a discrete computation from a truly continuous one. Namely, in order to detect the difference between these two kind of computations, we would have to push our investigations down to the atomic level. Here, however, we would encounter quantum uncertainty effects, preventing any possibility of reaching a definitive answer to our question. Besides, it is to be stressed that the computation itself of these uncertainty effects would be based, in turn, on analogic computations (such as those using a Schrödinger equation) in which the continuous nature of the involved quantities is accepted *a priori*.

The other class of alternative approaches are based upon computational models not inspired by logical operations but rather by physical and biological behavior. These approaches share the conviction that meaning and cognition are something emergent, at a macroscopic level, from a substratum of interacting computational entities (operating in an analogic way). And, in principle, emergence cannot be reduced to computation (be it Turing or hyper-Turing), even though computational models are useful in grasping some of its features. These approaches include:

- the *connectionist* approach (McClelland and Rumelhart, 1986);
- the *dynamicist* approach (Port and Van Gelder, 1995; Tschacher and Dauwalder, 1999; Van Gelder, 1998; 1999);
- the *embodied cognition* approach (Varela *et al.*, 1991);
- the approach based on *Artificial Life* (Langton, 1989).

There are considerable overlaps between these different approaches, as well as a number of divergences. Here, for lack of space, we do not enter into details about all the problems with each individual approach, limiting ourselves to remark that:

- the main problem of each approach is to prove that its models give rise to emergent phenomena;
- often these approaches make use of non-ideal models of emergence, as defined in the Chapter 4 of this book;

- all these approaches favour the adoption of a DYSAM-like approach; namely, they acknowledge the limited range of validity of each non-ideal model and, therefore, accept that no single model can account for a wide range of experimental data; such a circumstance has led to a significant development of architectural models, in which individual models are embedded within a complex structure of reciprocal interconnections, often undergoing changes as a function of interactions with a suitable environment (see, amongst others, Caelli *et al.*, 1999; Yao, 1999; Ito and Kaneko, 2000; García-Pedrajas *et al.*, 2002; Sutton and Jamieson, 2002; Tani, 2003).

To illustrate the problems arising within this class of approaches, we refer briefly to the connectionist approach, whose main modeling tool consists of neural networks, already described in subsection 5.5.1 of Chapter 5. Its main principles can, in their essence, be reduced to the following two statements:

- Two different levels of description of cognitive processing exist: the *macroscopic* one, in which cognitive processing is studied on the observational scale of traditional experimental psychology, and the *microscopic* one, in which we focus upon the operation of simple detectors of single features of input patterns or of complex cognitive constructs;
- the cognitive processing features observed at the macroscopic level *emerge*, as a sort of collective effect, from the (cooperative) interactions between the operations of the single detectors working at the microscopic level.

Usually behavior at the microscopic level is modelled through neural networks. However, as already described in Chapter 5, it is very difficult to prove that macroscopic features of neural dynamics emerge, as collective effects, from the interactions between single neurons. This means that the traditional connectionist approach is unable to prove that macroscopic cognitive behavior is a collective effect emerging from the interactions between the microcognitive units. This inability is due to the lack of a reliable theory of emergence and, in turn, is precisely the cause of a number of problems (still unsolved) which plague the connectionist approach including:

- ***The Catastrophic Interference Problem***, consisting in the fact that the learning of new input-output relationships gives rise to a dramatic decay of performance relative to previously learned relationships (see McCloskey & Cohen, 1989; Ratcliff, 1990);
- ***The inability to represent complex symbolic structures***, due to the fact that the vectors used to represent activation patterns of neural units are unsuitable for encoding the complex structural relationships occurring

within a data set (or within human knowledge) such as, for example, tree structures;

- ***The Binding Problem***, stemming from the fact that a neural network cannot learn from experience alone to correctly bind into a unique, holistic and global, entity a number of different local features *independently* from the very nature of the features themselves (see Hummel and Biederman, 1992);

- ***The Grounding Problem***, consisting of the fact that a neural network is unable to connect higher-level symbolic representations with lower-level space-time distributions of physical signals coming from the environment (see Harnad, 1990).

Despite these intrinsic difficulties, over the past few years all these problems have been dealt with and partly solved in a satisfactory way. Lack of space prevents us even from citing the relevant literature (reviews include those by French, 1999, on catastrophic interference, Wermter and Sun, 2000,, as well as Zhou *et al.*, 2003, on symbolic structure processing, Singer, 1999, on the binding problem, and Pessa and Terenzi, 2002, on the grounding problem). Neural network models have even been proposed to deal with emotional processes (for cognitive models of this kind, see the review by Arbib and Fellous, 2004) and psychopathologies (Reggia *et al.*, 1996; 1998; Aakerlund and Hemmingsen, 1998; Stein and Ludik, 1998). However, most solutions are based upon suitable arrangements of neural network modules within suitable architectures, generally designed to process particular kinds of data. So far no solution has made explicit use of concepts and methods taken from theories of emergence, described above in Chapters 4 and 5 (see Pessa, 2004). We can therefore assert that the connectionist approach, as well as the other approaches belonging to the same class, is unable to prove the validity of the principles themselves on which it is based. In other words, the traditional symbolic approach still cannot be completely ruled out, as non-ideal models (for instance, those based on neural networks) describe intrinsically disordered systems whose mathematical analysis is still nigh on impossible.

9.3 Theories of consciousness

The fact that most solutions of the problems afflicting connectionist models do not use methods and constructs taken from the theories of emergence prevents most of them from accounting for a number of experimental findings showing that emergent phenomena take place within both the brain and cognitive systems. These include:

     f.1) the existence of *long-range correlations* (both of a temporal and spatial nature) and of large-scale coherence of electroencephalographic signals (Freeman, 2000; Nunez, 2000; for evidence of a phase transition within the brain, see Freeman, 1999), showing the integration of various cognitive (Classen *et al.*, 1998; Sarnthein *et al.*, 1998) as well as affective (Hinrichs and Machleidt, 1992; Nielsen and Chénier, 1999) processes in all mental activity;

f.2) the existence of *long-range correlations* between the activities of different neuronal groups, sometimes interpreted as providing evidence for a synchronization of neuronal activities (Rodriguez *et al.*, 1999; see also the critique by Van der Velde and De Kamps, 2002);

f.3) the existence of (typically *middle-range*) correlations between different stimulation elements, as shown by the celebrated Gestalt effects in visual perception, as well as by a number of other effects, characterizing visual attention and evidencing in turn the prevalence of global views over local ones;

f.4) the existence of a number of experimental effects, in the psychology of language, of learning and of memory, demonstrating how holistic features can influence local ones; sometimes these effects are interpreted as showing the importance of the role played by context in driving cognitive processes.

Most cognitive scientists, however, choose to work in a non-systemic way, avoiding any reference to these effects, focusing their attention upon very specific experimental situations and restricted domains. Such an approach, however, is impossible when cognitive scientists deal with fundamental problems, such as accounting for "intelligence" or "consciousness". Without entering here into the details of the philosophical debate about the nature of consciousness (see, for instance, Searle, 1999; Globus *et al.*, 2004) we limit ourselves to stressing the fact that the phenomenology on which these constructs are based calls out for a suitable generalization of the concept of emergence used in physical domains. Recalling Chapters 4 and 5, we stressed that emergence needed three ingredients: bifurcation, spatial extent and fluctuations. This last ingredient, however, deserves special attention. Namely, in most cases of physical interest, fluctuations consist of noise which can be represented as a Markovian process without memory, that is where every state transition at a given time instant cannot be influenced by the states occurring at previous time instants. This often drives the system towards a final equilibrium state, corresponding to maximum entropy, reached through an irreversible dynamics. However, this is precisely what does *not* occur in most phenomena characterizing the biological world and especially in phenomena calling for intelligence and consciousness. This suggests that, in order to generalize the methods and concepts of emergence theory to the biological and cognitive domains, we should first resort to fluctuations represented by stochastic processes more general than Markovian ones without memory.

This can be done in a virtually unlimited number of different ways, for instance by exploiting the possibilities of nonlinear dynamical systems theory (an attempt was made by Zak, 2000). Here we mention the proposals for a theory of biological emergence and of consciousness based on Quantum Field Theory. Since the 1980s the Quantum Field Theory-based approach to living matter has been an active research domain (the best review is by Vitiello, 2001, Chapters 3-5). While it is impossible here to review all models and relevant experimental evidence, we stress that such a theory solves three main problems arising in the physical study of biological systems:

- biological processes usually require a very small amount of energy, often only slightly larger than the average energy of a thermal fluctuation; this makes biological systems highly sensitive even to very small inputs; how can such a small amount of energy produce such complex organizational effects?

- most biological processes require a non-dissipative and very efficient transmission of energy and information; how can this occur, despite the ubiquitous presence of dissipation sources in biological matter?

- most biological processes display a strong coherence; how can this be kept despite the ridiculously small decoherence time (of the order of 10^{-40} s) associated with the physical features of biological molecules and the temperatures characterizing biochemical reactions (see Tegmark, 2000; Alfinito *et al.*, 2001)?

It can be shown that, if we take into account the fact that most molecular components of living matter carry an electric dipole moment, we can introduce quantum fields describing the dipole excitations. In the case of water (the principal component of biological systems) both the equations describing these fields and their solutions are invariant with respect to rotations in 3-dimensional space. Water, however, embeds quasi-mono-dimensional macromolecular chains, such as proteins. In the 1970s Davydov showed how a small energy release at one end of such chains, produced typically by ATP hydrolysis, gives rise to a collective mode (that is to a collective motion of atoms belonging to the macromolecular chain) propagating along the chain itself under the form of a solitary wave. This non-dissipative form of energy transport (possible only within a quantum framework) triggers, in turn, a spontaneous symmetry breaking in the water molecules surrounding the chain. This means that the solutions of the equations describing dipole excitations are no longer invariant with respect to 3-dimensional rotations. However, as shown in Chapter 5, in Quantum Field Theory a spontaneous symmetry breaking is always associated with the occurrence of Goldstone bosons carrying long-range interactions. The effect of the latter is to induce long-range ordering within the surrounding water which, in turn, supports coherent phenomena in distant parts of the same biological system. This description is further complicated by the existence of finite volume effects and by the presence of electromagnetic long range forces. These latter imply the existence of different and, more complex, organizational levels beyond the simple one so far described. Their exploration is still in progress, mainly with reference to brain dynamics (Jibu and Yasue, 1995; Vitiello, 1995; for a recent review, see Jibu and Yasue, 2004; recent advances are contained in Pessa and Vitiello, 2004a; 2004b).

Summarizing the above discussion, we can say that models of physical emergence based on Quantum Field Theory appear as the only ones able to solve the problems mentioned above. Namely:

- they do not require large amounts of energy, as a collective process is based on the intrinsic features of dynamical laws fulfilled by quantum fields, and not on energy;
- collective processes, owing to their intrinsic coherence, allow non-dissipative energy transport (as occurs for laser light);
- coherence is maintained owing to the existence of Goldstone modes (possible only within Quantum Field Theory) which counteract every action tending to destroy it.

We can thus conclude that a Quantum Field Theory-based approach to living matter shows how suitable models of physical emergence can explain the main features of biological emergence. This has stimulated a number of proposals for models of consciousness based on these ideas (see Penrose, 1994; Hameroff and Penrose, 1996; Hameroff et al., 1996; 1998). On these topics there is a lively debate in progress and most technical problems are still unsolved, but now the road towards a theory of cognitive processes and consciousness based on the concept of emergence has been found.

9.4 Embodied Cognition

A seemingly different view on cognitive system modeling has been adopted so far by the proposers of the so-called *embodied cognition* (Varela et al., 1991). Within this context the reference is to *reactive robotics* (Brooks, 1991; Beer, 1995), to Artificial Life and to emergent computation, discussed above in Chapter 5. So far, the models of cognitive processing proposed within this approach are based only upon analogies drawn from collective behavior observed in Artificial Life models or from autonomous robot behaviors. While these topics are fascinating, however, it must be noted that these models suffer from a number of drawbacks, the main ones being:

- the difficulty in predicting, controlling, detecting and monitoring emergent phenomena, owing to the non-ideal nature of the models themselves;
- the difficulty in accounting for high-level cognitive processing.

With regard to the first problem, we recall that in Chapter 5 we showed how some non-ideal models of Artificial Life could be cast in the form of suitable ideal models, having at disposal enough mathematical machinery. We stress, however, that the latter is not a luxury, as it is the only means to achieve prediction, control and monitoring. In any case, the examples of model reduction so far known imply that the first problem can, in principle, be solved.

The second problem, on the contrary, is more serious and highlights the inability of current theoretical approaches to account for the high level of complexity inherent in organism-environment interactions. This inability stems from two circumstances:

- we lack the tools necessary for describing the environment in an acceptable way; this is due to the fact that western science has developed, from the beginning, in a non-systemic way, taking into consideration only isolated (or, at most, closed) systems; thus, even today, the description of open systems and of their interactions with the environment is still rudimentary, as no conceptual apparatus has been developed during the development of science to deal with these problems; this applies even to Quantum Field Theory which avoids the description of external evironments by resorting to infinite volume limits;
- even ideal models have been built to describe situations (common in a number of physical cases) which deal with only *two* different levels of description: the microscopic and the macroscopic; however, this is never the case in the biological realm (and, *a fortiori*, in the cognitive one) in which we observe a huge number of different levels of descriptions, often hierarchically interrelated: from macromolecules to cells, organs, organisms, minds and even societies, comprising complex chains of different forms of emergence which cannot be captured by the rudimentary tools (even though mathematically sophisticated) of ferromagnetism or of superconductivity theory.

These remarks point to a number of problems (even of a technical nature) which must absolutely be solved if the embodied cognition concept is not to remain a chimera. On the other hand, only the development of a new, more general, theory of emergence, invoked here, will allow a further development of a theoretical and conceptual apparatus for dealing with Collective Beings. Otherwise we will remain in the current situation, depicted in this book, in which, while conscious of the new needs of systemics for a generalization of its conceptual tools to deal with the increasing complexity of actual world, we can do nothing but introduce new ideas, such as DYSAM and ergodicity, which, in the absence of a substantial tradition of applications to concrete problems, are still in a primitive stage. On the other hand, however, we have a whole range of concepts, mathematical tools and theories which are expressed in a language which is difficult, if not impossible, to understand for most scientists (except for some mathematicians and some theoretical physicists), including many followers of the systemic approach. This gap is dangerous for the future development of systemics and the aim of this book is to show how, when facing complexity, we must go beyond this situation and work in a collaborative and transdisciplinary way to build a new systemics, without gaps and endowed with more powerful conceptual tools.

References

Aakerlund, L. and Hemmingsen, R., 1998. Neural networks as models of Psychopathology. *Biological Psychiatry*, **43**: 471-482.

Alfinito, E., Viglione, R.G. and Vitiello, G., 2001. The decoherence criterion. *Modern Physics Letters B*, **15**: 127-136.

Arbib, M.A. and Fellous, J.-M., 2004. Emotions: from brain to robot. *Trends in Cognitive Sciences*, **8**: 554-561.

Barsalou, L.N., 1992. *Cognitive Science: An overview for cognitive scientists*. Erlbaum, Hillsdale, NJ.

Bechtel, W. and Richardson, R., 1993. *Discovering complexity: Decomposition and localization as strategies in scientific research*. Princeton University Press, Princeton, NJ.

Beer, R.D., 1995. A dynamical systems perspective on agent-environment interaction. *Journal of Artificial Intelligence*, **72**: 173-215.

Bringsjord, S., 1992. *What Robots can and can't be*. Kluwer, Dordrecht.

Bringsjord, S. and Zenzen, M., 2003. *Superminds: People harness hypercomputation and more*. Kluwer, Dordrecht.

Brooks, R.A., 1991. Intelligence without representation. *Artificial Intelligence*, **47**: 139-159.

Caelli, T., Guan, L. and Wen, W., 1999. Modularity in neural computing. *Proceedings of the IEEE*, **87**: 1497-1518.

Classen, J., Gerloff, C., Honda, M. and Hallet, M., 1998. Integrative visuomotor behavior is associated with interregionally coherent oscillations in the human brain. *Journal of Neurophysiology*, **79**: 1567-1573.

Collins, A.M., 1977. Why Cognitive Science. *Cognitive Science*, **1**: 1-2.

Fetzer, J.H., 2001. *Computers and cognition: Why minds are not machines*. Kluwer, Dordrecht.

Fodor, J.A., 1981. *Representations: Philosophical essays on the foundations of Cognitive Science*. MIT Press, Cambridge, MA.

Fodor, J.A., 1983. *The modularity of mind*. MIT Press, Cambridge, MA.

Fodor, J.A., 1987. *Psychosemantics. The problem of meaning in the philosophy of mind*. MIT Press, Cambridge, MA.

Freeman, W.J., 1999. Noise-induced first-order phase transitions in chaotic brain activity. *International Journal of Bifurcation and Chaos*, **9**: 2215-2218.

Freeman, W.J., 2000. *Neurodynamics: An exploration of mesoscopic brain dynamics.* Springer, Berlin.

French, R., 1999. Catastrophic forgetting in connectionist networks. *Trends in Cognitive Sciences*, **3**: 128-135.

García-Pedrajas, N., Hervás-Martínez, C. and Muñoz-Pérez, J., 2002. Multi-objective cooperative coevolution of artificial neural networks (multi-objective cooperative networks). *Neural Networks*, **15**: 1259-1278.

Gazzaniga, M.S. (Ed.), 2004. *The Cognitive Neurosciences III: Third Edition.* MIT Press, Cambridge, MA.

Globus, G.G., Pribram, K.H. and Vitiello, G. (Eds.), 2004. *Brain and being. At the boundary between science, philosophy, language and arts.* Benjamins, Amsterdam.

Hameroff, S.R., Kaszniak, A.W. and Scott, A.C. (Eds.), 1996. *Toward a science of consciousness I. The first Tucson discussions and debates.* MIT Press, Cambridge, MA.

Hameroff, S.R., Kaszniak, A.W. and Scott, A.C. (Eds.), 1998. *Toward a science of consciousness II. The second Tucson discussions and debates.* MIT Press, Cambridge, MA.

Hameroff, S.R. and Penrose, R., 1996. Conscious events as orchestrated space-time selections. *Journal of Consciousness Studies*, **3**: 36-53.

Harnad, S., 1990. The symbol grounding problem. *Physica D*, **42**: 335-346.

Hauser, L., 1997. Searle's Chinese Box: Debunking the Chinese Room Argument. *Minds and Machines*, **7**: 199-226.

Hinrichs, H. and Machleidt, W., 1992. Basic emotions reflected in EEG-coherence. *International Journal of Psychophysiology*, **13**: 225-232.

Hummel, J.E. and Biederman, I., 1992. Dynamic binding in a neural network for shape recognition. *Psychological Review*, **99**: 480-517.

Ito, J. and Kaneko, K., 2000. Self-organized hierarchical structure in a plastic network of chaotic units. *Neural Networks*, **13**: 275-281.

Jibu, M. and Yasue, K., 1995. *Quantum Brain Dynamics and Consciousness: An Introduction.* Benjamins, Amsterdam.

Jibu, M. and Yasue, K., 2004. Quantum brain dynamics and Quantum Field Theory. In G.G.Globus, K.H.Pribram and G.Vitiello (Eds.). *Brain and being. At the boundary between science, philosophy, language and arts* (pp. 267-290). Benjamins, Amsterdam.

Langton, C.G. (Ed.), 1989. *Artificial Life.* Addison-Wesley, Reading, MA.

Laurence, S. and Margolis, E., 2002. Radical concept nativism. *Cognition*, **86**: 22-55.

Lindsay, P.H. and Norman, D.A., 1972. *Human Information Processing.* Academic Press, New York.

MacLennan, B.J., 2001. 'Transcending Turing computability'. Technical Report UT-CS-01-473, Department of Computer Science, University of Tennessee, Knoxville, TE.

McClelland, J.L. and Rumelhart, D.E. (Eds.), 1986. *Parallel Distributed Processing. Explorations in the microstructure of cognition*, 2 voll. MIT Press, Cambridge, MA.

McCloskey, M. and Cohen, N.J., 1989. Catastrophic interference in connectionist networks: The sequential learning problem. In G.H.Bower (Ed.). *The psychology of learning and motivation. Advances in research and theory* (vol. 24, pp. 109-165). Academic Press, San Diego, CA.

Newell, A. and Simon, H.A., 1976. Computer Science as Empirical Inquiry: Symbols and Search. *Communications of the ACM*, **19**: 113-126.

Nielsen, T.A. and Chénier, V., 1999. Variations in EEG coherence as an index of the affective content of dreams from REM sleep: Relationship with face imagery. *Brain and Cognition*, **41**: 200-212.

Norman, D.A., 1980. Twelve issues for Cognitive Science. *Cognitive Science*, **4**: 1-32.

Nunez, P.L., 2000. Toward a quantitative description of large scale neocortical dynamic function and EEG. *Behavioral and Brain Sciences*, **23**: 371-437.

Palmer, S.E. and Kimchi, R., 1986. The information processing approach to cognition. In T.J.Knapp and L.C.Robertson (Eds.). *Approaches to cognition: Contrasts and controversies* (pp. 37-77). Erlbaum, Hillsdale, NJ.

Penrose, R., 1994. *Shadows of the mind*. Oxford University Press, Oxford, UK.

Pessa, E., 2004. Quantum connectionism and the emergence of cognition. In G.G.Globus, K.H.Pribram and G.Vitiello (Eds.). *Brain and being. At the boundary between science, philosophy, language and arts* (pp. 127-145). Benjamins, Amsterdam.

Pessa, E. and Terenzi, G., 2002. A Neural Solution to Symbol Grounding Problem. In M.Marinaro and R.Tagliaferri (Eds.), *Neural Nets – WIRN 01* (pp.248-255). Springer, London.

Pessa, E. and Vitiello, G., 2004a. Quantum noise, entanglement and chaos in the Quantum Field Theory of Mind/Brain states. *Mind and Matter*, **1**: 59-79.

Pessa, E. and Vitiello, G., 2004b. Quantum noise induced entanglement and chaos in the dissipative quantum model of brain. *International Journal of Modern Physics B*. **18**: 841-858.

Piattelli-Palmarini, M., Piaget, J, and Chomsky, N. (Eds.), 1980. *Language and learning: The debate between Jean Piaget and Noam Chomsky*. Harvard University Press, Cambridge, MA.

Port, R. and Van Gelder, T.J. (Eds.), 1995. *Mind as motion: Explorations in the dynamics of cognition*. MIT Press, Cambridge, MA.

Pour-El, M.B. and Richards, I., 1983. Computability and noncomputability in classical analysis. *Transactions of American Mathematical Society*, **275**: 539-560.

Pour-El, M.B. and Richards, I., 1989. *Computability in analysis and physics*. Springer, Berlin-Heidelberg-New York.

Preston, J. and Bishop, M. (Eds.), 2002. *Views into the Chinese Room: New essays on Searle and Artificial Intelligence*. Oxford University Press, Oxford, UK.

Pylyshyn, Z.W., 1984. *Computation and Cognition*. MIT Press, Cambridge, MA.

Ratcliff, R., 1990. Connectionist models of recognition memory: Constraints imposed by learning and forgetting functions. *Psychological Review*, **97**: 285-308.

Reed, S.K., 1988. *Cognition. Theory and application*. Brooks/Cole, Pacific Grove, CA.

Reggia, J.A., Ruppin, E. and Berndt, R. (Eds.), 1996. *Neural modeling of brain and cognitive disorders*. World Scientific, Singapore.

Reggia, J.A., Ruppin, E. and Glanzman, D.L. (Eds.), 1999. *Disorders of brain, behavior, and cognition: The neurocomputational perspective*. Elsevier, New York.

Rodriguez, E., George, N., Lachaux, J.P., Martinerie, J., Renault, B. and Varela, F.J., 1999. Perception's shadow: Long-distance synchronization of human brain activity. *Nature*, **397**: 430-433.

Sarnthein, J., Petsche, H., Rappelsberger, P., Shaw, G.L. and Von Stein, A., 1998. Synchronization between prefrontal and posterior association cortex during human working memory. *Proceedings of the National Academy of Sciences USA*, **95**: 7092-7096.

Searle, J.R., 1980. Minds, brains, and programs. *Behavioral and Brain Sciences*, **3**: 417-458.

Searle, J.R., 1999. *The Mystery of Consciousness*. A New York Review Book, New York.

Siegelmann, H.T., 1999. *Neural networks and analog computation: Beyond the Turing limit.* Birkhäuser, Boston, MA.

Singer, W., 1999. Neuronal synchrony: A versatile code for the definition of relations? *Neuron,* **24**: 49-65.

Sloman, A., 2002. The irrelevance of Turing machines to AI. In M.Scheutz (Ed.). *Computationalism: New directions* (pp.87-127). MIT Press, Cambridge, MA.

Spelke, E.S., Breinlinger, K., Macombe, J. and Jacobson, K., 1992. Origins of knowledge. *Psychological Review,* **99**: 605-632.

Stannett, M., 2001. 'An introduction to post-Newtonian and non-Turing computation'. Technical Report CS-91-02. Department of Computer Science, Sheffield University, Sheffield, UK.

Stein, D. and Ludik, J. (Eds.), 1998. *Neural networks and Psychopathology.* Cambridge University Press, Cambridge, UK.

Stillings, N.A., Weisler, S.E., Chase, C.H., Feinstein, M.H., Garfield, J.L. and Rissland, E.L., 1995. *Cognitive Science. An introduction.* MIT Press, Cambridge, MA.

Sutton, J.P. and Jamieson, I.M.D., 2002. Reconfigurable networking for coordinated multi-agent sensing and communications. *Information Sciences,* **148**: 103-111.

Tani, J., 2003. Learning to generate articulated behavior through the bottom-up and the top-down interaction processes. *Neural Networks,* **16**: 11-23.

Tegmark, M., 2000. Why the brain is probably not a quantum computer. *Information Sciences,* **128**: 155-179.

Tschacher, W. and Dauwalder, J.-P. (Eds.), 1999. *Dynamics, Synergetics, Autonomous Agents: Nonlinear systems approaches to Cognitive Psychology and Cognitive Science.* World Scientific, Singapore.

Tulving, E., 1983. *Elements of episodic memory.* Oxford University Press, New York.

Van der Velde, F. and De Kamps, M., 2002. Synchrony in the eye of the beholder: An analysis of the role of neural synchronization in cognitive processes. *Brain and Mind,* **3**: 291-312.

Van Gelder, T.J., 1998. The dynamical hypothesis in Cognitive Science. *Behavioral and Brain Sciences,* **21**: 615-665.

Van Gelder, T.J., 1999. Dynamic approaches to cognition. In R.Wilson and F.Keil (Eds.). *The MIT Encyclopedia of Cognitive Science* (pp. 244-246). MIT Press, Cambridge, MA.

Varela, F., Thompson, E. and Rosch, E., 1991. *The Embodied Mind: Cognitive Science and Human Experience.* MIT Press, Cambridge, MA.

Vitiello, G., 1995. Dissipation and memory capacity in the quantum brain model. *International Journal of Modern Physics B.* **9**, 973-989.

Vitiello, G., 2001. *My double unveiled.* Benjamins, Amsterdam.

Yao, X., 1999. Evolving artificial neural networks. *Proceedings of the IEEE,* **87**: 1423-1447.

Wermter, S. and Sun, R., 2000. *Hybrid neural systems.* Springer, Berlin-Heidelberg-New York.

Zak, M., 2000. Dynamics of intelligent systems. *International Journal of Theoretical Physics,* **39**: 2107-2140.

Zhou, Z.-H., Jiang, Y. and Chen, S,-F., 2003. Extracting symbolic rules from trained neural network ensembles. *AI Communications* , **16**: 3-15.

Appendix 1

SOME SYSTEMIC PROPERTIES

This appendix presents a list of items related to system *properties* which, all being emergent, are indissolubly linked with the theoretical role of the observer as introduced in the book.

An overview of the wide variety of possible systemic characteristics may be found in Arecchi and Farini (Arecchi and Farini, 1996) as well in Von Bertalanffy (Von Bertalanffy, 1968; 1952). Reference websites are given in Principia Cybernetica Web (Heylighen *et al.*, 2002) and the Encyclopaedia Autopoietica (Whitaker, 2003). Shorter reviews include that by Banzhaf (Banzhaf, 2001). It must be remembered that terms such as *properties* and *characteristics* seem to be based on *objectiveness*, intended to be peculiar to the considered item, whereas our approach also refers to the observer's cognitive models and knowledge. The list of systemic properties described here include the following terms:

Adaptive
Allopoietic
Anticipatory
Autonomous
Autopoietic
Chaotic
Complex
Connectionist
Deterministic
Dissipative
Equifinal
Emergent
Ergodic
Far from equilibrium
Growing vs. developing
Heterogeneous
Heuristic
Hierarchical
Homeostatic
Non-dissipative
Non-ergodic

Open and Closed
Self-organized
Structures and systems
Systemic characteristics
Systemic equilibrium
Systemic structure and organization

1) Adaptive

This attribute applies to systems able to change their internal states as a function of changes occurring in their operational environment. A software system, for instance, is said to be *adaptive* if, during its standard activity it can decide upon actions to be taken based upon modelling the current environmental states and relationships, an activity which, in turn, depends upon the output of a suitable environment monitoring device. Such a system, of course, must operate by applying one or more (see *DYSAM*, chapter 2) strategies. For instance, software learning systems are adaptive systems (McTear, 1993). Another example is given by those *human machine interfaces* which adapt themselves to the characteristics of their user. The actions in an adaptive system often affect the environment, and hence such a system belongs to a constant feedback loop with its environment.

The ability to adapt, in the case of software systems, characterizes so-called 'intelligent' systems, and machine learning processes. It is a typical property of most complex systems, consisting of interacting autonomous agents, such as ant or bee societies, and, more in general, communities within ecological systems. The concept of adaptive system was introduced for the first time in 1947 by the British psychiatrist and cybernetician W. R. Ashby in a celebrated paper (Ashby, 1947). The same paper also introduced for the first time the expression "Self-organizing system".

2) Allopoietic

This is the property of a system which renders it incapable of producing its own composing elements and processes. Such a feature pertains to non-autonomous systems requiring an external intervention for their activation. Allopoiesis, therefore, refers to processes allowing an organization to produce another, different, organization. One example of an allopoietic system is an assembly line. A *allopoietic production process* produces entities different from those required to produce themselves: a ship-yard, for instance, produces ships but not the machinery necessary to produce them. The reproduction process in biology is often allopoietic because, for instance, babies are materially distinct from parental organisms: *reproduction is not self-production*. The concept of allopoiesis is opposite to that of autopoiesis (Varela, 1980; 1991).

3) Anticipatory

The attribute 'anticipatory' as well as the definition of anticipatory systems were introduced by Robert Rosen, a mathematical biologist (Rosen, 1985), as illustrated in Section. 1.4.2.

An anticipatory system is a system provided with a predictive model of itself and of its environment, enabling it to change its state in the successive step according to the predictions formulated by the model in the previous step. Due to the fact that they are provided with a predictive model of themselves and of their environment, *anticipatory systems behave as if they already know their future.* The book cited above refers to different kinds of anticipatory systems and their characteristics, by introducing examples in physical and biological systems. Systems regulated and controlled by feedback loops are based on corrective actions, which in turn depend upon predefined states to be reached, whereas anticipatory systems are regulated by processing pre-established models, provided by the systems themselves, able to anticipate future states.

4) Autonomous

The attribute 'autonomous' has several different meanings, even when restricting ourselves to systemic properties. One meaning derives from the mathematical theory of systems of differential equations (the area which saw the birth of General System Theory). Within it a system of differential equations is called *autonomous* if it lacks any explicit external input, that is, in more technical terms, when there is no explicit functional dependence of the dependent variables on the independent variables. An example of an autonomous system of differential equations regarding two dependent variables, here denoted as x and y, in turn depending upon only one independent variable t, is given by:

$$dx/dt = f(x, y) \quad ; \quad dy/dt = g(x, y) ,$$

where both functions f and g cannot contain t in an explicit way. An example of a non-autonomous system of differential equations regarding the same variables could be:

$$dx/dt = f(x, y) \quad ; \quad dy/dt = g(x, y) + h(t) ,$$

where the symbol $h(t)$ denotes an explicit function of the independent variable t. Therefore, an autonomous system of differential equations is a system lacking any external input and can be considered as a prototype for the model of a *closed* system. Despite this circumstance, the mathematical theory of autonomous systems is a well developed branch of mathematics (having given birth to systemic concepts such as 'equilibrium' or 'attractor'). On the contrary, the theory of non-autonomous systems is still in its infancy, owing to a number of considerable technical difficulties, notwithstanding its foremost importance in most applications including, for example, the theory of automatic control.

Another meaning of the attribute 'autonomous', completely opposite to the previous one, refers to systems, implemented through technical devices, which have some form of onboard 'intelligence', as well as computational capabilities, sensors and motors, a store or a production device for energy, standalone and communication capabilities. In a word, these systems can behave like an apparently 'intelligent' system behaving in the environment without the need for any explicit external control (to make a comparison, an automobile, despite its name, is not autonomous in this sense, as it needs an external driver). Typical examples of these autonomous systems are given by some kinds of robots (keeping in mind, of course, that not all robots are autonomous). In this case the attribute 'autonomous' denotes a system whose behavior follows its own internal model.

To be more precise, however, we have to describe what we mean by 'internal model'. It is possible to conceive very simple systems, endowed only with a set of inputs and a set of outputs, as well as a set of pre-stated relationships specifying how the presence of a given input can give rise to a given output. Such systems, which we will call input-output systems, or stimulus-response automata, were introduced many years ago by behavioral psychologists to describe the observed behaviors of human and animal subjects. In a very restricted sense, a system of input-output relationships, embedded within a given stimulus-response automaton, could be identified with its 'internal model'. On reflecting, however, such an identification is incorrect. Namely, the very concept of 'internal model' refers to an inner representation of the external state of affairs, coded through some form of internal state. It is, now, possible to show that it is mathematically impossible to represent any kind of automaton defined in terms of its input, its output and its internal state, through a behaviourally *equivalent* stimulus-response automaton. This means that the variety of possible behaviors will never be captured by resorting only to the latter kind of automata. The first proof of this was given by Arbib in 1969 (Arbib, 1969). In the same year, however, the celebrated book *Perceptrons*, by Marvin Minsky and Seymour Papert, appeared in which it was proved that a neural-like version of a simple stimulus-response device, known as a 'Perceptron', was unable even to learn to implement very simple input-output relationships, such as the Boolean function XOR (Minsky and Papert, 1969).

Finally, in 1989 a theorem stating that, to approximate with any level of precision any form of input-output relationship, at least a three-layered system of non-linear units are necessary (Hornik *et al.*, 1989). Three-layered means that there is an input layer, an intermediate (often called 'hidden') layer, and an output layer. The unavoidable existence of the intermediate layer, processing the input information in a non-trivial way, points to the need for at least a 'mimimal' internal representation to account for the complexity of the observed behaviors. It is just such a form of inner representation which we will identify with the 'internal model' previously introduced.

5) Autopoietic

This attribute denotes a process, or a system, able to produce itself. Literally the word means 'self-reproduction'. *Autopoiesis is a network of production processes, where the function of each component is to contribute to the production or transformation of other network components* (Maturana and Varela, 1973; Maturana and Varela, 1980; Capra, 1996). What must be noted is that autopoiesis may be considered as an organization common to all living beings. Even more, it may be intended, according to Maturana and Varela, as *the* characteristic of living systems.

In an autopoietic organization the components recursively[5] generate, through their mutual interactions, the same network producing themselves. It must be noted that *in such a view* an autopoietic system is *operationally* closed, having a predefined structure. Examples of autopoietic systems are given by cells, living organisms, corporations; social systems. Another example is given by production systems able to produce not only the products that they are designed to produce (such as automobiles), but even the production machinery itself.

6) Chaotic

Even this system attribute is endowed with a number of different meanings. Here two of them are cited: *deterministic chaos* (Lorenz, 1963) and *stochastic chaos* (Freeman *et al.*, 2001). Deterministic chaos is associated with time behaviors characterized by:

long-term unpredictability;
sensitivity to initial conditions.

Property a) is typical even of behaviors which are not deterministically chaotic (such as *noise*), whereas property b) is typical only of deterministically chaotic ones. To ascertain the presence of a) and b) within a given model system there are a number of quantitative indices (such as those stemming from Fourier analysis, Lorenz plots, Lyapunov exponents),

·

---

5 A recursive program calls itself a finite numbers of times during a computation. Consider, for instance, a program P able to add one unit to a parameter given in input, i.e., able to compute $P(X)=X+1$. To compute $X+2$ the same program can be applied to the previous result: $P(X+1)=X+2$. It is interesting to note how the concept of recursive program contributes to define the concept of effective algorithm and of computability. On the other hand, the concept of iteration consists in repeating an operation for a non-limited number of times; recursion is not expected to lead to convergence, whereas convergence is expected for iteration.

which can be computed with a satisfactory approximation once the evolution equations of the model are known in advance. Things are, however, very different for ascertaining the presence of deterministic chaos in experimentally observed behaviors for which no model is available. In the latter case, it is very difficult (and often impossible) to obtain reliable estimates of the indices characterizing the presence of deterministically chaotic behavior. This has induced many scientists to claim that deterministic chaos is only an invention of mathematicians and does not really exist in nature. The first discovery of deterministically chaotic phenomena did, in fact, occur within the domain of abstract mathematical models, as for *quasi-periodic* behaviors, at the end of the nineteenth century in systems subjected to forced oscillations whose period had an incommensurable ratio with the period of the self-oscillations of the system itself. And even the Lorenz system, described by equations (1.12) in Section 1.4.4, is nothing but a very abstract and artificial description of a complex natural system (the terrestrial atmosphere). After all, it is very difficult to believe that natural systems are correctly described by strictly deterministic models: noise is ubiquitous in nature and the quantum-like behavior of microscopic phenomena seems to exclude the plausibility of a purely deterministic model (we recall, on this point, that a truly quantum system can never be chaotic). Over the past few years, however, deterministic chaos has regained its conceptual importance. In particular, a number of papers (see, for example, Pessa and Vitiello, 2004) have shown how, at the starting point of a phase transition, the behavior of a system is really deterministically chaotic, and it is just such a circumstance which allows an understanding of the formation of domains, constraints, defects and all other structures which characterize,and drivefurther evolution following a phase transition.

Shortcomings in the concept of deterministic chaos led to the introduction of the second form of chaos cited above, *stochastic chaos*. The latter occurs within systems endowed with noise, deriving both from inner sources and the external environment. Deterministic chaos is characterized by a low dimensionality of attractors and the fact that time is the only independent variable, whereas stochastic chaos is associated with attractors of high dimensionality, with the presence of noise-created and noise-sustained structures, with spatio-temporal unpredictability and whence to the presence also of spatially independent variables. Although some mathematical tools exist for investigating deterministic chaos, there is still no satisfactory framework for dealing with stochastic chaos which, so far, can be observed mainly through computer simulations and experimental observations.

There is an abundance of examples showing the importance of deterministic (or stochastic) chaos. Climate is a classical case, with the problem of medium-long range weather forecasting. This brings to mind the well-known *butterfly effect* (mentioned in Chapter 1) in Lorenz's model, referring to the sensitivity of a system to initial conditions. Other examples are population fluctuations, fluid turbulence, electrical activities of the brain and heart. Even in the classical *Three Body Problem* (Barrow-Green, 1997), also introduced in Chapter 1 and consisting of determining the general behavior of a system of three bodies under gravitational interaction, chaotic solutions can occur for specific ranges of the ratios between their masses (Hénon, 1983). Finally, the simplest example of a deterministic chaotic system, which can even be studied using a pocket calculator, is the so-called *logistic map*:

$$x_{n+1} = 4 \lambda x_n (1 - x_n)$$

where x_n denotes the value of the map at the n-th iteration, and λ is a parameter whose value must lie between 0 and 1. It can be shown that, when this parameter increases above a critical value, approximately 0.936..., the behavior of the number sequence generated by this map becomes deterministically chaotic.

7) Complex

The concept of *complex* system (Flood and Carson, 1988; Serra *et al.*, 1986; Serra and Zanarini, 1990; Gregersen, 2003) is related to that of complexity (cfr. Casti, 1994), already introduced in Chapter 1. Many different definitions of complexity have been proposed. They can be subdivided into two main categories: those relating complexity to some form of *structural* organization of a given system or of its dynamics, and those relating complexity to the *functional* aspects connected to operation of the system under study within a suitable environment. Of course, both kinds of definitions refer to features which can be detected only in relation to a given observer, equipped with a cognitive system, goals, and suitable measuring devices.

Definitions of complexity related to structural features, very popular within the abstract theory of computation and, more in general, within computer science, include the following (see also Papadimitriou, 1993):

Spatial or *Kolmogorov complexity* (Kolmogorov, 1965; Chaitin, 1990), for a given system, is the minimum possible length of a computer program able to reproduce all behavioural features of the system itself. The solar system, for instance, is characterized by a very low spatial complexity and even a small computer program, including the Newtonian gravitational interactions between the planets and between the planets and the sun, suffices to reproduce all planetary orbits for every choice of evolution time. On the other hand, a social system is generally associated with a very high level of spatial complexity, as a computer program able to reproduce all observed behaviors of its members would be (if it existed) of prohibitive length;

Temporal (sometimes called *computational*) *complexity*, is defined, for a given system, as the minimum time needed by a computer program, already having the minimum possible length, to reproduce the behavioural features of the system. The temporal and spatial complexities can, in principle, be different from one another. For instance, although the solar system is characterized both by a low spatial complexity and a low temporal complexity, a deterministically chaotic system has a low spatial complexity but a very high temporal complexity. That is, although the totality of its behaviors can be generated through a very simple recursive formula (as in the logistic map mentioned above), the time needed to generate a very long sequence of such behaviors is at least directly proportional to the length of the sequence itself. Thus, on increasing this length in an unlimited manner, the temporal complexity will also increase in an unlimited manner. Similar considerations hold for many other algorithmic procedures, such as, for example, that used for computing the n-th figure of π;

Logical depth (sometimes confused with *computational complexity*; see Bennett, 1985; 1988) is defined, for a given system, as the minimum amount of computational resources (not only time but also, for instance, memory) needed by a computer program of the minimum possible length, to reproduce the system behavioural features.

Shifting now to definitions of complexity related to functional features, it must be admitted that they are almost entirely lacking or, at the most, rather vague. In general, however, a system can be considered as complex (in the functional sense) when phenomena of emergence take place or, in principle, could occur within it. All models of complex systems of this kind built so far comprise a large number of interacting entities, agents and processes behaving according to dynamical non-linear rules, often simulated through techniques such as those used in Neural Networks, Genetic Algorithms, Cellular Automata, and so on. A number of these systems appear to be endowed with a hierarchical self-organization and the fact that a system lies in a situation far from thermodynamic equilibrium can also characterize such systems.

A typical disciplinary context of complex systems is Biocomputing (Lundh *et al.,* 1997) dealing with, amongst others, biological regulatory mechanisms. Often a natural system is described as complex, referring in an implicit way to its functional features, when it comprises a large number of elements, all differing from one another. Typical examples of complex systems of this kind include the brain and large social organizations. So far, however, there is no mathematical proof that such systems show some form of emergence but there are many observations which can be accounted for only by formulating such a hypothesis.

8) Connectionist

In cognitive science the term *Connectionism* denotes a particular approach to the study of cognitive processes and is based on two main hypotheses:

Every cognitive process can be studied (and described) by resorting to (at least) two different levels of description: the *microscopic* (or *microcognitive*) level, and the *macroscopic* one. Cognitive processes, as observed within laboratory experiments and in everyday life, pertain only to the macroscopic level and can be detected only at this level;

Every macroscopic cognitive process is nothing but a collective effect emerging from the (cooperative) interaction between suitable microcognitive units, each of which are able to detect suitable microcognitive features underlying the process itself.

Sometimes these microcognitive features are qualified as *subsymbolic* but the distinction between symbolic and subsymbolic descriptions is in part illusory. Namely, a connectionist model, generally formulated in terms of artificial neural networks, is represented in a symbolic way, as is every other algorithmic model, independently of the approach adopted.

The main advantages of the connectionist approach lies in the possibility of accounting for very complex macroscopic cognitive features by resorting only to very simple microcognitive units and rules. This includes the possibility of directly relating the biological activities of the brain with high-level mental functions relying, of course, on the possibility of observing (and of controlling in some way) the emergence postulated in assumption b) made above. At present, such observations are not so clear and there is currently active debate on this approach.

The word 'connectionism', first introduced in 1921 by the psychologist E. K. Thorndike to denote a theory of human and animal behaviour based on (modifiable) 'connections' between stimuli and responses (Thorndike, 1921), owes its modern popularity to the fact that the microcognitive units are described as neural-like devices, interacting through suitable reciprocal 'connections', resembling the synaptic connections existing within biological neural networks. Each connection is normally associated with a *weight*, characterized by a numerical value. The effect of an input arriving at a neural-like microcognitive unit is mediated by the weight associated with its input line. Nowadays, the well-established technology of *Neural Networks* (Haykin, 1998) implementing connectionist models has wide applications in many domains of science and engineering (see, for example, Bishop, 1995; Rojas, 1996).

Connectionist systems could, in principle, process information in a parallel way, even though this possibility is really utilized only in hardware applications and seldom in software simulations (except for parallel computers). Their operation can also be interpreted through explicit logical rules, although such an interpretation is not necessary for designing or controlling them. The connectionist approach thus appears as being directly inspired by the neurophysiology of the brain. Eight properties are considered as essential for building a connectionist model, once known as a *Parallel Distributed Processing* (PDP) model (Rumelhart *et al.*, 1986; Rumelhart and

McClelland, 1986), to emphasize the possibility of parallel processing by the microcognitive units:

The existence of a *set of processing units*

The fact that each unit has a *state of activation*

The existence of an *output function* for each unit

The existence of a *pattern of connectivity* amongst units

The existence of a *propagation rule* for propagating patterns of activities through the network of connections

The existence of an *activation rule* for combining the inputs impinging on a unit with the current state of that unit to produce a new level of activation for the unit.

The existence of a *learning rule* whereby patterns of connectivity are modified by experience

The existence of an *environment* within which the system is assumed to operate

9) Deterministic

A system is deterministic (or rather its dynamics are deterministic) when a knowledge of the values of its state variables (or features) at a given instant in time enables us to compute, in principle with any given precision, the values of its state variables in any other previous or subsequent time instant. A system of this kind can be considered as a special subcase of a probabilistic system, corresponding to the fact that all probabilities can only be 0 or 1. It is possible to introduce a distinction between *globally deterministic* and *locally deterministic* systems. In the former, computation of the values of the state variables at time t_1, starting from a knowledge of their values at time t_0, requires a time which is independent of the difference $t_1 - t_0$. In locally deterministic systems, a knowledge of the values of the state variables at time t_0 only allows computation of the state variables at time $t_0 + dt$, where dt is the minimum time unit (in analytical mathematics this value is infinitesimal). Every globally deterministic system is also locally deterministic, but the converse is not true. Deterministically chaotic systems are only locally, but not globally, deterministic and this makes them unpredictable over long time scales.

10) Dissipative

Ilya Prigogine (Prigogine and Nicolis, 1967; Prigogine 1981; 1998) introduced the term "dissipative structures" referring to situations of **coexistence of change and stability** (see Chapter 1 and item 23 below). A frequently observed dissipative non-living structure is a vortex in a flux of running water: water continuously flows through the vortex but its characteristic funnel shape shrinking into the spiral remains. The same kinds of structure exist in atmospheric phenomena such as hurricanes. It is interesting to note how an established dissipative system needs a constant flux of matter from outside. Analogously, a living dissipative structure needs a constant flow of matter, such as air, water, food, and, in certain cases, light. Networks of metabolic processes keep such systems far from thermodynamic equilibrium, i.e. thermodynamic death.

This attribute describes a system where energy dissipation, concomitant with non-equilibrium conditions, allows the emergence of ordered structures. The stability of dissipative structures does not derive from low entropy production (intended as an index characterizing microscopic disorder of a system: entropy growth indicates a trend towards a more disordered phase – e.g., from solid to liquid, to gas), but from the ability of the system to transfer a large amount of entropy to its environment.

As these systems are far from thermodynamic equilibrium, they are able to dissipate the heat generated to support themselves, leading to the emergence of ordered configurations, i.e., to allow processes of self-organization (Von Foerster and Zopf, 1962). Such systems also contain continuously fluctuating subsystems able to give rise to new organizations: such processes are said to correspond to *bifurcation* points (see item 28 below) and it is impossible to predict whether such a system will degenerate into a chaotic situation or reach a higher organizational level. In the latter case *dissipative structures* are established. The attribute is related to the fact that they need more energy to continue to exist that their existence is limited precisely by their ability to dissipate heat. In a sense, a dissipative structure arises as an exact balance between the dissipation (e.g., by diffusion) and the nonlinearity enhancing its inner fluctuations. The concept is introduced in Chapter 1.

11) Equifinal
Finality has long been discussed in philosophy and science and the subject was also described in Bertalanffy's important book (Von Bertalanffy, 1968). Bertalanffy discussed three kinds of finalities associated, respectively, with the following situations:

The dynamical evolution of a system reaches a stationary state asymptotically over time;
The dynamical evolution never reaches this state;
The dynamical evolution is characterized by periodic oscillations.

In the first case variations in the values of the state variables may be expressed as a function of their distance from the stationary state. System changes may be described as if they depended on a future final state. Such a circumstance could be related to a teleological view expressed, for instance, by *minimum or maximum principles* (of a local or global nature). Bertalanffy noticed how this form of description is nothing but a *different expression of causality*: the final state simply corresponds to a limiting condition of the differential equations governing the dynamical evolution. Such a condition, however, could also be considered as describing a particular kind of finality, i.e., the so-called *equifinality*. This is a characteristic of dynamical systems able to reach the same final state independently of initial conditions and input.

12) Emergent

Emergence is a process of self-organization whereby interacting elements give rise to profound changes in system structures (*intrinsic emergence*, as in phase transitions and the arising of order from disorder) or to the establishment of properties (systemic properties, such as those listed in this Appendix). Such emergence is unpredictable by considering the properties of the constituent elements and requires the observer to adopt a new cognitive model, a new level of description with reference that used for the elements. The role of the observer in a process of emergence may be considered as that of a person who decodes, and understands a message received in an unknown code. The "message" in this case is given by the changes in system structure or by the appearance of system properties. System properties are all emergent in principle (see Chapter 3).

13) Ergodic

Chapter 5 includes a discussion about the historical origin of this term. Here we recall that this attribute characterizes a stochastic process in which all ensemble averages are equal to all temporal averages.

A collection of elements is said to form an ergodic set if:

- the behavior manifested by any element of the set over a given time interval has similarities with the behavior manifested by the same element over other different time intervals;

- the behavior of any element selected at random is similar to that one of any other element.

In practice, *identical* behavior is not strictly required, but there should be strong similarity over time and in the numerical average values considered.

In an ergodic population any single individual represents, over time and with reference to the statistical parameters taken into consideration, the entire population. The salient characteristics of an individual are identical to those of any other member of the considered set.

The concept may be extended to behaviors regarding equilibrium states and related transition probabilities. Besides, every ergodic process must also be a stationary process. The reverse, in general, is not true. While referring to the Chapter 5 for the technical details relating to ergodicity, we stress here once again that the role of the observer is crucial in order to detect ergodicity: **the latter is a property of the whole system "observed set plus the observer".** *Ergodicity refers to the cognitive model used by the observer and in this way it may be intended more as a characteristic of the observer, and the model used, than of the observed.* The crucial role of the observer relates to the fact that **a system may be ergodic and not ergodic at one and** the same time, as a function of the time and space scale adopted by the observer.

14) Far from equilibrium

Generally this attribute applies to open dynamical systems lying in a stationary state which is far from thermodynamical equilibrium. As the conditions for the occurrence of the latter are no longer satisfied, a situation far from equilibrium can be sustained only through suitable interactions with the external environment, often associated with some form of dissipation. Typically these states are reached due to the existence of suitable fluctuations within the dynamics of the system itself which, in the presence of favourable conditions, can be amplified so as to reach the far from equilibrium condition. Such a condition can be considered as a metastable state, with a finite lifetime, surviving as long as the influence of the environment is counterbalanced by an inner mechanism of amplification of the fluctuations. Once this balance is broken, two things can occur: either a new stationary metastable state is reached (when the amplification exceeds the influence of the environment), or the system falls into a thermodynamical stable equilibrium state (when the environmental influence exceeds the amplification mechanism). The most celebrated examples of systems lying in far from equilibrium states is given by the *dissipative structures* introduced by Prigogine and his school.

15) Growing *vs.* developing

The distinction between growth and development is of foremost importance within the biological, social and economics realms. In these contexts a system (a company, for instance) may grow (by increasing, for example, its employees) without undergoing development (when the quality of its product, for instance, is decreasing). On the other hand it could undergo development (new technology is implemented) without growth (the number of employees remains unchanged). The usefulness of the concept of 'development', however, is related to processes which are rather more complex and interesting that that used in the simple example above. More precisely, the concept is needed for processes in which there is a structural change, occurring within the considered system, giving rise to a more effective adaptation between the system and its environment. Thus, it is natural to speak of development when, for instance, the organization of a company changes to obtain a higher quality and better distribution of its product, while improving, at the same time, both the position of the company in the market, and the effects the company has on the environment. Such development must be distinguished from mere 'growth', consisting in the increase in employees or in the amount of goods produced which, in many cases, would not involve development of the kind described above.

Once this distinction (as already discussed in Chapter 6) has been introduced in an intuitive way, it should be specified in a more quantitative or formal way. This, however, appears to be difficult, due to the fact that the systems to which it applies are often complex systems. In any case, the similarities between this distinction and that holding in thermodynamics between *extensive* and *intensive* quantities should be stressed. As is well known, an extensive quantity is a quantity whose value varies by varying the volume of the system under consideration, or the number of its components. Examples include the total internal energy or total entropy. On the contrary, intensive quantities do not depend upon the volume nor on the number of components. Such quantities include density, temperature and magnetization per unit volume. Within a thermodynamical description, growth refers to increases in extensive quantities and development (a more appropriate term would be 'structural change') to intensive quantities. For instance, when the temperature of a liquid decreases, there is an increase in the 'organization' of its components until, at a critical temperature, there is a phase transition, to a new structural arrangement of the components, detected macroscopically as the formation of a solid phase. Even the phase transition is signalled by an intensive quantity, the so-called order parameter, which macroscopically measures the degree of ordering taking place after the crossing of the critical point of the transition under study (in the case of liquid-solid transition the order parameter is the sample density).

It would be wonderful if the concepts used within the theory of phase transitions could be translated, *tout court*, to the processes of development occurring in biological or social systems. Unfortunately the complexity of the latter have so far prevented such a translation. However, the analogy presented above could be a starting point for a more profound analysis of the concepts of growth and development.

16) Heterogeneous

In the literature the attribute 'heterogeneous' is currently used in many different ways. For the sake of brevity, we mention only three of the most important: *temporal* heterogeneity, *spatial* heterogeneity and *phase* heterogeneity. Temporal heterogeneity characterizes systems whose properties are explicit functions of time. For instance, when the time evolution of a system crosses a critical point corresponding to a process of emergence (e.g., a phase transition), the temporal behavior of the system *before* the crossing is very different from its behavior *after* the crossing. Therefore the dynamics of such a system are temporally heterogeneous. Of course, not all systems are characterized by temporally heterogeneous dynamics. For example, to a first approximation, the dynamics of the solar system are temporally homogeneous: even by shifting the time coordinate by a billion years, the planets would, more or less, follow the same orbits as they do now, with the same temporal features. Within classical mechanics, the dynamics of all systems, while being characterized by temporal heterogeneity, are invariant with respect to a change in the sign of the time coordinate (the so-called inversion of the arrow of time). Such an invariance is often referred to as a homogeneity with respect to time reversal. Besides, when classical systems fall into an equilibrium state, their dynamics are characterized by true temporal homogeneity. Clearly, systems following irreversible dynamics are necessarily associated with temporal heterogeneity.

The concept of spatial heterogeneity, on the other hand, was introduced because many systems live within a spatial domain and thus their features can, in principle, depend explicitly on the spatial coordinates. A force field (gravity, for instance) is a typical system characterized by spatial heterogeneity, the force varying from one position to another within the domain in which the field is defined. Spatial and temporal heterogeneity are mutually independent. For instance, there are systems whose evolution is temporally heterogeneous but not spatially heterogeneous. This is the case for chemical reactions taking place between two reactants in a vessel being stirred so as to avoid spatial fluctuations in their concentrations. On the other hand, there are systems with spatial but not temporal heterogeneity. One example is a liquid-vapour system in equilibrium at constant temperature and volume. In thermodynamics such systems are said to be in a *stationary* state (spatially heterogeneous, of course). Most dissipative structures, mentioned in this book, including those produced by the Beloussov-Zhabotinskij reaction, are stationary states of this kind.

Finally, phase heterogeneity denotes systems in which there is the simultaneous presence of at least two different phases (here the word 'phase' means a state of organization of the constituents of the system in the thermodynamical sense, such as the 'solid phase', or the 'liquid phase'). Often, a closed system containing at least two phases is called a *heterogeneous system*. A heterogeneous system is said to be in an *equilibrium condition* when there is no net change among the spatial distribution of the phases.

17) Heuristic

This word comes from the Greek verb εὑρίσχω (heuriskein), meaning 'to find', 'to discover'. The attribute 'heuristic' denotes a problem-solving or goal-finding procedure based on trials which follow a given strategy. This feature makes heuristic procedures profoundly different from 'algorithmic' ones, in which one follows precise algorithmic rules to reach the required solution. By extension, the attribute 'heuristic' is used to denote any system or any kind of conceptual device designed to follow procedures and methods based on trials and strategies.

Heuristic procedures for decision-making or optimisation are very popular in Artificial Intelligence, as well as in Mathematics, Physics and, in general, every scientific discipline. They are performed using computational tools such as Neural Networks and reasoning-based Expert Systems.

Generally speaking, a heuristic idea is a guideline to discover, or to drive empirical research.

18) Hierarchical

A hierarchy may be intended as a sequence of ranked, asymmetrical relationships between elements (or sets of elements). Typical examples of hierarchies have their elements arranged on *levels*. The concept may be extended to *towers* of dependent concepts and abstractions.

Often a system is qualified as hierarchical when it consists of structured subsystems (the word *structure* here denotes a set of stable, functional, relationships among components and among interactions; see also point 26), each of which is also hierarchical until the level of analysis reached encounters non-structured subsystems.

Examples of hierarchical systems are social systems, such as military, religious and bureaucratic organisations, security systems, information processing systems having different levels of authorized tasks. Recursive functional hierarchies are often considered in mathematics.

The theory of hierarchy has received considerable attention within systemics. It originated from a paper by H. A. Simon in 1962 (Simon, 1962). Other important contributions were made by A. Koestler (Koestler, 1967), who introduced the notion of 'holon', which denotes an entity in a hierarchy that is at one and the same time a whole and a part. Thus a holon operates as a quasi-autonomous whole which integrates its parts, while working to integrate itself into an upper level purpose or role. More recent contributions include those of Pattee (Pattee, 1973), Salthe (Salthe, 1985) and Ahl and Allen (Ahl and Allen, 1996).

19) Homeostatic

This is a quality of an open system whose inner state does not change while its components and its environment are changing. Such a system keeps its inner state in a changing environment through internal adjustments and regulation. The process of homeostasis in living systems is performed by a number of self-regulating subsystems, such as the nervous and the endocrine systems.

20) Non-dissipative

In physics this attribute refers to an ideal system in which total energy is conserved during its time evolution. One typical example is a pendulum without friction. This theoretical simplification is useful in order to predict the behavior of systems whose total energy is approximately constant because this constancy adds a further constraint (sometimes called a *first integral*) which simplifies the process of solution of the equations of motion. Such systems, with perfect energy conservation, do not obey the *second principle of thermodynamics* (one of the many ways of expressing this principle is: *a thermal machine integrally transforming heat into work can not exist*) and their time evolution is perfectly reversible. A modern mathematical generalization of a non-dissipative system is given by the concept of *integrable system* (see, e.g., Hitchin *et al.*, 1999).

21) Non-ergodic

This refers to systems in which the behavioural observations do not follow the law of large numbers, in the sense that the sampling space is always different from the true phase space of the system under study. A typical example of a non-ergodic system is given by a glass. Namely, within it the long-term behaviour of any element depends in a crucial way on its location within the glass and, of course, on the particular history of the cooling process giving rise to the glass itself. In other words, there is no possibility of a general law underlying long-term behaviour, independent of the initial conditions. Clearly, therefore, non-ergodic behaviors lack

repetition (manifesting, for instance, only unique transition phases);

stability (for instance when transition probabilities are so variable that observations are not sufficient to model it).

Biological evolution and social processes generating structural changes are typically non-ergodic.

22) Open and Closed

Systems considered as isolated from their environment and reach an equilibrium state, that is a final state unequivocally determined by initial conditions are **closed** systems.

In **open** systems, on the contrary, stationary equilibrium states, where system composition is constant in spite of continuous exchange of its component parts, can be established (Von Bertalanffy, 1950; Prigogine and Nicolis, 1967; Nicolis and Prigogine, 1977). An open system tends to resist perturbations which tend to move it away from its evolution process, while this is not true for closed systems. Systems may be closed to matter/energy flows (autarkic), closed to information flows (independent), closed to organization. Within standard macroscopic thermodynamics a system is considered closed if it is able to emit and absorb energy and information, but not matter. On the other hand, for **open systems**, it is possible to prove that the same final state may be reached in different ways and starting from different initial states. Open systems may reach stationary states manifesting constant composition in spite of continuous exchange of its component parts with the environment. In open systems there is permeability between them and the environment, due to the fact that there is matter exchange, as typically happens with living systems (Miller, 1978).

However, the reader is warned against relying on the difference between closed and open systems only as specified by standard macroscopic thermodynamics.

On this point, it should first be stressed that it is possible to use the attribute of open and closed by referring to specific aspects of a system: so a system could be simultaneously open and closed. For instance, systems consisting of agents, each of which, in turn, is equipped with a cognitive system, may be considered from different points of view (by referring physical, thermodynamic or cognitive aspects) and can therefore be open from some aspects and closed for others (Minati *et al.*, 1996; 1998).

23) Self-organized

A system is called a *self-organizing system* if it can change its internal structure and its mode of operation in response to external influences (Banzhaf, 2001). The concept of self-organization developed during the first years of Cybernetics (Ashby, 1956). The study of binary networks, where every element could assume a state of *on* or *off* (such as light bulbs) depending upon previous ones and upon other nodes according to commutation rules, brought self-organization processes to light in the networks themselves, no longer flashing in a random manner, but in an ordered way.

Descriptions of self-organization processes were introduced by I. Prigogine (Prigogine and Nicolis, 1967; Nicolis and Prigogine, 1977) in his theory of "dissipative structures". The focus of his research was on the ability of systems far from thermodynamic equilibrium to manifest stability under non-equilibrium conditions through the formation of stationary states far from equilibrium.

At least three classical examples of this kind of phenomenon may be recalled.

- *Bénard instability*, produced, for instance, by *uniformly* warming from the bottom a thin layer of liquid, until a certain critical difference of temperature between inferior and upper surfaces is established, and heat is transferred by the coherent motion of a large quantity of molecules forming a structure of hexagonal, honeycomb-like cells in the liquid,. Far from being only a laboratory curiosity, the same effect may even be detected in warm air flows rising from the Earth's external atmospheric layers, producing vortices of circulating air leaving the marks of their behavior on Arctic snow or on desert sand. The so-called Bénard's rollers are an expression of the same effect. In this case in a layer of liquid, uniformly warmed from the bottom as in the previous experiment, when reaching a critical temperature, rollers of molecules of liquid are established (see also *bifurcation* in point 28).
- Chemical oscillating phenomena generating remarkable chromatic changes such as the famous *Belousov-Zhabotinsky reaction* (Belousov, 1959). Because such reactions take place at regular intervals they can be considered as "chemical clocks". Change of color in a uniform way for the entire chemical system in which the reaction is taking place implies that the reaction must act globally, by using a high level of organization, of order, for each molecule composing the system. This effect occurs spontaneously and is an expression of a stationary state far from equilibrium. For the story of the research leading to the *Belousov-Zhabotinsky* reaction see Winfree (Winfree, 1984) and for *periodic Liquid Phase Reactions* see Zhabotinsky (Zhabotinsky, 1964; Zaikin and

Zhabotinsky, 1970). An introduction to *Biological and Biochemical Oscillators* is also available (Goldbeter, 1996; Chance *et al.*, 1973).

• Another phenomenon refers to the emergence of *coherence in light emission*, typical of the *laser* effect. The first laser was introduced in 1958 and, at the same time as Prigogine's work in the early sixties, physicists like H. Haken realized how this coordinated light emission, giving rise to spontaneous coherence, should be considered as a *self-organized process*. Haken established a new discipline devoted to study these phenomena called "Synergetics" (Haken, 1979; 1983) or "Science of combined effects". Self-organization may also be identified when collective spontaneous rhythms are established, when one of these spontaneously becomes the order parameter, then that is the dominant one.

Processes of self-organization take place in the emergence of *collective phenomena* as introduced in Chapter 3.

24) Structures and systems

Usually people consider the *properties of elements* and *processes* (called *behaviors* in some contexts) as conceptually equivalent. Namely, within an 'Aristotelian' framework, a *behavior may be considered as a property* of a given element or entity (e. g., the behavior of fluid having a particular viscosity in a given context is intended as its property) and a *property may be considered as a behavior* (e. g., the property of a solid of burning in a particular way is intended as describing its behavior under particular conditions).

Such an equivalence, however, is misleading when dealing with complex systems (and not only). Therefore, a distinction will be made between the concepts of the *property of an element* and a *process* (or *behavior)*. More precisely, we refer to two different frameworks:

1. One is based upon the existence of primeval entities, called *elements*, each defined through a suitable set of *properties*, which act as the source of *relationships* with other elements, relationships whose features depend only upon the properties of the elements themselves. The latter can be viewed as irreducible to further analysis, at least at a given basic level. When considering a set of elements, together with their properties, the ensuing relationships define, in an unambiguous way, a *structure*. For instance, we can describe a crystal as a set of irreducible elements (its component atoms), each endowed with a set of properties (location, charge, mass and so on), from which we can derive in a unique way the relationships holding between the elements (the electrostatic forces acting between the atoms which keep the spatial configuration of the crystal lattice stable). Within such a framework we thus have a precise description of crystal *structure*. Some authors (e.g., Campbell and Bickhard, 2004; Brown and Harré, 1988) speak of this framework as *particle metaphysics*;

2. The second is based on *processes* (or *behaviors*) viewed as coherent wholes where the focus is on the conditions which lead to the emergence and duration/retention of such coherence. Although it is sometimes possible to describe these conditions in terms of suitable interactions between given elements, these descriptions are always partial. Often a process can allow for a number of different descriptions of this kind, none of them being exhaustive, and each of them being irreducible to every other of the same kind. Thus we start from the beginning due to the fact that the process implies a *system*, whereas the *structure* of the latter, provided it can be defined, is not so essential for understanding why the system is just *that* system. For instance, resorting again to the previous example, the crystal is now viewed as a whole, as an organized configuration of the electrostatic field (neglecting fluctuations). It is, of course, possible to interpret the locations where this field is particularly strong as if they were particles (or atoms), but such an interpretation is used only as long as it is convenient for practical computations. The authors cited above speak of this framework as *field metaphysics* (or *process metaphysics*).

While it is clear that it is possible to relate the two frameworks, it is equally clear that only within the second can we deal with the problem of understanding the emergence of systems as systems.

In order to introduce more precise definitions, it must first be stressed that *Structures*, i.e. *structured sets,* are established by *relationships* between *elements* or, better, between *properties of elements.* Each element is considered to be *coupled* through its relationship with others. Structures are *networked* elements.

This general concept of structure is related to that used in mathematics, where an algebraic structure is given by a set J having an internal composition rule, say T. Usually the properties of *structures* are different from the properties of their component elements. The *properties of a structure* may be found by knowing the *properties of its elements* **and** the *relationships* between them.

There are examples of structures in many disciplines such as engineering, biology, linguistics, sociology, and information sciences. The concept is so powerful and general as to have a very wide range of applications.

On the other hand, **Systems** are established by *relationships* between *processes* (or *behaviors*), i.e., *interactions*.

In his historic book (Von Bertalanffy, 1968) introduces a system S as characterized by suitable state variables Q_1, Q_2, . . . , Q_n, whose instantaneous values specify the state of the system. In most cases the associated process consists in the time evolution of the state variables, ruled by a system of *ordinary differential equations*, as introduced in Chapter 1.

It should be recalled that processes and behaviors may be described only with reference to a context, even if the latter consists only of the observer itself.

Besides, as a system is a whole entity, it is also possible to describe it by resorting to the first framework (particle metaphysics) and to introduce the concept of *property of a system*, considered as an element, precisely because of its wholeness. It is then possible to recognize that the *properties* of *systems* are different from the properties of their component elements (given that it is possible to identify them). A cue for a system being emergent as a whole is provided by the fact that its *properties* (when it is viewed as a single element, by adopting particle metaphysics) can *not* be found by knowing the *properties of its elements* and the rules describing their mutual *interactions* (resorting here to particle metaphysics for the component elements).

Interaction processes render emergent, properties which are not deducible from a knowledge of the properties of component elements nor from the rules of their interaction. *Emergence* is not a process established *by definition* like structural properties in simplified configurations, but generated by complexity, non-linearity, fluctuations and the fundamental role of the observer, using suitable levels of description and cognitive models. The *calculi of emergence* are, of course, much more difficult to implement, owing to the fact that the theoretical apparatus of modern science is still based upon particle metaphysics. On this point, the last few years have seen the introduction of new tools such as, for instance, Neural Networks, Cellular Automata, models used for simulations in Artificial Life and for Autonomous Robots.

We further stress that, while *structures and properties* are such by definition, *systems and processes* (or *behaviors*) are such as a function of the level of description and of the cognitive model used by the observer, being an integral part of the process of emergence.

The difference between the concepts of *interaction* and *composition* are discussed in Chapter 5.

25) Systemic characteristics
As the characteristics of systems are different from those typical of their elements, nor reducible to them, the following Table A1.2 highlights these differences through examples of such characteristics.

Table A1.2. Characteristics of elements and of generated systems.

Football players	Team
Weight, age	*Harmony, game strategy*
Cells	Living being
Function	*Behavior*
Students	School
Numbers	*Collective ability to learn*
Brain components	Memory, Intelligence
Configuration	*Processing capabilities*
Single musician	Orchestra
Instrument played	*Polyphony*
Words	A poem, a book, a story
Correct Grammar	*Meaning*
Musical notes	Music
Correctness in the score	*Harmony*
Couple	Family
Synchronization of interests	*Emergence of roles*
Soldiers	Army
Single abilities	*Ability to apply military strategies*
Workers	Corporation
Quantity	*Value*
Animals	Herds, swarms, flocks, packs
Quantity, single behavior	*Collective behavior*

The crucial aspect common to systems thinking and to recent evolution of scientific thinking is the role of the observer. The theoretical role of the observer in Systemics, with particular reference to emergence is discussed in Chapters 2 and 3.

26) Systemic equilibrium

The problem of Systemic equilibrium is a very complex one. It involves different systemic aspects such as: thermodynamic equilibrium, dynamic equilibrium, stationary states far from equilibrium, phase space and attractors, bifurcation, symmetry breaking.

a) Thermodynamic equilibrium

In terms of macroscopic thermodynamics, a system is described by its state variables (such as the pressure, volume and temperature of a gas). A *state* of the system is defined by a set of values, one for each state variable. Clearly, the concept of state is related to the observability of the values of those state variables, which is lacking in many practical cases (e.g., fast evolution, perturbation induced by the act of observation, presence of noise, etc.). Thus when speaking of the states of a system, we have to assume that their changes are very slow (thus reducing the thermodynamic description of those states to a limiting case and only approximating the transformations taking place in the real world). Each state can then be associated with given rates of change in the state variables. A state is then called a *thermodynamic equilibrium state* if the rates of change associated with it are all vanishing.

The class of thermodynamic equilibrium states can be further subdivided into two sub-categories: those which are *stable* and those which are *unstable*. Roughly speaking, a thermodynamic equilibrium state is considered stable if, when a small perturbation forces the system to leave that state, once the perturbation has ceased the system returns to its previous equilibrium state. *Thermodynamic stability conditions* (without entering into detail) provide for the stability of a given equilibrium state.

The concept of thermodynamic equilibrium state has been further generalized within the theory of differential equations. For instance, when dealing with an autonomous system of ordinary differential equations, characterized by suitable state variables $Q_1, Q_2, \ldots, Q_n$ (this time devoid of any thermodynamical interpretation) such as:

$$\begin{cases} dQ_1 / dt = f_1 (Q_1, Q_2, ..., Q_n) \\ dQ_2 / dt = f_2 (Q_1, Q_2, ..., Q_n) \\ \cdots\cdots\cdots\cdots\cdots\cdots\cdots\cdots \\ dQ_n / dt = f_n (Q_1, Q_2, ..., Q_n) \end{cases}$$

the equilibrium state is defined by the condition:

$$dQ_1/dt = 0, dQ_2/dt = 0, \ldots, dQ_n/dt = 0.$$

Unlike thermodynamic equilibrium states, however, characterization of the stability of equilibrium states in systems of differential equations is much more difficult, being associated with a very rich phenomenology of possible cases. This domain forms the subject matter of *Stability Theory*, for which the reader is referred to standard texts (e.g., Guckenheimer and Holmes, 1983; Glendinning, 1994).

We limit ourselves to observing that, in the case of systems of partial differential equations, containing rates of change of the state variables even with respect to space coordinates (and not only with respect to the time coordinate), different definitions of equilibrium state are possible. More precisely, when the rates of change with respect to all independent variables (both spatial and temporal) all tend to zero, we speak of *homogeneous stationary equilibrium states*. When, instead, only the rates of change with respect to time tend to zero, we speak of *non-homogeneous stationary equilibrium states*. There is also a stability theory for these generalizations of the concept of equilibrium state. Such a theory, which is a fundamental tool in the theory of pattern formation and of dissipative structures, is far more complicated than the corresponding theory for equilibrium states of ordinary differential equations.

b) Dynamic equilibrium

The concept of equilibrium state introduced above can be further generalized to *dynamic equilibrium*. Consider the solution of a system of differential equations, expressing state variables as a function of time, and denoted by:

$$Q = f(t, Q_0)$$

where Q denotes the vector of the values of the state variables, f is a vector function of time t and Q_0 the vector of the initial values of the state variables. Let us then consider another vector function:

$$Q = g(t) .$$

This is known as a *dynamical equilibrium state* for the system described by the differential equations above if, when t tends to infinity, the difference between the functions f and g tends to zero for a suitable choice of initial values within a given interval. In other words, here the equilibrium situation is given not by a static state but by a behavior. Needless to say, the search for dynamical equilibrium states is far more difficult than that for normal equilibrium states. However, such dynamical equilibrium states allow a wider range of possibilities for the modeller.

c) Stationary states far from equilibrium

The concept of stationary state far from equilibrium is applied to systems characterized by many different equilibrium states. In this case, every equilibrium state, when stable in the sense described above, is always stable only with respect to perturbations below a given amplitude. In other words, each stable equilibrium state is associated with a given *basin of attraction*, such that, when a perturbation leads the system to a state located within the basin of attraction of a given equilibrium state, the system will unavoidably return to that equilibrium state. It is clear that in such a situation the distinction between different equilibrium states cannot be based only upon their stability with respect to perturbations of a given amplitude. Generally, in this case, an external criterion is introduced which allows one to privilege, among the various possible stable equilibrium states, that endowed with a particular property. Often such a criterion consists in selecting the *global equilibrium states*, that is those states in which some quantity, defined in terms of the state variables, reaches its absolute minimum. In physics it is common to identify such a quantity with the total energy, but not always: sometimes the free energy is preferred (as in the case of dissipative structures). Such a choice, of course, depends upon technical considerations as well as the purposes of the modeller. There are, amongst others, a large number of models where the choice of such a quantity is still impossible (this is the case, for instance, of non-integrable systems). In any case, once a global equilibrium state has been identified (in principle there could be various states of this kind), in some contexts known as a *ground state*, all other equilibrium states are called *metastable equilibrium states*. This latter term refers to the fact that all these states, in the presence of fluctuations, have a finite lifetime, as there will be a non-zero probability of a fluctuation occurring of such amplitude that it will push the system outside the basin of attraction of the metastable equilibrium state, leading it to fall into the global equilibrium state. If the latter is unique, its lifetime in the presence of fluctuations will be infinite, because any fluctuation, even if it takes the system temporarily into the basin of attraction of a metastable equilibrium state, will be sooner or later counterbalanced by another fluctuation leading the system from the metastable situation back down to the global equilibrium state. For this reason metastable equilibrium states are also called *far from equilibrium stationary states*.

A typical example of a stationary state far from equilibrium is given by the case of Bénard cells. It was introduced in the previous section referring to self-organization (see point 25). When the system considered gradually moves away from equilibrium (equilibrium in this case is a uniform temperature throughout the whole liquid) it reaches a critical instability point where the so-called Bénard cells, ordered hexagonal, honeycomb-like cells emerge. This instability is an example of self-organization, of a non-equilibrium state sustained by a continuous heat flow. It is important to note that the behavior of a dissipative structure, **maintained in a stationary state far from equilibrium**, is not assumed to follow universal laws, but is typical of that system. It has been shown how, when the system is close to equilibrium, the phenomena follow universal laws. When far from equilibrium, the trend is from universal validity toward particular cases. The same occurs for living systems trying to stay as long as possible in a stationary state, characterized by continuous changes in their metabolism and flows of matter, involving a very large number of chemical reactions. Equilibrium is established when all chemical reactions stop. When approaching equilibrium conditions, a system manifests weak fluctuations, evolving towards a state of minimum entropy. One measure of this is the tendency towards the reduction of temperature produced by heat dissipation, considering temperature as a measure of the molecular dynamics of the system. On the other hand, when a system moves away from equilibrium, entropy production increases. The system no longer tends towards equilibrium; various stationary points can be reached. **Bifurcation** points characterize the situations where the system may take very different directions (Scott Kelso, 1995).

Another example of a system in a far from equilibrium stationary state is the atmosphere of a planet, as described by Lovelock (Lovelock, 1988). He pointed out how, from an analysis of the atmosphere of a planet, it was possible to obtain information about the existence of life upon it. For instance, by examining the atmosphere of Mars, he noted little oxygen, carbon dioxide and the absence of methane. On a planet without life, **all** physico-chemical reactions between gases essential for life, if present, are assumed to have already taken place. On Earth, however, gases such as methane and oxygen, notwithstanding their high probability of reaction, *coexist*, constituting a mixture far from thermodynamic equilibrium. It is possible to assume that this is caused simply by the existence of life on Earth. Living organisms produce and use gases, such as oxygen, and their continuous production and use takes place. In this way, the planet Earth appears as an open system, far from equilibrium, subject to a constant flow of matter and energy. See also the *Gaia theory* (Lovelock, 1988; Hamilton and Lenton, 1998).

d)   Phase space and attractors

Phase space is an abstract space where each state variable of the system is associated with a coordinate axis. It is, of course, possible to represent in a graphical way such *n*-dimensional space (here *n* is the number of state variables) only in the particular cases of 2 or 3 dimensions. **System behavior can be depicted as the motion of a point along a trajectory in the phase space.**

If a system approaches a given domain of its phase space as time tends to infinity, this domain is called an *attractor* of the system dynamics. There are three kinds of attractors:

• *fixed points*, consisting of isolated equilibrium states;
• *periodic attractors*, typical of systems exhibiting periodic oscillations in the long term;
• *strange attractors*, corresponding to fractal domains of phase space, usually present in chaotic systems.

The analysis of non-linear systems is simply based upon an analysis of the topological characteristics of their attractors and known as *qualitative analysis*. Attractors provide information about the kind of system behavior.

e) Bifurcation

The concept of bifurcation is applied to systems whose evolution equations contain, besides its dependent state variables and independent variables (such as time and spatial coordinates), also a number of *parameters*, that is, quantities whose value can vary, not as an effect of the state of the system itself, but only under the influence of an external environment. For instance, the motion of a point particle in 1-dimensional space is governed by Newton's celebrated Second Law:

$$F(x) = m \, d^2x/dt^2$$

where x denotes the position of the particle and plays the role of state variable, whereas $F(x)$ describes the force field. This evolution equation, however, also contains a parameter: the mass m, whose value depends neither upon x nor upon t. In a sense, the mass represents the action of the external environment on the system under study, an action which cannot in any way be counterbalanced by the system itself. We could, for example, consider the environment as the modeller freely choosing the value of m for its own purposes, or as Nature which, through its own laws, fixes the value of m. Whatever our opinion, what matters is that a parameter is a kind of variable which is different from the usual dependent and independent variables in the evolution equations. Of course, except for a very small number of models, designed only for mathematical commodity, all models of evolutionary systems must contain some kind of parameter.

Within systems containing parameters, then, the main question is: what happens when the value of a parameter undergoes a change? The word 'happens' here refers to the 'qualitative' structure of the behaviors of the system as described by the evolution equations being studied and of their solutions. Such a 'qualitative' structure can be described in precise terms when the evolution equations, for instance, coincide with differential equations, and consist of the number and the kind of attractors of system dynamics. Two possibilities then exist: 1) the number and the kind of attractors remain unchanged when a parameter value is varied; 2) the number and/or the kind of attractors changes as the value of a parameter crosses a *critical* value. In the second case we speak of a *bifurcation* occurring in correspondence to the critical value.

A simple example is that of a system described by an algebraic equation (here any evolution is absent) for a single state variable x having the form:

$$x^3 - \mu x = 0$$

Here μ denotes a parameter (sometimes called, when we focus our attention on its variation, a *control parameter*). It is easy to see that, if we identify the behavior of our system with the real part of the solution of the equation, when the control parameter $\square$ is positive we have three different possible behaviors (i.e., solutions, given by $x = 0$, $x = \pm \sqrt{\mu}$), whereas, when the control parameter is negative, we have only one possible behavior (given by the solution $x = 0$). Thus the critical value $\mu = 0$ marks a bifurcation point, and upon crossing it the very nature of the solutions changes.

There is a consistent body of theories on bifurcation, not only in differential equations, but also in other kinds of evolution equations. Such theories, known collectively as *Bifurcation Theory*, use the most advanced tools of mathematics and constitute the technical core of modelling of phenomena such as morphogenesis, pattern formation, emergence, and so on. For more technical details the reader is referred to standard textbooks on this subject (e.g., Sattinger, 1973; Guckenheimer and Holmes, 1983; Glendinning, 1994).

f) Symmetry breaking

The concept of symmetry breaking first arose in Theoretical Physics. Historically, it is very difficult to attribute its introduction to any individual physicist. However, three important contributors should be mentioned:

- Pierre Curie - at the beginning of the Twentieth Century he remarked upon the role played by symmetry considerations within physical transitions;
- Emmy Noether - in 1915 she proved her celebrated theorem connecting invariance in symmetry transformations with the conservation of suitable physical quantities during the evolution of a physical system;
- Lev Davidovic Landau - starting in the 1930s, he based his whole theory of phase transitions upon the concept of symmetry breaking.

In technical terms the expression *symmetry transformation* denotes a transformation of suitable variables in the evolution equations of a given system (for a bibliography on these topics see, for example, Itzykson and Zuber, 1986; Moriyasu, 1983; Olver, 1995; Sewell, 1986; Umezawa, 1993). Such a transformation can act both upon the form of these equations, as well as the form of their solutions. We thus have *symmetry breaking* when a symmetry transformation leaves the form of the evolution equations invariant, but changes the form of their solutions. A typical example is that of a sample of matter consisting of atoms which, at a given temperature, is paramagnetic. The form of the equations describing the atomic motion is invariant with respect to particular symmetry transformations constituted by space rotations around a given axis. The solutions of these equations also possess the same invariance. As a matter of fact, if the sample is exposed to an external magnetic field, whatever its direction, this will give rise, within the material, to an induced field exactly aligned with the external one. However, when the temperature is decreased, there is a critical point (the so-called *Curie point*) where a transition from the paramagnetic to the ferromagnetic phase occurs. This gives rise to an internal magnetic field of macroscopic dimensions, deriving from the alignment of the magnetic fields of the individual atoms due to their interactions. Besides the formation of North and South magnetic poles within the sample, the presence of such a field leads to the existence of a preferred direction: that of the internal magnetic field. Thus, although the form of the equations describing the motions of the atoms continues to be invariant with respect to the symmetry transformations constituted by spatial rotations, their solutions are not, as the preferred direction breaks such invariance. This phase transition (and, according to Landau, *every* phase transition) can thus be associated with a breaking of symmetry.

Other interesting examples of symmetry breaking occur within the field theories of fundamental interactions, especially those based upon the so-called *gauge theories*. The mechanism underlying these forms of symmetry breaking is based, this time, upon symmetry transformations acting, not upon independent variables, but on the field variables themselves. More precisely, the form of the evolution equations is invariant with respect to suitable symmetry transformations acting upon the dependent variables. However, these equations contain one or more parameters such that, when their value is below a critical value, even the solutions of the equations themselves are invariant with respect to these transformations. Such invariance occurs because the equations are associated with a unique state of minimal energy, the so-called *ground state*, so that all behaviours will, in the long term, unavoidably reach this state. However, when such parameters cross a critical value, the equations, while retaining invariant their form, allow for the existence of a multiplicity of different ground states, all associated with the same value of the energy minimum. Thus, by acting, through the symmetry transformations, upon the state variables, we can, in principle, transform the evolution towards a particular ground state into an evolution towards another, different, ground state. The solutions are therefore no longer invariant with respect to the symmetry transformations and we have a breaking of symmetry. Such a mechanism is, in fact, acting even in the ferromagnetic case introduced above, as we have two possible, conceptually equivalent ground states,: that with all atoms 'up' and that with all atoms 'down'. The fact that in real ferromagnets only one of these ground states is actually observed is due to the fact that the physical extension of the ferromagnet is finite. It should be stressed that Landau, by starting from his theory of phase transitions, introduced this second form of symmetry breaking.

27) Systemic structure and organization

Authors such as Maturana and Varela (Maturana and Varela, 1973) introduced this kind of distinction. When dealing with a living, and thus autopoietic, system, it is possible to conceive its **organization** as being defined through a set of relationships and interactions amongst its components. Such a set identifies the system as belonging to a class (animal, vegetal, organs, virus, etc.). There is, however, no reference to the particular nature of the components. Like autopoiesis, this is to be intended as a general organizational scheme which is common, for instance, to all living beings. In other words, the organization is independent of the characteristics and properties of the system components. **Structure,** on the other hand, is to be intended as an organization applied *in practice* to specific cases, to components having characteristics and properties to which the general organizational schema is applied.

One can say that, when dealing with *organization*, the reference is to networks of relations with undefined parameters, whereas in the case of *structure* the reference is to networks having well-defined parameters. An example of the difference between organization and structure is given by the existence of two different ways of describing a Neural Network: either as a system with *n inputs*, *m hidden layers* and *s outputs*, or as a network with precise values of connection weights and well-defined transfer functions associated with the single neurons. Some authors speak of the former description as a specification of network *architecture*.

References

Arbib, M. A., 1969, Memory Limitations of Stimulus-Response Models, *Psychological Review*, **76**:507-510.

Ahl, V., and Allen, T. F. H.,. 1996, *Hierarchy theory, a vision, vocabulary and epistemology.* Columbia University Press, New York.

Arecchi, F. T., and Farini, A., 1996, *Lexicon of Complexity.* Studio Editoriale Fiorentino, Florence, Italy.

Ashby, W. R., 1947, Principles of the Self-Organizing Dynamic System, *Journal of General Psychology* **37**:125-128.

Ashby, W. R., 1956, *An Introduction to Cybernetics.* Wiley, New York

Banzhaf, N., 2001, Self-Organizing Systems. In: *Encyclopaedia of Physical Science and Technology*, 3rd edition, vol. 15, (R. A. Meyers, ed.), Academic Press, New York, pp. 589-598.

Barrow-Green, J., 1997, *Poincaré and the Three Body Problem.* American Mathematical Society,. Providence, RI.

Belousov, B. P., 1959, A periodic chemical reaction and its mechanism. *Sbornik Referatoo po Radiatsionnoi Meditsine*, Medgiz, Moscow, 145-147.

Bennett, C. H.,1988, Logical Depth and Physical Complexity. In *The Universal Turing Machine- a Half Century Survey* (R. Herken, ed.), Oxford University Press, New York. pp.227-257.

Bennett, C. H., 1985, Information, Dissipation, and Definition of Organization. In *Emerging Syntheses in Science* (P. Pines, ed.), Santa Fe Institute, Santa Fe, NM, pp. 297-313.

Bishop, C. M.,1995, *Neural Networks for Pattern Recognition.* Oxford University Press, Oxford, UK.

Brown, H. R. and Harré, R., 1988, *Philosophical Foundations of Quantum Field Theory.* Oxford University Press, Oxford, UK.

Capra, F. ,1996, *The Web of life.* Doubleday/Doubleday, New York.

Casti, J. L., 1994, *Complexification. Explaining a Paradoxical World through the Science of Surprise*. Harper Collins, New York.

Chaitin, G., 1990. *Information, Randomness, and Incompleteness*. Word Scientific, Singapore.

Chance B., Pye, E. K., Ghosh, A. K., and Hess, B., (eds.), 1973, *Biological and Biochemical Oscillators*. Academic Press, New York.

Flood, R., and Carson E., 1988, *Dealing with Complexity. An Introduction to the Theory and Application of Systems Science*. Kluwer, New York

Freeman, W. J., Kozma, R., and Werbos, P., 2001, Adaptive behavior in complex stochastic dynamical systems, *BioSystems* 59:109-123.

Glendinning P., 1994, *Stability, Instability and Chaos: An introduction to the theory of Nonlinear Differential Equations,* Cambridge University Press, Cambridge, UK.

Goldbeter, A., 1996, *Biochemical Oscillations and Biological Rhythms*. Cambridge University Press, Cambridge, UK.

Gregersen, N. H., (ed.), 2003, *From Complexity to Life: on the Emergence of Life and Meaning*. Oxford University Press, New York.

Guckenheimer J., and Holmes P., 1983, *Nonlinear oscillations, dynamical systems and bifurcation of vector fields,* Springer, Berlin.

Haken, H.,1979, Pattern formation and pattern recognition - an attempt at a synthesis. In *Pattern formation by dynamical systems and pattern recognition* (H. Haken ed.), Springer, Heidelberg.

Haken, H., 1983, *Synergetics, an Introduction: Nonequilibrium Phase Transitions and Self-Organization in Physics, Chemistry, and Biology.* Springer, New York.

Hamilton, W. D., and Lenton, T., M.,1998, Spora and Gaia: how microbes fly with their clouds, *Ethology, Ecology & Evolution* 10:1-16.

Haykin, S., 1998, *Neural networks; A comprehensive foundation* (2nd edition). Prentice Hall, Upper Saddle River, NJ.

Hénon, M., 1983, Numerical explorations of Hamiltonian systems. In *Chaotic Behavior of Deterministic Systems* (G. Iooss, R. H. G. Helleman, and R. Stora, eds.), North-Holland, Amsterdam, pp. 53-170.

Hitchin, N. J., Segal, G. B., and Ward, R. S., 1999, *Integrable Systems*. Oxford University Press, Oxford, UK.

Hornik, K., Stinchombe, M., White, H., 1989, Multi-layer Feedforward Networks are Universal Approximators, *Neural Networks* 2:359 – 366.

Itzykson, C., Zuber, J. B., 1986, *Quantum Field Theory,* McGraw-Hill, Singapore.

Koestler, A., 1967, *The ghost in the machine*. Macmillan, New York

Kolmogorov, A. N., 1965, Three Approaches to the Quantitative Definition of Information, *Problems in Information Transmission* 1:3-11.

Lorenz, E., 1963, Deterministic Non Period Flow, *Journal Of The Atmospheric Sciences* 20: 130-141.

Lovelock, J. E., 1988, *The ages of Gaia: a biography of our living Earth.* W. W. Norton, New York.

Lundh, D., Olsson B., and Narayanan, A., (eds.), 1997, *Proceedings of BCEC97. BioComputing and Emergent Computation*. World Scientific, Singapore.

Maturana, H., and Varela, F., 1973, *Autopoiesis: The Organization of the Living*. -original edition- De Maquinas y Seres Vivos, Editorial Universitaria, Santiago, Chile.

Maturana, H., and Varela, F., 1980, *Autopoiesis and Cognition*. Reidel, Dordrecht.

McTear, M., 1993, User modeling for adaptive computer systems: A survey of recent developments, *Artificial Intelligence Review* 7:157-184.

Miller, J. G., 1978, *Living Systems.* McGraw Hill, New York.

Minati, G., Penna, M. P., and Pessa, E., 1996, Towards a general theory of logically open systems. In: *3rd Systems Science European Congress*, (E. Pessa, M. P. Penna, A. Montesanto, eds.), Kappa, Rome, Italy, pp. 957-960

Minati, G., Penna, M. P., and Pessa, E., 1998, Thermodynamic and Logical Openness in General Systems, *Systems Research and Behavioral Science* 15:131-145.

Minsky, M., and Papert, S., 1969, *Perceptrons*, MIT Press, Cambridge, MA.

Moriyasu, K., 1983, *An elementary primer for gauge theory,* World Scientific, Singapore

Nicolis, G., and Prigogine, I., 1977, *Self-Organization in Nonequilibrium Systems: From Dissipative Structures to Order through Fluctuations.* Wiley, New York.

Olver, P. J., 1995, *Equivalence, invariants, and symmetry,* Cambridge University Press, Cambridge, UK.

Papadimitriou, C. H., 1993, Computational Complexity, Addison Wesley, Redwood City, CA.

Pattee, H.. H., (ed.), 1973, *Hierarchy theory: the challenge or complex systems.* Braziller, New York.

Pessa, E., and Vitiello, G., 2004, Quantum noise, entanglement and chaos in the quantum theory of mind/brain states, *Mind and Matter* 1:59-79.

Prigogine, I., 1981, *From Being to Becoming: Time and Complexity in the Physical Sciences.* W. H. Freeman & Co., New York.

Prigogine, I., and Nicolis, G, 1967, On Symmetry-Breaking Instabilities in Dissipative Systems, *The Journal of Chemical Physics* 46:3542-3550.

Prigogine, I.,1998, *The End of Certainty: Time, Chaos, and the New Laws of Nature.* The Free Press, New York.

Rojas, R., 1996, *Neural networks. A systematic introduction.* Springer, Berlin-Heidelberg-New York.

Rosen, R., 1985, *Anticipatory Systems.* Pergamon Press, New York.

Rumelhart, D. E., Hinton, G. E., and Williams, R. J., 1986, Learning representations by back-propagating errors, *Nature* 323:533-536.

Rumelhart, D. E., and McClelland, J. L., 1986, PDP models and general issues in cognitive science. In: *Parallel Distributed Processing: Explorations in the Microstructure of Cognition, Vol. 1: Foundations.* (D. E. Rumelhart, J. L. McClelland and the PDP Research Group, eds.), MIT Press, Cambridge, MA, pp. 3-45.

Salthe, S., 1985, *Evolving Hierarchical Systems: their structure and representation.* Columbia University Press, New York.

Sattinger, D. H., 1973, *Topics in stability and bifurcation theory.* Springer, Berlin.

Scott Kelso, J. A , 1995, *Dynamic Patterns.* MIT Press, Cambridge, MA.

Serra, R., and Zanarini G., 1990, *Complex systems and cognitive processes*, Springer, Heidelberg.

Serra, R., Zanarini G., Andretta, M., and Compiani M.,1986, *Introduction to the physics of complex systems.* Pergamon Books, Oxford, UK.

Simon, H.. A., 1962, The architecture of complexity. In: *Proceedings of the American philosophical society* 106: 467-482.

Thorndike, E. K., 1921, *The Teacher's World Book.* Teachers College, New York.

Umezawa, H., 1993, *Advanced Field Theory. Micro. Macro, and Thermal Physics.* American Institute of Physics, New York.

Varela, F., 1980, Describing the logic of the living. In *Autopoeisis: A theory of the living organization* (M. Zeleny, ed.), North-Holland, New York., pp.36-48.

Varela, F., 1991, Organism: A meshwork of selfless selves, *Revue Européenne Sciences Social* **29**:173-198

Von Bertalanffy, L. ,1968, *General Systems Theory*, Braziller, New York.

Von Bertalanffy, L., 1950, The Theory of Open Systems in Physics and Biology, *Science* **111**:23-29.

Von Bertalanffy, L., 1952, *Problems of Life. An Evaluation of Modern Biological and Scientific Thought.* Harper & Brothers, New York.

Von Foerster, H., and Zopf, G. W., (eds.),1962, *Principles of Self-Organization.* Pergamon, New York.

Winfree, A. T.,1984, The prehistory of the Belousov-Zhabotinsky oscillator, *Journal of Chemical Education* **61**:661-663.

Zaikin, A. N., and Zhabotinsky, A. M. , 1970, Concentration wave propagation in two-dimensional liquid phase self oscillating system, *Nature* **225**:535-537.

Zhabotinsky, A. M., 1964, Periodic liquid phase reactions, In: Proceedings of the Academy of Sciences, USSR, **157**: 392-395.

Web Resources
(Note: in some Internet sources the date of publication is not indicated)

Campbell, R.J. and Bickhard, M.H., 2004,"Physicalism, Emergence and Downward Causation", http://www.lehigh.edu/~mhb0/pubspage.html.

Heylighen, F., Joslyn, C., and Turchin, V., (eds), 2002, *Principia Cybernetica Web* (Principia Cybernetica, Brussels), http://pespmc1.vub.ac.be

Whitaker, R., 2003, Encyclopaedia Autopoietica, http://www.enolagaia.com/EA.html

Appendix 2

SOME QUESTIONS AND ANSWERS ABOUT *SYSTEMICS*

1. What is intended by the term *Systemics*?
2. Is it possible to falsify Systemics?
3. What does it mean to act *systemically*?
4. What are the advantages of acting systemically?
5. What are the disadvantages and major critiques of Systemics?
6. Why is Systemics not regularly taught at school?
7. What are the limits of Systemics?
8. What relationships exist between systems and music?

1. What is intended by the term *Systemics*?

The term Systemics refers to:

- a framework within which to work for recognising systemic properties and invariants, as introduced in Chapter 1;
- a general strategy for dealing with problems, by focussing not only on the elements of a system but also on their mutual interactions, *represented* by relationships between behaviors (see Appendix 1, point 26), and on the aspects of emergence, as introduced in Chapter 3;
- a methodology which stresses the need for viewing events, processes or complex entities in a global manner.

Systemics, intended as a more general (not *generic*) cultural approach, as outlined in Chapter 1, derives from, amongst others, Wiener and Ashby's *Cybernetics*, Von Foerster's *Second Cybernetics*, Bertalanffy's *General Systems Theory*, Prigogine's works in the field of self-organization theory and Haken's Synergetics. Fundamentally, Systemics proposes the use of alternative approaches to those based upon strategies which only identify and study single component behaviours. It is also based upon theoretical and experimental evidence that the latter approaches are absolutely ineffective in dealing with complex problems.

Often the term Systemics or even systemic concepts are mentioned briefly and generally used with implicit, intuitive references, thus diffusing ambiguity and imprecision. This allows, for example, concepts such as system, set, organism, device and references to theories, including *systems theory* and *control theory*, to overlap in a confused manner.

More specifically *Systemics* has both *disciplinary* and *interdisciplinary* applications.

a) *Disciplinary applications* examine, from the point of view of a given
   discipline, the properties of the global, emergent behaviour of
   configurations of interacting elements which give rise to *systemic
   properties* (see those listed in Chapter 2) not present at the level of
   system elements. Within this conceptual framework different approaches
   and models have been used in various disciplines including physics,
   biology, economics and management.
   All too often, within philosophical, and sometimes scientific,
   investigations the Systemic approach is opposed by modern
   reductionism, still based on a mechanistic approach, which assumes that
   any phenomenon may be conceptually reconstructed from its component
   parts. The latter, although recognising the need for taking into account
   interactions between individual components, refutes the need for
   adopting a new conceptual paradigm to explain emergent properties.
   Edward O. Wilson, an exponent of this way of thinking, states(Wilson,
   1998): "Nature is organized by simple universal laws to which all other
   laws and principles can be reduced" (p.55). "... all tangible phenomena,
   from the birth of stars to the working of social institutions, are based on
   material processes that are ultimately reducible, however long and
   tortuous the sequences, to the laws of physics" (p. 226). Another
   exponent of reductionism, Francis Crick, states in (Crick, 1994): "...
   while the whole may not be the simple sum of its separate parts, its
   behaviour can, at least in principle, be *understood* from the nature and
   behaviour of its parts *plus* the knowledge of how all these parts interact"
   (p.11). The above example refers to how in chemistry elements are
   combined in different ways producing different substances with specific
   properties which are very different from those of the component
   elements. This may be explained in terms of quantum mechanics. "It is
   curious that nobody derives some mystical satisfaction by saying 'the
   benzene molecule is more than the sum of its parts' ...". The reference is
   to the possibility of establishing a new holism avoiding what Wilson calls
   the mysticism of past holists, such as that of Morgan mentioned in
   Chapter 3. Another example of this nature would be the production of
   green water by *mixing* yellow and blue water.
   The effectiveness of the arguments put forward by authors such as
   Wilson and Crick relies almost exclusively on the adoption of what, in
   point 24) of Appendix 1, we called *particle metaphysics*. This consists in
   assuming that every system can be described, in ultimate analysis, as
   consisting of a set of irreducible (and unchangeable both in number and
   properties) elements (its 'atoms'), reciprocally interacting in a suitable
   way. Such a framework, amongst others, coincides with that traditionally
   adopted in modern mathematics, where each formal structure is reduced
   into a set-theoretic form, where the concept of 'set', of course, refers to
   the existence of a predefined and fixed number of 'elements', whose
   nature is given once and for all. It is clear, then, how, once particle

metaphysics has been adopted (a mathematician could, perhaps, call it *set metaphysics*), the conclusion of reductionists such as Wilson and Crick immediately follows. Namely, if the description of a system can, even though in some tortuous and non-linear way, be reduced to a set of elements plus their interactions, why not rely only on the latter by eliminating all 'systemic' properties? After all, Occam's razor states that *entia non sunt moltiplicanda praeter necessitatem*.

In point 24) of Appendix 1 above it was stressed, however, that modern physics (and even modern mathematics) allows for the introduction of a better framework, called *field metaphysics* (or *process metaphysics*). Within it (see Campbell and Bickhard, 2004) systems are conceived as networks of processes, rather than as sets of interacting elements. Typically a process is described through a function (or, more generally, a functional). As a consequence the concept of 'element', rather than being primitive and irreducible, is a derived one. In other words, it is simply a temporary *label* introduced to shorten and summarize certain descriptions of processes under suitable conditions. It would be more appropriate to speak, in general terms, of an 'effective' element, rather than simply of an element. This entails, of course, that both the number and the properties of the elements can change over time (or with other independent variables). Field metaphysics can be formalized in various ways, but the best known one is Quantum Field Theory (QFT). Without entering into detail (see Itzykson and Zuber, 1986; Umezawa, 1993), we limit ourselves to recalling that it can be mathematically proven that within QFT there are an infinite number of different *representations*, physically not reciprocally equivalent, of the same process (Haag, 1961; Hepp, 1972). Thus it can never occur that a system description be reducible *only* to a particular set of interacting elements, simply because the same system allows for infinite other, physically not equivalent, descriptions. Such a circumstance leads to a number of consequences such as, for instance, that for each system, there are a number of *different* particle-like interpretations (Clifton and Halvorson, 2001). But the most important of these consequences is that, at least for suitable systems, the 'systemic' properties can never be reduced to the properties of elements.

The *opportunity* for a paradigm shift when dealing with phenomena characterized by properties which individual components do not possess was introduced in Chapter 1. The term *opportunity* should be stressed: the strategy is not based on objectivism, on searching for the truth, for reality. A very good domain for understanding the nature of the controversy between reductionism and the systemic view is *linguistics*. The meaning of a statement is not the sum of the meanings of the different words used, nor a more complex non-linear function of them. Neither is the meaning of a book the sum of the meanings of the different statements, the 'elements', within it. Here the reductionist approach has no sense.

b) *Interdisciplinary applications* of Systemics take place when a single, specific discipline is able to represent and use models, problems, approaches, inferences and solutions of another discipline whilst using its own language.

Interdisciplinary approaches can be carried out in various ways.

A book is said to be interdisciplinary when the same subject is discussed in different disciplinary contexts such as growth, for example, in economics, biology, chemistry and evolutionary studies.

An educational process is said to be interdisciplinary when it is not based on teaching separate, individual disciplines, but on teaching various disciplines at one and the same time dealing, for example, with mathematics whilst teaching economics, geography whilst teaching history, and so on.

Interdisciplinary ways of studying, carrying out scientific research, and practicing professions are based on taking into account multiple simultaneous perspectives and on using different simultaneous knowledge and approaches. Typical examples of *interdisciplinary problems*, which can be dealt with *only* by interdisciplinary approaches, include the environmental and cognitive sciences which are interdisciplinary *in principle* because *based on*, and not only *involving*, different disciplines.

Physicians and managers in the health sector, for instance, are supposed to tailor and to adapt their actions according to the context by using different approaches and not by applying *the right one*. Results from applying one approach based on specific knowledge influence successive actions. The idea is to apply the *same strategy* by using different approaches and resources.

Following the description of interdisciplinary applications mentioned above, careful attention should be paid in the possible misuse of *metaphors and analogies* because the latter offer no guarantee of rigour in transposing results, effects and conclusions. Their usage should be considered only as representational, as attempts to merely make generic, able to *induce* but *not substitute* autonomous ways of reasoning. Namely correspondences and analogies should be introduced between different *autonomous* disciplinary applications only by referring to *complete* processes of reasoning.

Concepts, reasoning or conclusions reached within a given discipline by *analogy* to those used in another discipline are not considered as complete, and neither robust nor rigorous.

Interdisciplinary applications take place when the *same* models are applied to different problems represented in the *same* way (these applications are typically *quantitative*, and mathematical models include variables representing different kinds of phenomena).

Understanding *interdisciplinarity* as equivalent to applying the same rules in different contexts, such as arithmetic for counting apples, money and people, is completely wrong because the idea of *interdisciplinary applications* relates to generalizing not rules (such as computing), but models, simulations and representations endowed with the same meaning, the same effects, the same behaviour, the same effects taking place in different contexts even when using different variables referring to different processes and phenomena. Similar mathematical models may be used to describe population dynamics, markets, ecological processes or physical phenomena.

Another example of *interdisciplinary application* is the opposite process: which no longer resorts to the same model for different phenomena but relies on multi-modeling, as introduced with *DYSAM* (see Chapter 2), which simultaneously uses various models from different approaches (competitive, evolutionary, collective, rule-based, optimisation-based) for studying a single, specific phenomenon.

It is also possible to apply Systemics to frameworks other than disciplines and not in a metaphorical sense, but with reference to specific processes of knowledge representation. Consider *languages*, for instance. Interdisciplinary processes in this case are *represented* by the process of *translation* taking place when at least one person has a knowledge of both languages. Translating allows different levels of performance. The maximum is represented by reciprocal, interactive representations and another level is the reading of translations. Peculiar to the process of translating is the need for the translator. The theoretical impracticability of the old dream of *automatic translation* is conceptually very important. Translating is not possible without reference to a human *translator*. Although it may be possible to *simulate* the translation process by using technologies which can create a translator profile, different translators generate different versions of the same text. Translation processes can even be carried out from and to the same language, such as when a student is asked to repeat a lesson using his/her own words. As for interdisciplinary applications it is the representation of the same ideas in different ways.

Knowledge, disciplines and even interdisciplinary applications, are represented by using languages. When referring to languages we relate to what they actually represent. The process of translating generates new meanings when we deal with knowledge representations starting from different disciplines. This could also be the case for *religions*. A very innovative approach would be to compare the different representations of the same concepts (such as God, good and bad, life, safety, sanctity, etc.) over various religions. A good start would be to organize synoptic correspondences between different religions.

Transdisciplinary inquiry within Systemics takes place when systemic properties, experienced and dealt with by disciplinary and interdisciplinary approaches, are studied *per se*. Research on trans-disciplinarity relates to architectures of networked correspondences and analogies, abstractions, methodologies, logic, models and representations of systemic properties. Research at the trans-disciplinary level relates to a more general and higher level of abstraction than disciplinary and inter-disciplinary research, *but not independent* from the latter. This is a very important and crucial point. If this level is considered independent, then it is assumed as autonomous, independent of applications, results, experiences and models developed at disciplinary and interdisciplinary levels, constituting the so-called *Galilean telescope* for trans-disciplinarity (Minati, 2004). It is important to underline that trans-disciplinary approaches relate to the establishment of *robust theoretical generalizations*, for instance, by modelling, representing, through simulation processes and not by a *metaphorical* usage of systemic knowledge. The process of *generalizing* asks for a crucial theoretical effort, while making generic and metaphoric allows an extension of the usage of concept trading with less rigour, less specificity, and at a lower theoretical level. The common misunderstanding of considering making things generic and metaphoric as equivalent to generalizing must be avoided.

2. Is it possible to falsify Systemics?

First of all, it should be recalled that the *falsification principle* was introduced, in opposition to the *verification principle*, by the "Vienna circle". This neo-positivistic principle regards the problem of excluding from scientific discussion all scientific laws because, due to their general applicability, they cannot be verified by a finite series of observations (whether it be for the law of gravitation or for observations such as "all rocks are black"). According to K. R. Popper (1902-1994), the main exponent of an approach based upon falsifying, any scientific system cannot be selected once and for all but it must be possible to confute it through experience. The success of a critical confuting experiment is sufficient to refute, to invalidate, the hypothesis forming the basis of a scientific theory.

It is therefore possible to falsify a predictive theory which has general applicability such as occurs for scientific laws (regarding gravitation, mechanics, hydrodynamics, etc.). Falsification could represent at least one cornerstone of this general approach.

Let us now consider the concept of system. Its definition is based on the concept of interacting components which lead to the emergence of a new reality endowed with characteristics and properties different from those of its component parts. Within the literature, as shown in Chapter 1, we often find the statement that the condition of *interaction* of composing elements (i.e. one's behaviour affecting another's) is *necessary* to establish (that is, to create the emergence of) a system.

Falsification may then be precisely applied to this situation. An important warning has to be made, however: a falsification process is always dependent upon the features of the observer's cognitive system. In other words, it is not endowed with an absolute, objective, meaning but the interpretation of its outcomes lies in the cognitive schemata of the observer or, better, in the interaction of the cognitive schemata of many different observers. This remark is very important especially when dealing with systemic concepts, which are always defined with relation to an observer.

Once this necessary caution is taken into account, a critical experiment, the falsifying phenomenon, could then consist in detecting emergence, that is the establishment of systemic behaviour from *non*-interacting elements. Or the retention of systemic behaviour by a system whose components have ceased to interact amongst themselves. Both detection and interaction are, of course, relative to a given observer.

Examples of such a situation are:

– Individual musicians playing alone. Subsequently, their performances are artificially superimposed and orchestral effects take place without any actual interaction. A situation may also occur where all orchestral components *minus one* play together, adapting to a pre-recorded individual performance, integrating their playing and harmonizing with it. *In this case all the others are playing for the individual.*

– A set behaving as a system, taking on typical systemic properties, in spite of any interaction, based for instance on communicating.

When looking at these examples, it is possible to notice (always referring to a given observer) that the absence of interaction prevents the very existence of the system itself. Thus, we could say that so far the falsification experiment gave a negativeresult. In principle, however, there is nothing preventing a future positive outcome relative to some observer and to some observational level.

3. What does it mean to act *systemically*?

As described above, *Systemics* is based on the concept of *system*, identifiable in various fields and disciplines, as a main conceptual tool to identify, induce, model emergence, and to find correspondences. In turn, correspondences among models and methods in different disciplines are realized, induced, when applied to systems.

First of all *non-systemic* problems are effectively dealt with by using non-systemic approaches. Non-systemic problems (of course, the attributes 'systemic' and 'non-systemic' are always relative to a given observer) are those with a single or only a few linearly inter-related variables; they are associated with modeling problems where the purpose is to deal with single non-interacting behaviors. Often this schema which does not require a systemic approach may be a *simplification* of the effective configuration to be modeled. The level of the problem is decided by the observer by deciding the level of description, by keeping in mind the levels of effectiveness and the results expected. There are many examples of this, for instance, in physics. For historical reasons the very first *laws* in physics were non-systemic or expressed in a non-systemic way because the first approach was to simplify, to reduce, making a problem understandable and manageable by using the mathematical tools then available.

The need to deal with a problem systemically depends upon its level of description and on the type of problem as mentioned above.

The availability of new scientific tools (especially mathematical ones) has provided the possibility of formulating problems in a more sophisticated way, suitable for systemic descriptions. In short, there are now more sophisticated languages for describing problems in more global and effective ways, one of these is the systemic one.

Systemic problems are those described and considered by an observer adopting a systemic level of description. The latter can be adopted because it is very appropriate for the problem in hand or because it is congenial to the observer or because of the type of results expected.

The classical analytical approach is based on decomposing problems into different parts which are assumed   to form, when composed of an appropriate configuration of real and objective relationships, the overall problem. In such an approach the configuration of relationships, also representing interactions, looks like a network of causes-and-effects assumed to be more controllable and manageable the *more* initial conditions and details are known.

In a systemic approach first of all the suitability of the level of description and of the threshold levels considered significant in reference to the observer's requirements are explored. One often takes into account the difference between *microscopic, macroscopic and mesoscopic* descriptions (Serra *et al.,* 1986; Serra and Zanarini, 1990). The three levels of description mentioned above are very appropriate for describing, for instance, chaotic systems, such as the atmosphere . In this case systemic approaches have to be adopted for weather forecasting. When dealing with chaotic systems, linear approaches, based only upon looking for cause-effect relationships and on continuously refining data knowledge by assuming that it is an effective strategy to reach a better knowledge of the whole system, are ineffective.

4. What are the advantages of acting systemically?

Some problems, such as those considered in this book, are dealt with effectively by using systemic approaches while little success is obtained using reductionist approaches.

The great advantage of using systemic approaches is to avoid the use of resources for dealing with single elements, at the *microscopic level*, while the problem to be dealt with is the *emergent* one. The crucial point relates to the ability to *recognize* when a problem is systemic, that is to identify the related level of description.

A typical behavior when facing an emergent problem is to react by focusing on *effects*, on *symptoms*, that is without paying attention to the fact that a symptom may be just the very first manifestation, at a simpler level of description, of a more complex problem. By the way, a problem is *complex* when it is described by non-linearity, as systemic, emergent, resulting from many interacting components. Whereas, a problem is *complicated* for the observer with reference to the problem itself, i.e., it requires approaches, tools, conceptual schemas and previous knowledge in order to understand the description and to process solutions. A problem is intended to be *difficult* for an agent with reference to the agent's specific abilities, resources, especially the cognitive ones.

The extremely critical issue is to adopt a level of description suitable for showing interactions, collective effects and emergent phenomena.

A systemic approach can take advantage of schema, models, methods and solutions developed in a specific field and using them in another. Even the mere metaphorical use of schema and models belonging to a context may induce analogies and approaches in different fields. It is possible to take advantage of knowledge developed in various fields, different from the original one, using *inter-disciplinary applications*.

5. What are the disadvantages and major critiques of Systemics?

It is almost always possible to use such a level of description for a problem that the systemic approach seems appropriate. In this way the adoption of the systemic approach does not depend upon an evaluation of the most suitable approach but as *ideologically opposed* to the reductionist approach.

It may be difficult, when adopting a systemic approach, to realize when a problem is better dealt with by using a non-systemic approach. This may occur when problems are always *described* in a systemic way, but in some cases this may not be the most effective approach. Namely, the reductionist approach may be very effective for some problems, at certain levels of description.

Thus, using the systemic approach on principle may lead one to overlook some very effective actions which can be taken on individual elements, such as optimizing single processes. It is not a valid approach to expect systemics to be effective when non-systemic approaches are more suitable and appropriate. Often a problem is considered as more *complex*, more *complicated* as a way to *escape* non-systemic solutions in a specific discipline because of a lack of knowledge and competence in that discipline.

First of all it must be shown that the choice to act systemically is a cultural and scientific one, searching for effectiveness and appropriateness, and *not* an *ideological* one. The dichotomy *Systemics-is-good, Non-Systemics-is-bad* must be rejected because it is non-effective, rigid and dogmatic. Nevertheless such a *simplification* is based upon the characteristics of the systemic approach, as a cultural framework, to guarantee in some way the refusal of disciplinary rigidity and of dividing knowledge and competence.

The effectiveness and significance of the systemic approach is *necessarily* based on knowledge and competence in a given discipline, although this is not a *sufficient* condition. Systemics is not a short cut to escape studying specific disciplines: on the contrary, the depth of knowledge must be sufficient to be able to apply a systemic approach. Systemics may be intended as a bridge between disciplines: strong, solid riverbanks are required to build a robust and resistant bridge.

It is ineffective and perhaps a source of error to act systemically in a not well-known context by simply trying to replicate other findings and to identify invariants, correspondences and analogies, without an adequate knowledge of the context under consideration. One risk of the systemic approach is to hurriedly produce analogies, correspondences and relationships amongst concepts, events and phenomena not adequately studied from a disciplinary point of view.

6. Why is Systemics not regularly taught at school?

Referring to point 1 of this Appendix, Systemics should be intended as an approach, a strategy to deal with levels of descriptions of problems, using resources together with a theoretical framework allowing the introduction of specific, systemic, disciplinary tools (as introduced in chapter 4). An example of education in a given discipline using Systemics is the teaching of control theory *only* in engineering.

Systemics may not be explicitly taught as a specific discipline, but introduced *while* teaching disciplines such as physics, chemistry or economics. It is impossible to teach, to introduce, Systemics without teaching a discipline. Systemics is an *added value* to disciplines. Disciplines taught without reference to Systemics form the basis for reductionism.

There are now established fields of research, such as the cognitive and environmental sciences, which cannot, on principle, be dealt with by single disciplines, that is by reductionist approaches, as already mentioned in point 1 above.

A greater systemic content may perhaps be found in pre-school teaching and in elementary schools for children. During these formative years, in fact, it is not a matter of developing specific competences but rather of harmony among still unbonded valences.

The possibility of establishing a single, disciplinary, separated educational process has been intended as maturing from the evolutionary age, when pupils may focus and concentrate on specific subjects, disciplines, thus becoming *real* students. To this end, educational institutions have been established only for teaching single, individual disciplines, knowledge and competence (Banathy, 1991).

Systemics should be not introduced and taught *after* developing disciplinary knowledge, but *during* this process, as a methodology not in addition to, but integrated into teaching. Besides, Systemics may be taught *not* as a specific discipline, but at a disciplinary, inter-disciplinary and trans-disciplinary level.

Usually, knowledge production, such as that from publishers, research and educational activities, is *organized* and based on disciplinary distinctions and single subject competence. On the other hand, Systemics cannot be considered *alone*, autonomously, without disciplinary or interdisciplinary applications.

7. What are the limits of Systemics?

a) The most important current limit of Systemics is the lack of a *general* theory of emergence, *General Systems Theory* overlapping with a *Theory of Emergence* (Minati, 2004). We refer to the concept of emergence as in the literature and as introduced in this book. Briefly, a theory of emergence should scientifically explain in general the process of establishment of the properties of a system which the individual components do not possess.

There are already some very powerful *aspects* of such a theory. When dealing, for instance, with organized systems performing functions; with synergic effects as introduced by Synergetics; with processes of self-organization and dissipative structures; with the establishment of collective behaviors. This book also discusses emergence for social systems.

This kind of a *general theory* is particularly lacking for studying living systems, the process of evolution and consciousness (Vitiello, 2001).

In philosophy this problem relates to the classical question of the transformation of *quantity in quality* dealt with by many philosophies such as *Dialectics* and so-called *Historical Materialism* (Engels, 1947).

b) Because of this limit and notwithstanding the many well-established academic publications, the eminent scholars involved, and both disciplinary and interdisciplinary applications, Systemics does not receive a warm welcome in specific disciplinary research contexts. This is both because of assuming that the approach of focusing on single applications is more effective, leaving interdisciplinary applications as secondary interests (like popularizing), and because of considering systemic approaches as additions providing little or no theoretical contributions, assumed to exist *only* in detailed mono-disciplinary studies. This view assumes that Systemics takes advantage of mono-disciplinary knowledge rather than supporting the development of interdisciplinary knowledge which can, , in turn, be used in disciplinary applications. This is not so much a *limit* of Systemics as a limit of a persisting idealistic approach and understanding of science, which is less and less effective when dealing with complex problems. Processes of knowledge production and education in particular are still influenced by this traditional approach. The required change is not related so much to the acceptance of Systemics but to the effective processes of knowledge management for dealing with complex systems.

c) Another limit refers to the unavailability of application-, discipline-independent ways for evaluating the effectiveness and appropriateness of interdisciplinary approaches. It is very difficult to evaluate interdisciplinary methodologies *in general*, without referring to the disciplines expected to have interdisciplinary applications. **The availability of a suitable language** to *autonomously* **(that is, without reference to disciplinary contexts) describe interdisciplinarity is related to trans-disciplinarity and that is the real** *dream* **of Systemics.**

Behind Systemics there is the implicit hope of the unity of knowledge, of the existence of only one language, valid for any disciplinary context, able to unitarily express any event, level of description and any process of becoming. The breaking up of such a language, its splitting into a multitude of different languages (i.e., disciplines), is well illustrated by the **Tower of Babel** *in the first book of the Torah (Genesis).*

d) With reference to language, a limit of systemic representations can occur due to an imbalance between the language used and meaning to be represented. Interdisciplinary applications, and the possibility of *translating* concepts, invariants, characteristics, methods, properties, problems and solutions, between disciplines requires processes which relate different languages. In order to make this possible they must have comparable representional *semantic power*. *Metaphorical correspondences* may be typically used between contexts using languages which have very different *semantic power*.

e) Moreover, the *scientific dimension* of Systemics can be a limitation: when it is applied only to scientific disciplines and interdisciplinary applications. Here the reference is to the (usually only metaphorical) correspondences between scientific and non-scientific contexts, such as artistic ones.

Examples of correspondences between artistic contexts include setting music to a movie or illustrating a book or organizing a stage design for a ballet.

We cannot avoid mentioning how *The Creation of Man* in the *Sistine Chapel* in Rome, painted in 1510 by Michelangelo (1475 - 1564), the backdrop behind God whose finger has been interpreted as transmitting life to Adam, has the shape of the *human brain* (Figure A2.1). In 1990 the neurologist Frank L. Meshberger published a paper illustrating how the Sistine Chapel panel titled "The Creation of Adam" shows an anatomically accurate representation of the brain (Meshberger, 1990).

It is known that Michelangelo performed anatomical dissections. Many interpretations have been given and many others are possible. The traditional interpretation focused upon God giving *life* to Adam. Michelangelo's intention may, however, have been to portray God as giving *intellect* to Human Beings, through the image of the human brain. This is an example of one kind of knowledge portraying a very powerful message referring to another kind of knowledge: in this case a very powerful message to future generations, by assuming that only future knowledge becoming generally available will make Michelangelo's message clear.

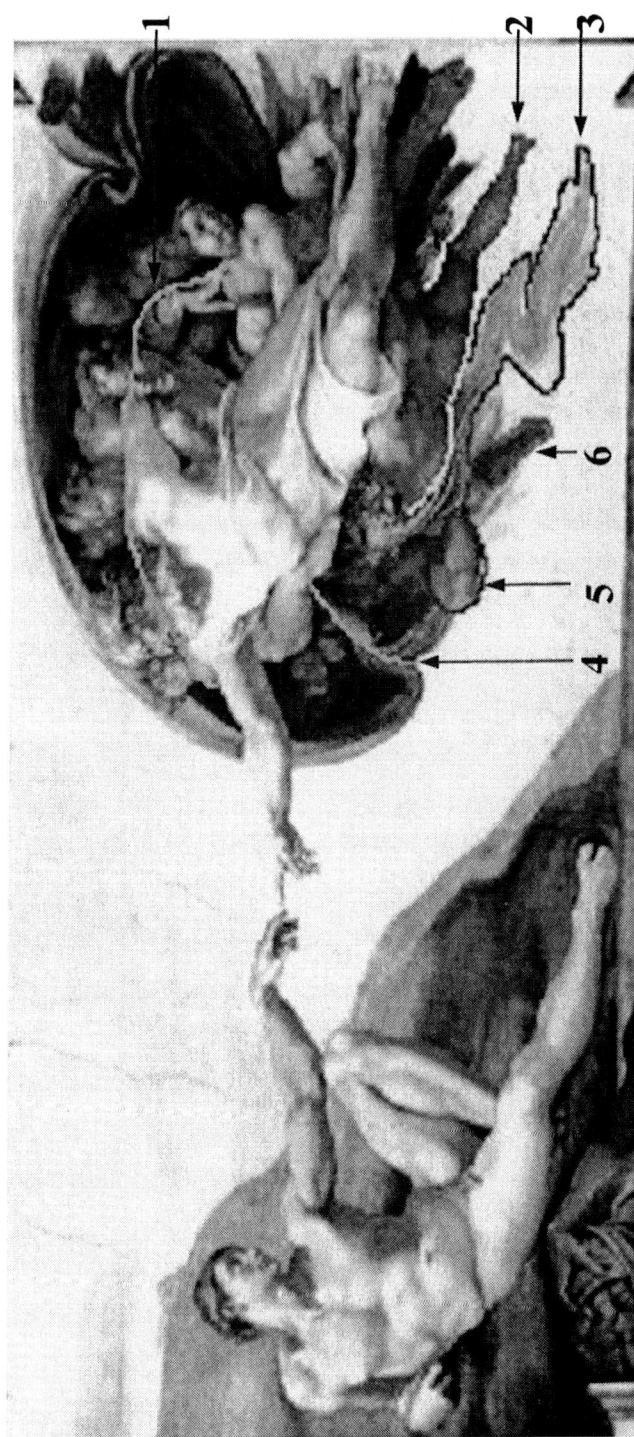

Figure A2.1. The brain with anatomic details as painted by Michelangelo in The Creation of Man in the Sistine Chapel. **1** - cingulate gyrus, cingulate convolution (gyrus cinguli); **2** - brain axis, brainstem and spinal marrow, spinal cord (medulla spinalis); **3** - vertebral artery (arteria vertebralis); **4** - fissure of Sylvius, lateral cerebral fissure; **5** - pituitary gland, hypophysis (glandula pituitaria); **6** - optic chiasm, nerve and tract.

The need to *understand in order to produce artistic beauty*, as in painting and sculpturing was peculiar to the *Renaissance period* (1450-1600). Leonardo da Vinci's (1452-1519) drawings are the finest examples of the intimate combination of the need to understand (for instance, how birds move their wings to fly, how muscles of the human body allow movements) not only for artificially reproducing the same effects (dams to make rivers navigable, rotating wings to to allow human-made flying machines), but also to produce artistic beauty. (Romei, 1994).

8. What relationships exist between systems and music?

Processes of representing - to be used by cognitive processes - concepts, ideas, information, knowledge and emotions in different languages are based, for instance, on words, symbols, numbers and pictures. Sentences of natural languages may be considered as *systemic* compositions of elements (words) having their own meaning, whereas music, due to its high level of abstraction, can be considered as consisting of ratios, structures and relationships (established by the theory of music) among single, individual sounds without meaning. This is the basis of the difference between *nice sounds and music*. Music is *man-made*. Human beings needed to *invent* musical instruments to create music, i.e., *architectures of structures of ratios*, and not just to produce sounds. **Musical instruments may be considered as *devices* for generating physical (acoustical) entities in such a flexible way that they can be organized into dynamic structures using (musical) grammar and syntax** (Teperley, 2001). Phrases of music are discourses made of ratios and not of elements like words.

Musical instruments have the specific purposes to produce effects (sounds) to be organized through the building of a *structure* among them. Musical instruments are generators of very dynamic and flexible physical quantities to be related in a very complex network of ratios (i.e., *musical structures*). The listener is invited to fill the *empty architecture* of music, "abstractions without contents", by thinking and making music *emergent*.

This is closely related to the effects of music on infants and children which implicitly induce effects on cognitive structures, as studied in the psychology of music and in cognitive science (Dowling *et al.*, 2002; Janata *et al.*, 2002; Pineau and Tillmann, 2001; Clynes, 1982; Tillmann and Bigand,1996).

Musical structures are also related to the *dimensionality* of music, as when considering the correspondence between the latter and *polyphonic (simultaneous* musical texture of several melodic vocal or instrumental lines) *counterpoint* (the art of combining simultaneously and subsequently independent melodies).

More specifically:

- *Polyphony*, "Polyphonos" (*many-voiced*) and "Polyphonia" occur in ancient Greek with no specific reference to a musical technique and the term is still used in a general sense. The term is "used to designate various important categories in music: namely, music in more than one part, music in many parts, and the style in which all or several of the parts move, to some extent, independently." (Sadie, 2001). The topic of Polyphony is a very important one in music and has been approached in many different ways. An overview of the many approaches may be found in Sadie (Sadie, 2001). Since the 1200's polyphony has been intended as: (1) multiplicity of parts, contrasting *dyaphonia* and *polyphonia*; (2) several parts of equal importance; (3) equal development of individual parts, distinguishing between polyphony and heterophony; and (4) simultaneous use of several structures. One of the first descriptions of this practice may be found in the *"Musica enchiriadis"* (late 9th cent.), and was later developed into freer forms of countermelody. The fugues and chorale settings of J. S. Bach (1685-1750) in the baroque era are considered by some as the epitome.

Boulez defined "polyphony as a combination of structures of which one is answerable by another" (Boulez, 1963). Polyphony has been considered by ethnomusicologists with reference to diaphony, plurivocal, polyvocal, polyphonic parallelism, multiphonic, multi-sonance. For others "true" polyphony is a procedure, which must be multi-part, simultaneous, hetero-rhythmic and non-parallel. To clarify the idea one can consider polyphonic music as *independent but harmonically related musical lines,* as a *simultaneous* musical texture of several melodic vocal or instrumental lines. In the polyphonic composition of autonomous musical lines it is possible to recognize, in this specific representation, the meaning of emergence processes (Minati, 2002), of the establishment of systems from interacting components.

The theory of *counterpoint* was formulated about 1330. Counterpoint is intended in music as the art of combining, simultaneously and subsequently, independent melodies. Drawing parallels with the linguistic domain, harmony may be considered as the grammar of music and counterpoint as the syntax. The term *counterpoint* was introduced by Giovanni de Muris in the 14th century and derives from the Latin *punctum contra punctum* (point against point), meaning note against note in the notation of plainsong. The academic study of counterpoint was long based upon *Gradus ad Parnassum* (1725) by Johann Joseph Fux (1660-1741), an Austrian theorist and composer. The study of counterpoint was subdivided into five species (note against note, two notes against one, four notes against one, syncopation and florid counterpoint) which combines the other species. In the 15th century the great Italian polyphonic composers introduced the free use of counterpoint (*ricercare, fantasia, a imitazioni*). Polyphonic forms were later given a brilliant and sophisticated expression during the baroque era in the works of J. S. Bach. "Counterpoint and polyphony are practically synonymous" (Sadie, 2001).

The more multi-dimensionality, the more *abstraction*, as intended in cognitive science and related to intelligence, is represented, i.e. simultaneous related, associated processes of thinking. Not *something* abstract. But *abstracting*, intended as a process. The process of transforming into a higher level of generalization and having the possibility to use, in many different contexts, all kinds of resources (for instance, the ability to measure, to use methodologies and approaches, to identify rules, etc.). The process of abstraction is related to the ability to move from *applying functions* to *invent uses, to create associations*. The process of abstraction is related to the concept of intelligence. In Cognitive Science *intelligence* is no longer intended as the ability to solve problems, but to make *correspondences*, to *associative thinking*, for learning, memory (rebuilding information and not just store-and-retrieve), decision making-processes, reasoning, making inferences and perceiving as expressed by the Cognitive System (Barsalou, 1992). Thought is related to cognitive processing as studied in Cognitive Science. Thinking refers to the process of creating relations and correspondences between items of perceived information and processed information.

Singing may be intended as a combination of the creation of musical structures by using elements possessing mutual ratios, but also with their own meaning, like the words of a natural language. A singer is a particular *musical device* able to produce sounds organized into a structure, with each element (word) having its own meaning. . There is the combination of musical meaning (from the musical structure) with the semantic meaning of a natural language.

The concept of *architecture* refers to organization, style, strategy, procedure and ways of using resources. The concept has many disciplinary applications: architecture intended as the art or science of planning and building structures, the method or style of building. The concept is used in Cognitive science, Software technology, Hardware technology, Linguistics, Laws, Engineering, Music, Chemistry, Biology and so on. *System architectures* make reference to *architecture, structures of elements, of relations and interactions among them* leading to the emergence of a system. The relation between systems and music is that music results from a process of *emergence*. As in emergence, the observer has a crucial role: this is evident by considering the difference between hearing and listening to music.

By using *musical structures* it is possible to organize architectures of them (as in polyphonic counterpoint music). Musical architectures are *structures of structures*.

One of the peculiarities of music is that it does not need to be *translated because it consists of architectures of structures based on ratios between physical entities (sounds).* Music may be arranged for different instruments and styles, but translation is not necessary. The same is true for images.

Music may be intended as a systemic *lab* where it is possible to produce, to represent in a very flexible way, architectures. Music using different architectures *represents* processes (thinking, intelligence and other mental processes) more than *accompanying* them (movies, theatre, ceremonies). This relates to knowledge representation in cognitive science. There are examples of correspondence and *continuity* within the same architecture, the same style of using resources, in different fields. For instance, baroque architecture, music, dress, behaving in general, ways of thinking and so on. **There is an important correspondence between polyphonic counterpoint architectures generating harmony among different simultaneous melodic lines and associative thinking, which lies at the basis of systems thinking** (Minati, 2002).

References

Banathy, B. H., 1991, *Systems Design of Education. A Journey to Create the Future.* Educational Technology Publications, Englewood Cliffs, NJ.

Barsalou, L. N., 1992, *Cognitive Science: An Overview for Cognitive Scientists.* Erlbaum, Hillsdale, NJ.

Boulez, P., 1963, *Penser la musique aujourd'hui.* Gonthier, Paris.

Clifton, R. K., and Halvorson, H. P., 2001, Entanglement and open systems in algebraic quantum field theory, *Studies in the History and Philosophy of Modern Physics* **32**:1-31.

Clynes, M., (ed.), 1982, *Music, mind and brain: the neuropsychology of music.* Kluwer, New York.

Crick, F., 1994, *The astonishing hypothesis: the scientific search for the soul.* Scribner, New York.

Dowling, W. J., Tillmann, B., and Ayers, D. 2002, Memory and the experience of hearing music, *Music Perception* **19**:249-276.

Engels, F., 1947, *Anti-Dühring. Herr Eugen Dühring's Revolution in Science.* Progress Publishers, Moscow, Russia (first published in 1877).

Haag, R., 1961, Canonical commutation relations in Quantum Field Theory and Functional Integration. In: *Lectures in Theoretical Physics,* (W. E. Brittin, B. W. Downs and J. Downs, eds.), Wiley, New York, vol. 3, pp. 353-381.

Hepp, K., 1972, Quantum theory of measurement and macroscopic observables, *Helvetica Physica Acta* **45**:237-248.

Itzykson, C., and Zuber, J. -B., 1986, *Quantum Field Theory.* McGraw-Hill, Singapore.

Janata, P., Tillmann, B., and Bharucha, J. J., 2002, Listening to polyphonic music recruits domain-general attention and working memory circuits, *Cognitive, Affective and Behavioral Neuroscience* **2**:121-140.

Meshberger, F. I., 1990, An Interpretation of Michelangelo's Creation of Adam Based on Neuroanatomy, *Journal of the American Medical Association* **264**:1837-1841.

Minati, G., 2004, Towards a Systemics of Emergence, In *Proceedings or the 48th Conference of the International Society for the Systems Sciences,* (J. K. Allen, J. Wilby, eds.), Asilomar, CA, 04-11, pp. 1-14.

Pineau, M., and Tillmann, B., 2001, *Perception des structures musicales.* L'Harmattan, Paris.

Romei, F., 1994, *Leonardo da Vinci: Artist, Inventor and Scientist of the Renaissance.* Bedrick, New York.

Sadie, S., (ed.), 2001, *The New Grove Dictionary of Music and Musicians.* Second Edition Macmillan, London.

Serra R., Zanarini G., Andretta M., and Compiani, M., 1986, *Introduction to the physics of - complex systems.* Pergamon Books, Oxford, UK.

Serra, R., and Zanarini, G., 1990, *Complex systems and cognitive processes*, Springer, Heidelberg.

Teperley, D., 2001, *The Cognition Of Basic Musical Structures.* MIT Press, Cambridge, MA.

Tillmann, B., and Bigand, E., 1996, Does formal structure influence perceived musical expressivity ? *Psychology of Music* **24**:3-17.

Umezawa, H., 1993, *Advanced Field Theory. Micro, Macro, and Thermal Physics.* American Institute of Physics, New York

Vitiello, G., 2001, *My double Unveiled: the dissipative quantum model of brain.* Benjamins, Amsterdam.

Wilson, E. O., 1998, *Consilience: the unity of knowledge.* Alfred A. Knopf, New York.

Web Resources

Campbell, R.J. and Bickhard, M. H., 2004, "Physicalism, Emergence and Downward Causation", http://www.lehigh.edu/~mhb0/physicalemergence.pdf

Minati, G., 2002, Music and systems architecture, In: Proceedings of the 5th European Systems Science Congress Hersonissos, Iraklion, Crete, Greece. http://www.afscet.asso.fr/resSystemica/Crete02/Minati.pdf,

Index